W9-BHI-890

SECOND EDITION

CASE FILES®
Neuroscience

Eugene C. Toy, MD
Vice Chair of Academic Affairs and
    Residency Program Director
Department of Obstetrics and Gynecology
Houston Methodist Hospital
John S. Dunn Sr. Academic Chair of
    Obstetrics and Gynecology
St. Joseph Medical Center
Houston, Texas
Clerkship Director and Clinical Professor
Department of Obstetrics and Gynecology
University of Texas Medical School at
    Houston
Houston, Texas

Evan Yale Snyder, MD, PhD
Professor and Director
Center for Stem Cells & Regenerative
    Medicine
Sanford-Burnham Medical Research
    Institute
Department of Pediatrics
University of California-San Diego
La Jolla, California

Josh Neman, PhD
Assistant Professor, Research
Department of Neurological Surgery
Keck School Medicine
University of Southern California
Los Angeles, California
Formerly:
Research Fellow
Division of Neurosurgery
City of Hope Comprehensive
    Cancer Center
Beckman Research Institute
Duarte, California

Rahul Jandial, MD, PhD
Assistant Professor, Neurosurgery
City of Hope Comprehensive
    Cancer Center
Beckman Research Institute
Duarte, California

Medical

New York Chicago San Francisco Athens London Madrid
Mexico City Milan New Delhi Singapore Sydney Toronto

Case Files®: Neuroscience, Second Edition

1 2 3 4 5 6 7 8 9 0   DOC/DOC   19 18 17 16 15 14

ISBN 978-0-07-179025-3
MHID 0-07-179025-X

Notice

Medicine is an ever-changing science. As new research and clinical experience broaden our knowledge, changes in treatment and drug therapy are required. The authors and the publisher of this work have checked with sources believed to be reliable in their efforts to provide information that is complete and generally in accord with the standard accepted at the time of publication. However, in view of the possibility of human error or changes in medical sciences, neither the editors nor the publisher nor any other party who has been involved in the preparation or publication of this work warrants that the information contained herein is in every respect accurate or complete, and they disclaim all responsibility for any errors or omissions or for the results obtained from use of the information contained in this work. Readers are encouraged to confirm the information contained herein with other sources. For example and in particular, readers are advised to check the product information sheet included in the package of each drug they plan to administer to be certain that the information contained in this work is accurate and that changes have not been made in the recommended dose or in the contraindications for administration. This recommendation is of particular importance in connection with new or infrequently used drugs.

This book was set in Goudy by Cenveo® Publisher Services.
The editors were Catherine A. Johnson and Cindy Yoo.
The production supervisor was Catherine Saggese.
Project management was provided by Anupriya Tyagi, Cenveo Publisher Services.
The cover designer was Thomas De Pierro.
RR Donnelley was printer and binder.

This book is printed on acid-free paper.

Library of Congress Cataloging-in-Publication Data

Toy, Eugene C., author.
 Case files. Neuroscience / Eugene C. Toy, Evan Yale Snyder, Josh Neman, Rahul Jandial. — Second edition.
  p. ; cm.
 Neuroscience
 Preceded by: Case files. Neuroscience / Eugene C.Toy ... [et al.]. 2009.
 Includes bibliographical references and index.
 ISBN 978-0-07-179488-6 (pbk. : alk. paper) — ISBN 0-07-179488-3
 I. Snyder, Evan Y., author. II. Neman, Josh, author. III. Jandial, Rahul, author. IV. Title.
 V. Title: Neuroscience.
 [DNLM: 1. Nervous System Diseases—Case Reports. 2. Nervous System Diseases—Problems and Exercises.
 3. Neurosciences—Case Reports. 4. Neurosciences—Problems and Exercises. WL 18.2]
 QP355.2
 612.8—dc23
                                                                              2014006911

McGraw-Hill Education books are available at special quantity discounts to use as premiums and sales promotions or for use in corporate training programs. To contact a representative, please visit the Contact Us pages at www.mhprofessional.com.

To my mentor, Dr. Patricia Butler who, as the Associate Dean of
Educational Programs at the University of Texas Medical School at
Houston, inspires us all to excellence in the dissemination of knowledge.
She has made a tremendous impact on countless medical students and faculty.

—ECT

My work on this book is dedicated to the memory of my father,
Harry Snyder, and in honor of my mother, Martha Snyder,
who inspired me to strive for excellence in helping others.
I also dedicate this to my sons, Ciaran and Madden, whom I hope, in turn,
to inspire in the same way. And I dedicate this to my wife, Angela,
who helps me to be the best person I can.

—EYS

To my greatest mentors—my parents Joseph and Farideh. Your guidance and
support have allowed me to always follow my passion and dreams.
For this, I am eternally grateful.

—JN

To my grandfather, Mani Ram Jandial, for creating a family foundation of
both loyalty and audacity, principles from which I continue to benefit and
hope to pass on to my own beloved sons, Ronak, Kai, and Zain.

—RJ

CONTENTS

CONTRIBUTORS

Athena R. Anderson, MS
Research Associate
Department of Neurosurgery
City of Hope
Duarte, California
Intracerebral Hemorrhage and Increased Intracranial Pressure

Cecilia Choy, MS
Doctoral Student, Class of 2015
Division of Neurosurgery
City of Hope
Duarte, California
Intracerebral Hemorrhage and Increased Intracranial Pressure

Tatsuhiro Fujii
Medical Student, Class of 2016
Keck School of Medicine
University of Southern California
Los Angeles, California
Basal Ganglia
Cell Types of the Nervous System
Cerebellum
CNS Development
Formation of Cerebral Cortex
The Myelin Sheath and Action Potential
Neural Synapses
The Neuron
Neurotransmitter Receptors

Michael W. Lew, MD
Professor and Chair
Department of Anesthesiology
City of Hope National Medical Center
Duarte, California
Intracerebral Hemorrhage and Increased Intracranial Pressure

Josh Neman, PhD
Assistant Professor, Research
Department of Neurological Surgery
Keck School Medicine
University of Southern California
Los Angeles, California
Formerly:
Research Fellow

Division of Neurosurgery
City of Hope Comprehensive Cancer Center
Beckman Research Institute
Duarte, California
Addiction
Alzheimer Disease
Approach to Learning Neurosciences
Audition
Axonal Injury
Brain Laterality
Breathing
Cell Fate Determination
Cell Types of Nervous System
CNS Development
Consciousness
Disconnection Syndromes
Electrical Properties of Nerves and Resting Membrane Potential
Executive Function
Eye Movements
Formation of the Cerebral Cortex
Hyperthermia
Language Disorders
Leep and Limbic System
Movement Control
Nerve Growth Factors
Neural Control
Neural Repair
Neural Stem Cells
Neuroendocrine Axis
Neurogenesis
Neurogenic Diabetes Insipidus
Neuromuscular Junction
Neuronal Migration
Neurotransmitter Release
Neurotransmitter Types
Neurulation
Nociception
Olfaction
Parasympathetic Nervous System
Peripheral Nervous System
Proprioception
Reticular-Activating System
Spatial Cognition
Spinothalamic Pathway
Sympathetic Nervous System
Synaptic Integration
Vision
Visual Perception

John L. Raytis, MD
Assistant Clinical Professor
Department of Anesthesiology
City of Hope National Medical Center
Duarte, California
Intracerebral Hemorrhage and Increased Intracranial Pressure

Allison L. Toy
Senior Nursing Student
Scott & White Nursing School
University of Mary Hardin-Baylor
Belton, Texas
Primary Manuscript Reviewer

FIRST EDITION CONTRIBUTORS

Lissa C. Baird, MD
Resident in Neurosurgery
Division of Neurosurgery
University of California, San Diego
Medical Center
San Diego, California
Audition
Basal Ganglia
Cerebellum
Eye Movements
Movement Control
Nociception
Olfaction
Proprioception
Spinothalamic Pathway
Vision

Melanie Hayden Gephart, MD, MAS
Resident Neurosurgeon
Department of Neurosurgery
Stanford University Hospital and Clinic
Stanford, California
Alzheimer Disease
Brain Laterality
Consciousness
Disconnection Syndromes
Executive Function
Language Disorders
Spatial Cognition
Visual Perception

Allen Ho, MD
Resident, Department of Neurosurgery
School of Medicine
Stanford University
Stanford, California
Axonal Injury
Nerve Growth Factors
Neural Repair
Neural Stem Cells

Andrew D. Nguyen, MD, PhD
Assistant Professor of Neurosurgery
University of California San Diego Health System
San Diego, California
Cell Fate Determination
CNS Development
Formation of the Cerebral Cortex
Neurogenesis
Neuronal Migration
Neurulation
Peripheral Nervous System

Evan Ou
Medical Student, Class of 2011
University of California, San Diego School of Medicine
San Diego, California
Consciousness
Disconnection
Executive Function
Syndromes

Min S. Park, MD
Division of Neurosurgery
Naval Medical Center San Diego
San Diego, California
Cell Types of Nervous System
Electrical Properties of Nerves and Resting Membrane Potential
The Myelin Sheath and Action Potentials
Neural Synapses
Neuromuscular Junction
The Neuron
Neurotransmitter Receptors
Neurotransmitter Release
Neurotransmitter Types
Synaptic Integration

Pooja Rani Patel, MD
Resident, Department of Obstetrics and Gynecology
Formerly Medical Student, Class of 2008
Baylor College of Medicine
Houston, Texas
Principal Student Reviewer and Manuscript Coordinator
Approach to Learning Neurosciences Spinothalamic Pathway

Brett Reichwage, MD
Resident Neurosurgeon
Division of Neurological Surgery
University of Florida
Gainesville, Florida
Addiction
Breathing, Neural Control
Hyperthermia
Neuroendocrine Axis
Neurogenic Diabetes Insipidus
Parasympathetic Nervous System
Reticular-Activating System
Sleep and Limbic System
Sympathetic Nervous System

Amol M. Shah
Medical Student, Class of 2011
University of California
San Diego School of Medicine
San Diego, California
Addiction
Breathing
Neural Control
Reticular-Activating System

Mary Kendall Thoman
Medical Student, Class of 2010
University of Texas Medical School at Houston
Houston, Texas
Axonal Injury
Nerve Growth Factors
Neural Repair
Neural Stem Cells

We appreciate all the kind remarks and suggestions from the many medical students over the past 5 years. Your positive reception has been an incredible encouragement, especially in light of the short life of the *Case Files*® series. In this second edition of *Case Files*®: *Neuroscience*, the basic format of the book has been retained. Improvements were made in updating many of the chapters including a new case correlation section, in which related cases are referenced to allow the student to review other neurobiological concepts or disease concepts. A new case of subarachnoid hemorrhage has been added. Expanded neuroscience pearls with clinical points highlighting common material covered in the USMLE step 1 have been included. The multiple-choice questions have been carefully reviewed and rewritten to ensure that they comply with the National Board and USMLE format. Through this second edition, we hope that the reader will continue to enjoy learning diagnosis and management through the simulated clinical cases. It certainly is a privilege to be teachers for so many enthusiastic and receptive students, and it is with humility that we present this second edition.

The Authors

ACKNOWLEDGMENTS

The inspiration for this basic science series occurred at an educational retreat led by Dr. Maximillian Buja, who at the time was the Dean of the medical school. Dr. Buja served as the Dean of the University of Texas Health Science Center at Houston Medical School from 1995 to 2003 before being appointed Executive Vice President for Academic Affairs. It has been such a joy to work together with Drs. Jandial and Snyder and more recently Josh Neman, who are brilliant neuroscientists and teachers.

I would like to thank McGraw-Hill for believing in the concept of teaching by clinical cases. I owe a great debt to Catherine Johnson, who has been a fantastically encouraging and enthusiastic editor. It has been amazing to work together with my daughter Allison, who is a senior nursing student at the Scott and White School of Nursing; she is an astute manuscript reviewer and already in her early career she has a good clinical acumen and a clear writing style. At St. Joseph Medical Center, I would like to recognize our outstanding administrators: Pat Mathews, and Drs. Thomas Taylor and John Bertini. At Methodist Hospital, I appreciate Drs. Mark Boom, Judy Paukert, and Alan Kaplan. and I appreciate Linda Bergstrom's advice and assistance. Without the help from my colleagues, Drs. Konrad Harms, Priti Schachel, and Gizelle Brooks-Carter, this manuscript could not have been written. Most importantly, I am humbled by the love, affection, and encouragement from my lovely wife Terri, and our children: Andy and his wife Anna, Michael, Allison, and Christina.

Eugene C. Toy, MD

Mastering the extensive and diverse areas of knowledge within a field as broad as neuroscience is a formidable task. It is even more difficult to draw on that knowledge, relate it to a clinical setting, and apply it to the context of the individual patient. To gain these skills, the student learns best with good models, appropriate guidance by experienced teachers, and inspiration toward self-directed, diligent reading. Clearly, there is no replacement for education at the bench. Even with accurate knowledge of the basic science, the application of that knowledge is not always easy. Thus, this collection of patient cases is designed to simulate the clinical approach and stress the clinical relevance to the neurosciences. However, it should also be remembered that, although we often talk about basic research as going "from bench to bedside", it is actually the inquisitiveness and insight and hypotheses of the clinician that drives basic researchers to ponder certain questions at the bench. Hence the pathway of "bedside-to-bench" is equally as powerful. It is that pathway that we hope this book might also stimulate among its readers.

Most importantly, the explanations for the cases emphasize the mechanisms and structure–function principles rather than merely rote questions and answers. This book is organized for versatility to allow the student "in a rush" to go quickly through the scenarios and check the corresponding answers or to consider the thought-provoking explanations. The answers are arranged from simple to complex: the bare answers, a clinical correlation of the case, an approach to the pertinent topic including objectives and definitions, a comprehension test at the end, neuroscience pearls for emphasis, and a list of references for further reading. A listing of cases is included in Section III to aid the student who desires to test his/her knowledge of a certain area or to review a topic including basic definitions. We intentionally used open-ended questions in the case scenarios to encourage the student to think through relations and mechanisms.

Approach to Learning Neurosciences

The Big Picture

Neuroscience is unique in that it incorporates an understanding of science on multiple levels, from a molecular understanding of events such as with receptors, at the synaptic level, to a global understanding of the sensory/motor tracts and their spatial interactions. It is by understanding all these concepts that a student can better grasp the clinical presentations of neurological disorders and the theory behind treatment options. The student should thus approach each neuroscience topic from both aspects if applicable. For example, when studying **multiple sclerosis** (MS), the student should understand that on the molecular level this disease involves destruction of **oligodendrocytes**, which are responsible for creating and maintaining **myelin sheaths** around axons of the central nervous system. The student should then review the **nodes of Ranvier** and concepts pertaining to **saltatory conduction**. Next, the student should take a step back and look at the condition from a neuroanatomic perspective. For example, if the patient with MS presents with left impaired adduction on right gaze, but has normal convergence, and normal left abduction on left gaze, not only should the student be able to diagnose the patient with a left **intranuclear ophthalmoplegia (INO)**, but should also understand that the lesion is in the left **medial longitudinal fasciculus (MLF)** and then proceed to review the anatomy of the MLF tract (ie, that the left MLF yokes the left cranial nucleus VI to the right cranial nucleus III). The student should strive for an understanding such that symptoms should make sense rather than rely on blind memorization!

Know the Tracts

There is no way to avoid it; the student has to memorize the various neural tracts (ie, spinothalamic tract, corticospinal tract, etc.) **forward and backward.** It is easier to first take each tract separately and memorize the exact pathway the neurons in that tract travel throughout the body, noting any decussations or synapses so that the student can determine whether lesions would have ipsilateral (same side as the lesion) or contralateral (opposite side of the lesion) symptoms, or which nuclei are involved. The **second step** would be to **synthesize this information** by taking various cross-sections of the nervous system (from the spinal cord to the brain) and being able to identify where each tract is, reviewing at the same time where that tract is coming from upstream and going to downstream. It is important to note that the terms *upstream* and *downstream* may be different spatially, depending on the tract referred to. For example, *upstream* means a cross-section *above* the one being studied when referring to the corticospinal tract since the tract travels caudally. However, *upstream* means a cross-section *below* the one being studied when referring to the spinothalamic tract since this tract travels cranially (from the peripheral inputs to the cortex). The third step involves knowing the cross-sections so thoroughly that the student can draw any cross-section, incorporating all the tracts and nuclei involved in that cross-section. Throughout the studying process, the students should be asking themselves, *if there is a lesion in this structure or at this level, what symptoms will be manifested?*

Understand the Terminology

Although it takes less effort to memorize medical terminology without understanding the origin of the term, it is much more effective in the long term to understand the reason behind the name of a structure or pathological condition. Going back to our example of MS, the *scleroses* refers to the plaques or lesions in the white matter, while the term *multiple* refers to variety in location and time. In other words, in order to diagnose MS, a patient must have at least two anatomically separate lesions occurring at two distinct time periods. Similarly, the student should not just simply memorize structures like the previously mentioned *spinothalamic tract* and *corticospinal tracts*. Rather, the student should understand that the *spinothalamic* tract receives input at the *spinal* cord level that travels to and synapses in *thalamic* nuclei. Likewise, the *corticospinal* tract sends information originating from cells in motor *cortex* to the *spinal cord*, which eventually coordinates muscular movement via the lower motor neurons.

NEUROSCIENCE PEARLS

▶ The student should seek to understand the neuroscience on a molecular level, synaptic level, and a higher level such as sensory/motor tract level.

▶ An understanding of the neural tracts should allow synthesis and drawing the cross-sections from brain to spinal cord.

▶ The student should strive to understand medical terminology rather than blindly memorizing.

Clinical Cases

A 53-year-old man presents to the emergency department following a new-onset generalized seizure. After recovery from the seizure period, he is alert and oriented to person, place, and time, although he has no specific memory of the seizure. There are no neurological deficits noted on the physical examination, but magnetic resonance imaging (MRI) and positron emission tomography (PET) scans indicate the presence of a sizable lesion. The patient undergoes surgery for resection of the malignancy and is diagnosed with a malignant primary brain tumor, which is identified as a glioblastoma multiforme (GBM). The neurosurgeon's plan is to place the patient on long-term seizure prophylaxis and steroids in combination with stereotactic radiotherapy.

▶ What other types of tumors are most closely related to this patient's malignancy?
▶ What are the imaging findings most characteristic of these types of tumors?
▶ What are the hallmark pathological findings of these types of tumors?

ANSWERS TO CASE 1:

Cell Types of the Nervous System

Summary: A 53-year-old man with new-onset seizure and a left parietal lobe mass undergoes surgery to resect a primary brain tumor, diagnosed as a GBM. He also receives adjuvant radiation treatment.

- **Related tumors:** Low-grade astrocytomas and anaplastic astrocytomas.

- **Imaging characteristics:** A ring-enhancing lesion in the left parietal lobe with surrounding edema and mass effect.

- **Neuropathological findings:** GBMs exhibit marked hypercellularity, nuclear pleomorphism, microvascular proliferation, and pseudopalisading of tumor cells around areas of necrosis.

CLINICAL CORRELATION

The clinical presentation of this patient is very typical for GBM, the most severe and malignant form of primary intracranial tumor.

Astrocytomas are a group of primary brain tumors derived from astrocytes, the star-shaped glial cells forming the latticework structure that supports neuron function (see Discussion). The World Health Organization (WHO) classifies astrocytomas on a continuum from grades I to IV, based on pathological findings. Grades I and II consist of low-grade astrocytomas, grade III consists of anaplastic astrocytomas, and grade IV includes GBMs, which represent 15%-20% of all primary brain cancers. Tumor types on the WHO continuum exhibit progressively more hypercellularity and nuclear pleomorphism (a nonhomogeneous group of cells with nuclei of various sizes and shapes). Proliferation of tumor cells around vascular structures along white matter tracts occurs in both anaplastic astrocytomas and GBMs. Necrosis with pseudopalisading of tumor cells is only found in GBMs.

A wide array of symptoms ranging from drowsiness and fatigue to motor and communication difficulties has been reported. Treatment consists of surgery, radiotherapy, and chemotherapy, either individually or in combination. Despite continued advances in therapeutic approach, the prognosis of GBM remains unfavorable, and median survival for patients has remained consistently low (Table 1-1). Due to the highly aggressive nature of GBMs and the frequency of tumor recurrence, regular

Table 1-1 • SURVIVAL ESTIMATES FOR ASTROCYTOMAS	
Astrocytoma Type (WHO Grade)	Median Survival
Low-grade astrocytomas (I)	8-10 years
Low-grade astrocytomas (II)	7-8 years
Anaplastic astrocytomas (III)	~2 years
Glioblastoma multiforme (IV)	<1 year

tumor surveillance through imaging and clinical evaluation remains vital for disease management.

Oligodendroglioma is another type of primary brain tumor that arises from the oligodendrocyte. This is the glial cell that provides the myelin sheaths around the axons of neurons in the central nervous system. Oligodendrogliomas generally affect adults in their fifth decade of life, developing in the frontal lobes and presenting with seizures. They have a characteristic appearance on radiographs and on microscopic examination, due to the calcifications pattern in the tumor. Similar to the management of astrocytomas, treatment is multimodal and consists of an appropriate surgical procedure followed by radiation and chemotherapy. The presence or absence of specific chromosomal deletions on the long arm of chromosome 1 or the short arm of chromosome 19 has been shown to affect the sensitivity of these tumors to chemotherapy. Oligodendrogliomas tend to have a better prognosis than astrocytomas, with a 10-year survival of 10%-30% reported in the literature.

Other tumors arising from cells of the nervous system include mixed tumors with oligodendral and astrocytic components, **schwannomas** arising from **Schwann cells** of the peripheral nervous system, and **gangliogliomas** arising from both glial cells and neurons. This list is not exhaustive but is representative of the variety of pathologies found in primary brain tumors.

APPROACH TO:
Cell Types of the Nervous System

OBJECTIVES

1. Differentiate between the central and peripheral nervous systems.

2. Know the names of each cell type in the nervous system.

3. Describe the role of each cell type within the nervous system.

4. Identify the components that make up the blood-brain barrier (BBB).

DEFINITIONS

MENINGES: A series of three membranes which encapsulates the central nervous system (CNS).

SUBARACHNOID SPACE: The space between the arachnoid and pia maters that contains delicate connective tissue, blood vessels, and cerebrospinal fluid (CSF).

GLIAL CELLS: Cells that support neurons and form the structural framework for the nervous system. These include astrocytes, oligodendrocytes, and microglia.

MYELIN: A phospholipid bilayer which insulates the axon and allows for faster propagation of an action potential, commonly referred to as saltatory conduction.

DISCUSSION

The human nervous system can be divided into two main components: the **central nervous system** (CNS) and the **peripheral nervous system** (PNS) (Figure 1-1). The CNS is contained entirely within the **meninges**, while the PNS is distributed outside of the meninges. The meningeal layers consist of the dura mater, pia mater, and the arachnoid mater. Rarely, certain types of tumors, referred to as **meningiomas,** can develop in the meninges.

There are two main types of cell within the nervous system: nerve cells (**neurons**) and glial cells (**glia**). The majority are glia, which form the structural framework of the nervous system and provide functional support for the neurons.

- **Oligodendrocytes,** found only in the central nervous system, and **Schwann cells,** found only in the peripheral nervous system, are glial cells that perform similar functions. They produce myelin, which forms the fatty sheath that insulates the axons of neurons for faster conduction of electrical signals. One oligodendrocyte can provide myelin sheaths for many axons, but a Schwann cell provides myelin for just one axon.

- **Microglia,** the phagocytes of the nervous system, are mobilized by insults to the CNS and remove debris following neuronal injury or death. They arise from

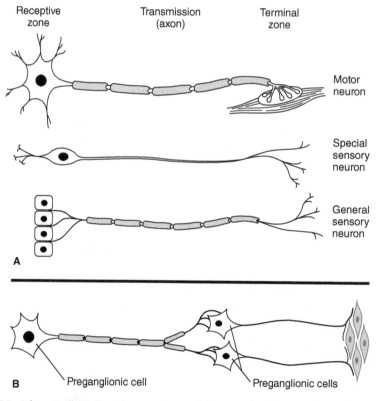

Figure 1-1. Schematic illustration of nerve cell types. **A:** Motor and sensory neurons. **B:** Preganglionic cells.

macrophages outside of the nervous system and are physiologically unrelated to other glial cells.

- **Astrocytes,** the most numerous type of glial cell, are star-shaped cells that fill the interneuronal space in the CNS. They provide structural support for the neurons in the CNS, insulate and separate neurons from one another, and help to regulate the potassium ion concentration in the extracellular space around neurons.

- **Tight junctions** between the foot processes of astrocytes and the endothelial membranes of blood vessels help to maintain the **blood-brain barrier,** an almost impermeable lining of the brain's capillaries and venules that prevents certain toxic substances in the blood from entering the brain. This protective mechanism can present an obstacle for efficacious delivery of therapeutic agents to the CNS.

- **Neurons** are the signaling unit within the nervous system and are the only cells in the nervous system involved with the conduction of electrical impulses. While there are a number of morphologically different types of neurons, they all share the same cellular phenotype consisting of a cell body or soma, multiple dendrites, axon(s), and a presynaptic terminal.

COMPREHENSION QUESTIONS

1.1 A 37-year-old woman comes to see you in your family medicine clinic because the medication she is taking for her moderately severe seasonal allergies makes her extremely drowsy. You advise her to switch from the current first-generation antihistamine she is taking to a newer second-generation antihistamine because you know the latter has less incidence of drowsiness as it cannot penetrate the BBB. Which cell type in the CNS is most responsible for forming the BBB?

A. Astrocyte

B. Microglia

C. Oligodendrocyte

D. Schwann cell

1.2 A 45-year-old man presents to his primary care physician with complaints of persistent headache for the past several months. After an extensive workup, he is found to have an intraparenchymal brain tumor. On biopsy, the pathologist reports that the tumor has myelin elements. From what cell type in the CNS did this tumor likely arise?

A. Astrocyte

B. Microglia

C. Schwann cell

D. Oligodendrocyte

1.3 A 21-year-old college student is brought into the emergency room complaining of the worst headache of his life. On examination he is noted to have a stiff neck and photophobia. As the physician on duty, you are concerned that this young man may have bleeding from a ruptured aneurysm. You perform a CT scan followed by a lumbar puncture which returns bloody cerebrospinal fluid (CSF), confirming your suspicion. Neurosurgery is called and the patient is rushed to surgery. Where was this bloody CSF obtained from?

A. Epidural space

B. Intraparenchymally

C. Subarachnoid space

D. Subdural space

ANSWERS

1.1 **A. Astrocytes** extend foot processes that, in conjunction with the endothelial cells of the cerebral capillaries, are a critical component of the BBB. The tight junctions of the endothelial cells are most important in maintaining the BBB. The BBB is a very important structure that, as the name implies, restricts access of molecules from the blood to the brain. This restriction is generally based on size and lipid solubility such that small, lipophilic molecules can cross, whereas larger, lipophobic molecules cannot. The BBB also helps to prevent bacteria and viruses from entering the brain. Oligodendrocytes are responsible for myelinating CNS axons, Schwann cells for myelinating PNS axons, and microglia are the phagocytes of the CNS.

1.2 **D.** Tumors arising from **oligodendrocytes** will contain myelin elements and are located in the CNS. In the analysis of tumors, it is important to remember that dysfunctional tumor cells arise from normal cells and thus express the same markers as the cells from which they arise. In this case, the tumor contains myelin elements and thus most likely arises from a cell that produces myelin: either an oligodendrocyte or a Schwann cell. Because the question refers specifically to the CNS, the correct answer is oligodendrocyte. The other cell types listed do not produce myelin.

1.3 **C.** Because CSF flows in the **subarachnoid space** between the arachnoid mater and the pia mater, bloody CSF indicates blood in the subarachnoid space. Bleeding into the CNS is an emergency situation with a grave prognosis despite appropriate and timely management. Epidural bleeds typically present acutely following trauma and are commonly associated with a laceration of the middle meningeal artery. Subdural bleeds often present subacutely several weeks following minor head trauma, especially in the elderly. Intraparenchymal bleeds present in a variety of ways, depending on the amount and location of the bleeding in the brain parenchyma.

NEUROSCIENCE PEARLS

▶ Glioblastomas are a form of astrocytoma, and are the most malignant forms of primary brain cancers.

▶ The most important structure creating the blood-brain barrier is the specialized tight junction of the endothelial cells.

▶ The central nervous system is contained within the meninges (consisting of the dura mater, pia mater, and arachnoid mater), while the peripheral nervous system is distributed outside the meninges.

▶ Neurons conduct electrical impulses throughout the nervous system.

▶ Glial cells support neurons and form the structural framework for the nervous system.

▶ Oligodendrocytes produce myelin in the CNS, while Schwann cells produce myelin in the PNS.

▶ Microglia arise from macrophages and phagocytose debris following neuronal injury or death.

▶ Astrocytes provide structural support for the neurons in the CNS, insulate and separate neurons from one another, and help to regulate the potassium ion concentration in the extracellular space around neurons.

▶ Tight junctions between the foot processes of astrocytes and the endothelial membrane of the vessels help to maintain the BBB.

REFERENCES

Bear MF, Connors B, Paradiso M, eds. *Neuroscience: Exploring the Brain*. 3rd ed. Baltimore, MD: Lippincott Williams & Wilkins; 2006.

Kandel E, Schwartz J, Jessell T, eds. *Principles of Neural Science*. 5th ed. New York, NY: McGraw-Hill; 2012.

Squire LR, Berg D, Bloom FE, du Lac, S, eds. *Fundamental Neuroscience*. 4th ed. San Diego, CA: Academic Press; 2012.

An otherwise healthy 54-year-old woman comes to the physician's office complaining of weakness, initially involving her hands and progressing to her legs over several months. She states that her hands feel stiff and clumsy and that she has difficulty buttoning her clothes or using her keys. She has fallen several times in the past couple of weeks when walking up and down stairs at home. Lately, both she and her husband have noticed that she will "stumble over words" when she talks. She denies having any funny sensations or numbness and has no family history of neurological disease.

On physical examination, she is alert, appropriate, and oriented to person, place, and time. Her cranial nerves are intact except for some subtle fasciculations (involuntary muscle twitches) of her tongue. Strength testing of her right arm and leg shows mild weakness with pronator drift. Muscle stretch reflexes are brisk in upper extremities but reduced in her right leg. She also has difficulty with rapid alternating movements of her right arm. There are no abnormal findings on sensory examination. The physician has a presumptive diagnosis of **amyotrophic lateral sclerosis** (ALS) and explains that this is an upper motor neuron disease.

▶ What part of the nervous system is likely to be involved?
▶ Explain the mechanism of the fasciculations in this disease.
▶ What are the pathological findings for this disease?

ANSWERS TO CASE 2:
The Neuron

Summary: A 54-year-old woman with progressive weakness of her arms and legs with subtle speech difficulties is diagnosed with ALS.

- **Part of nervous system affected:** The central nervous system (CNS), specifically the motor neurons.

- **Fasciculations:** ALS is a progressive, neurodegenerative disease affecting motor neurons, which are cells within the CNS that control voluntary muscle movement. The loss of signal communication to the muscles caused by the degeneration of these cells leads to muscle atrophy, which manifests itself in fasciculations.

- **Pathology:** Loss of motor neurons in the brainstem nuclei and the anterior horns of the spinal cord, along with degeneration of the corticospinal tracts in the spinal cord.

CLINICAL CORRELATION

ALS is a progressive degenerative disorder of the **motor neurons** of the spinal cord, brainstem nuclei, and corticospinal tracts. In its classical presentation, patients in their sixth decade initially complain of difficulty with fine finger movements and stiffness and weakness of the fingers and hands. The patients can also experience cramps or **fasciculations** of the muscles in the upper extremities. As time progresses, atrophic weakness involves both upper extremities, and spasticity and hyperreflexia develops in the lower extremities. The muscles of the neck, tongue, pharynx, and larynx may also become involved. Depending on the extent of involvement of the various motor neurons, **a mixed upper and lower motor neuron disease becomes evident.** There are no cognitive, sensory, or autonomic disturbances in this disease. An electromyelogram (EMG) obtained for confirmatory purposes reveals widespread fibrillations and fasciculations, evidence of active denervation and reinnervation of the muscles. Analysis of cerebrospinal fluid (CSF) may reveal normal or slightly elevated protein levels. Unfortunately, the etiology of ALS is unknown. As the disease reaches its ultimate conclusion, the patient is left with flaccid paralysis of all voluntary muscles with the exception of the extraocular and sphincter muscles. Treatment is directed toward minimizing the disability. For instance, medications such as baclofen and diazepam can be used to treat spasticity. The median survival from onset of symptoms is 3-4 years, although individuals can live a normal life span if proper care is available.

APPROACH TO:
The Neuron

OBJECTIVES

1. Describe the role and structure of a neuron.
2. Know the classification of neurons based upon morphology.
3. Know the classification of neurons based upon function.

DEFINITIONS

CYTOSKELETON: The internal network of fibers that form the framework inside a cell. In the neuron, the cytoskeleton consists of three types of filaments—microtubules, neurofilaments, and microfilaments—and determines the shape of the neuron, provides cell rigidity, and creates a mechanism for intracellular transport.

SOMA: The cell body of the neuron, which contains the organelles and acts as the metabolic center of the cell.

DENDRITES: Armlike extensions protruding from the cell membrane that increase the surface area of the neuron and receive signals from other neurons.

AXON: The main conducting unit for carrying electrical signals, known as action potentials, to other cells. Action potentials are formed at the axon hillock, the origin of the axon at the cell body, and are propagated down the axon in a rapid all-or-none transmission to the next cell.

PRESYNAPTIC TERMINAL: The end of an axon that forms a synapse or neurochemical cleft with the next cell (neural, muscular, etc). Signals are transmitted across the synapse via the release of neurotransmitter into the synaptic cleft that diffuse into the postsynaptic cell, chemically triggering a physiological response.

DISCUSSION

Morphological Features

Neurons, the basic functional components of the brain, are cells that act as the main signaling units of the nervous system. The cytoskeleton of the neuron is composed of three types of filaments, each constructed from different classes of proteins. Together they form the cytoskeletal structure of the neurons that determine the shape of the neuron, provide stiffness to the cell, and create a mechanism for intracellular transport.

- **Microtubules** (about 25 nm in diameter), the largest of the filaments, are constructed from 13 different types of protofilaments in a circular array and serve to maintain the neuron's processes. The protofilaments are formed by two subunits, alpha and beta tubulin, in an alternating pattern. Microtubules are also involved in axoplasmic transport, or movement of various cell components along an axon such as mitochondria, lipids, synaptic vesicles, proteins, and so on.

- **Neurofilaments** or intermediate filaments are approximately 10 nm in diameter and are more abundant than either microtubules or microfilaments. Neurofilaments are made from strands of proteins that pair into a helical structure. The pairs of proteins then twist together to form larger protofilaments. Four protofilaments combine to form the final neurofilament. Neurofilaments can form cross-links with microtubules to provide stiffness and shape to the cell. Within the context of Alzheimer disease, neurofilaments are modified to form pathological protein aggregates called neurofibrillary tangles.

- **Microfilaments** are, as the name implies, the thinnest of the fiber types at approximately 3-5 nm in diameter and are formed by the polymerization of the protein actin to form a double helical structure. Microfilaments are located at the periphery of the cell adjacent to the cytoplasm of the cell. Together with multiple types of actin-binding proteins, they assist in the dynamic functions of the cell.

  There are four distinct morphological features in a typical neuron.

- The **cell body** or **soma** acts as the metabolic center of the neuron and contains the ultrastructural organelles, such as the nucleus (DNA storage and ribosome production), endoplasmic reticulum (protein synthesis), mitochondria (energy through ATP production), lysosomes (waste disposal), and vesicles (packaging and transport) necessary to carry out basic metabolic functions.

- Multiple short **dendrites** branch out from the cell body in a treelike fashion. Dendrites function mainly to increase the surface area of the neurons and to receive signals from other neurons.

- A single **axon** arises from the cell body and serves as the main conducting unit for carrying electrical signals to other neurons and muscles. These electrical signals, called **action potentials**, are rapid all-or-none transmissions down the length of the axon. Action potentials are formed at the **axon hillock**, the origin of the axon from the cell body. To facilitate transmission of the electrical signal, axons may be wrapped in **myelin sheaths** produced by oligodendrocytes or Schwann cells.

- The **presynaptic terminal** at the end of the axon forms a synapse with the postsynaptic cell, that is, other neurons or muscle cells. When the action potential reaches the presynaptic terminal, a cascade of events occurs culminating in the release of neurotransmitters into the synaptic cleft. The neurotransmitters diffuse to the postsynaptic cell, where they trigger certain physiological responses.

  Santiago Ramon y Cajal, using silver staining techniques developed by Camillo Golgi, was first to notice that the shape of the neuron could be used for classification. Based upon the number of axons and dendrites which originate from the cell body, neurons can be classified into several broad groups.

- **Unipolar neurons** are the simplest nerve cells and have a single process emerging from the cell body. This process branches with one branch functioning as the axon and the other branches functioning as dendrites. They are found throughout the central nervous system of invertebrates and in the autonomic nervous system of vertebrates.

- **Bipolar neurons** have two processes attached to an oval-shaped cell body. One process functions as a dendrite and carries information from the periphery of the organism. The other process functions as the axon carrying information toward the central nervous system. Bipolar neurons are found in the retina of the eye and the olfactory epithelium of the nose.

- **Pseudo-unipolar neurons** are a variant of the bipolar neuron and have a single process which emerges from the cell body. This single process splits into two; one functions as a dendrite carrying information from the periphery to the cell body, and the other functions as the axon transmitting the information to the central nervous system. They are found in the dorsal root ganglia of the spinal cord and relay touch, pressure, and pain sensations from the extremities to the spinal cord.

- **Multipolar neurons** are the main neuron type in the mammalian central nervous system. These neurons have a single axon and multiple dendrites emerging from the cell body. The processes of multipolar neurons vary in number, diameter, and length, depending on the number of synaptic contacts they make with other neurons.

Functional Classification

Neurons can also be classified according to the function that they serve within the nervous system.

- **Sensory** or **afferent neurons** carry information from the body to the central nervous system. Sensory information can include pain, pressure, touch, and joint position sense, among others.

- **Motor neurons** carry commands from the brain and spinal cord to muscles and glands in the periphery.

- **Interneurons** comprise all neurons which do not fit into the previous two groups. Interneurons carry information within the nervous system and connect with neighboring or distant neurons.

All behaviors are mediated by a series of neurons that form signaling networks. For instance, mechanoreceptors carry information concerning joint position through sensory neurons to the spinal cord. This information is transmitted through various interneurons in the spinal cord and brain to allow the central nervous system to process and determine where an individual's limb is in relation to the body and space. Once a decision is made to move the extremity, a variety of interneurons transmits the information to motor neurons that relay the final command to the muscles. In this role, interneurons, through a variety of neurotransmitters, can act to either excite or inhibit other neurons.

Spinal Cord Neurons

Neurons within the spinal cord are organized into somatosensory and motor neurons. Somatosensory neurons form an ascending tract of neurons running up the spinal cord transmitting information about touch, proprioception (balance and self-awareness), vibration, pain, and temperature. Motor neurons form a descending tract of neurons running down the spinal cord to control muscle function throughout the body. The cord itself is divided into several sections axially: the dorsal, lateral, and ventral horns. Somatosensory neurons from the body run into the dorsal horn, where they connect with interneurons in the lateral horn that transmit information via the ascending tracts to the brain, or in the case of a spinal reflex, directly to motor neurons in the ventral horn. Motor neurons running out of the ventral horn are similarly controlled by direct reflex signaling from a sensory neuron or by efferent (motor) neurons traveling down the descending tract from the brain.

CASE CORRELATES

- See Cases 1-10 (cellular and molecular neuroscience cases).

COMPREHENSION QUESTIONS

2.1 A 35-year-old man is brought to the emergency room (ER) by the paramedics after having his left hand crushed in an industrial accident. He is in excruciating pain. What morphologic type of neuron is responsible for transmitting the pain signal from his mangled stump to his CNS?

A. Bipolar

B. Multipolar

C. Pseudo-unipolar

D. Unipolar

2.2 When an action potential reaches the presynaptic terminal, a set of complex cellular functions is set in motion, involving a wide variety of different proteins with different functions. How do the majority of the elements involved in this signaling cascade come to be found in the presynaptic terminal?

A. Diffusion

B. They are synthesized in the presynaptic terminal

C. They are transported there by kinesin interaction with microtubules

D. They are transported there by dynein interaction with microtubules

2.3 After a neurotransmitter is released from the presynaptic terminal, it interacts with a postsynaptic neuron to cause its effect. Which morphologic feature of the neuron is responsible for receiving synaptic transmissions and transmitting these postsynaptic signals to the rest of the cell?

A. Axon

B. Dendrite

C. Soma

D. Terminal bouton

ANSWERS

2.1 **C.** The neurons of the dorsal root ganglia are **pseudo-unipolar.** Pain is transmitted from sensory end organs to sensory neurons with cell bodies located in the dorsal root ganglia, which then transmit the signal to second-order sensory neurons in the spinal cord. In function they are quite similar to bipolar neurons, in that they have one large dendrite and one large axon, but these processes fuse prior to entering the cell body, so there is in fact only one process leaving the soma.

2.2 **C.** The interaction between the cytoskeletal microtubules and carrier molecules is responsible for fast axonal transport. **Kinesin** is responsible for anterograde (away from the cell body) transport and **dynein** is responsible for retrograde (toward the cell body) transport. Remember that almost all macromolecules and organelles found in the presynaptic terminal were originally synthesized in the cell body but that the axon is far too long to rely upon diffusion for the molecules to travel down the length of the axon.

2.3 **B. Dendrites** are responsible for transmitting signals toward the cell body. Axons are the set of neural processes which transmit signals away from the cell body. The soma is responsible for housing virtually all of the biomolecular machinery of the cell, and the terminal bouton houses the presynaptic neurotransmitter-releasing system.

NEUROSCIENCE PEARLS

▶ ALS is the most common motor neuron disease and affects both upper and lower motor neurons.

▶ ALS is associated with loss of neurons in the anterior horn (lower motor neuron), and degeneration of the lateral corticospinal tracts (upper motor neuron).

▶ The cytoskeleton of the neuron is composed of three types of filaments: microtubules, neurofilaments, and microfilaments.

▶ Microtubules are involved in the movement of various cell components along an axon.

▶ There are four distinct morphological features in a typical neuron: soma, dendrite, axon, and presynaptic terminal.

▶ Neurons can be classified as unipolar cells, bipolar cells, pseudounipolar neuron, and multipolar neurons according to the number of axons and dendrites, which originate from the cell body.

▶ Neurons can be classified as sensory/afferent neurons, motor/efferent neurons, or interneurons according to function.

▶ Interneurons connect with neighboring or distant neurons and can act either to excite or inhibit other neurons.

REFERENCES

Bear MF, Connors B, Paradiso M, eds. *Neuroscience: Exploring the Brain*. 3rd ed. Baltimore, MD: Lippincott Williams & Wilkins; 2006.

Kandel E, Schwartz J, Jessell T, eds. *Principles of Neural Science*. 5th ed. New York, NY: McGraw-Hill; 2012.

Purves D, Augustine GJ, Fitzpatrick D, Hall WC, eds. *Neuroscience*. 5th ed. Sunderland, MA: Sinauer Associates Inc; 2011.

A 39-year-old man on business in Japan was brought to the emergency room (ER) by paramedics with initial complaints of numbness and tingling around the face and mouth. His symptoms progressed to include lightheadedness and nausea/vomiting, which prompted the visit. He had just completed a successful business venture and had celebrated the occasion with dinner at a fine restaurant by feasting on some puffer fish, a Japanese delicacy. Upon his arrival, he rapidly became unable to move and had severe respiratory distress. Telemetry recordings demonstrated an irregular heart beat. He was quickly intubated and placed on mechanical ventilation. Unfortunately, he expired shortly after his presentation. Autopsy confirmed a diagnosis of tetrodotoxin (TTX) poisoning.

▶ What is the biochemical mechanism of this disease?
▶ How does TTX inhibit neural activity?
▶ What treatment options are available?

ANSWERS TO CASE 3:

Electrical Properties of Nerves and
Resting Membrane Potential

Summary: A 39-year-old businessman presents to the ER with complaints of numbness and tingling in his mouth which progresses to paralysis, respiratory failure, and death.

- **Mechanism:** TTX binds to the fast voltage-gated sodium channel at a site, which is located in the extracellular pore opening of the ion channel, known as *site 1*. This binding temporarily disables the function of the ion channel, resulting in death by paralysis of muscles.

- **Neural activity:** TTX is a potent neurotoxin with no known antidote that blocks action potentials in nerves by binding to the pores of the fast voltage-gated sodium channels in nerve cell membranes. Specifically, the action potential is inhibited during the rapid depolarization phase.

- **Treatment:** There is no known antidote for TTX poisoning. Emptying the patient's stomach, ingesting activated charcoal to bind the toxin, and supportive care with mechanical ventilation are the limits of current medical care.

CLINICAL CORRELATION

TTX is a potent neurotoxin commonly found in species of fish in the order Tetraodontiformes. TTX binds to the voltage-gated sodium channels in nerves and prevents the uptake of sodium ions by the cell. While it does not bind irreversibly to the channel, TTX has a very strong affinity for the channel and is not easily removed. Symptoms may begin with numbness and paresthesia of the face, mouth, and extremities and progress to lightheadedness or dizziness, nausea/vomiting, paralysis, and respiratory distress. For up to 50% of individuals with tetrodotoxicity, death will generally occur within 24 hours of ingestion. Because of the potential dangers inherent in the ingestion of puffer fish or fugu, only specially trained and licensed chefs are allowed to purchase and prepare this Japanese delicacy.

APPROACH TO:

Electrical Properties of Nerves and
Resting Membrane Potential

OBJECTIVES

1. Describe the signaling mechanisms used in the nervous system.

2. Describe the necessary elements to create an electrochemical gradient.

3. Describe how the membrane and channels work to create the environment necessary for signaling.

DEFINITIONS

ELECTROCHEMICAL GRADIENT: The gradient across a cell membrane created by the differential concentrations of charged ions on either side of the membrane (for example, the difference in concentrations of potassium and sodium ions inside and outside a neuron that when regulated by their respective ion channels allow for signaling down an axon to occur).

ION CHANNELS: Transmembrane proteins in the cell membrane which open and close to allow for the passage of ions.

GATING: The process by which channels undergo conformational changes to allow for the passage of ions.

RESTING MEMBRANE POTENTIAL: The potential created in a neuron at rest by resting channels and sodium-potassium pumps.

DISCUSSION

The nervous system depends on two types of signaling mechanisms, **electrical** and **chemical**, to propagate information throughout the nervous system. Rapid changes in the electrical potential across the neuronal cell membrane generate electrical signals that are transmitted down the length of the neuron. This system requires (1) an intact membrane to separate ions and maintain an **electrochemical gradient** and (2) **ion channels** to allow for the selective passage of ions of specific charges to generate the electrical signal.

The cell membrane of the neuron is formed by a lipid bilayer and is generally impermeable to charged particles. The double layer of phospholipids is hydrophobic. Charged ions are hydrophilic and as a result attract water molecules. This allows the neuronal cell membrane to separate charges across its surface to maintain the electrochemical gradient. However, to create and use the energy stored in the electrochemical gradient, structures must exist to allow for the passage of ions across this membrane. Ion channels, formed by transmembrane spanning proteins, serve that specific function within the neuron. The basic structure consists of transmembrane proteins with carbohydrate groups attached to their surface and a central pore-forming region to allow for the passage of ions. This pore-forming region spans the entirety of the membrane and is generally made up of two or more subunits.

Ion channels must also be selective for specific charged particles. One method by which channels select for specific ions is by size. Although the diameter of a potassium ion (K^+) is larger than the diameter of a sodium ion (Na^+), the Na^+ ions demonstrate a stronger electrostatic attraction for water molecules. Thus, in a solution the Na^+ ion has a larger shell of water than K^+ ions. Channels can therefore select for K^+ ions based upon the size differential in a solution. Other types of channels are selective for specific ions based upon the ion's electrical affinity to charged portions of the channel. The attraction between an ion and the channel must be sufficiently strong enough to overcome the hydrostatic attraction of the ion. Once the shell of water surrounding the ion is shed, the ion can diffuse through the channel.

The flow of ions through a channel is passive and governed by the electrochemical gradient. Some ion channels are highly selective for a specific anion or cation,

Table 3-1 • APPROXIMATE ION CONCENTRATION ACROSS NEURONAL CELL MEMBRANE

Ion	[Out][a] (mmol)	[In] (mmol)	Equilibrium Potential (mV)
Na+	150	15	+55
K+	5.5	150	−90
Cl−	125	9	−70
Organic anions	−	385	−

[a][Out] and [In] represent concentrations of ions outside and inside of the cell membrane, respectively.

while others are more indiscriminate. Ion channels also open and close based upon the needs of the neuron. This change in state requires a conformational change of the proteins that form the channel, a process called **gating**.

To understand the electrical properties of the neuron, we must have an understanding of the electrochemical gradient. Particular ions are distributed unequally across the cell membrane. Concentrations of Na+ and Cl− are greater on the outside of the cell, while concentrations of K+ and organic anions, such as charged amino acids and proteins, are greater on the inside of the cell (Table 3-1). The organic ions are incapable of passing across the cell membrane.

Because the cell membrane is essentially impermeable to charged particles, ions must rely on specific channels and transporters to gain entry or exit from the neuron. There are two general types of ion channels in the neuron, **resting** and **gated channels**. Resting channels are normally open, as their name implies, in the resting state of the neuron and are important for the establishment of the **resting membrane potential**. Gated channels are typically closed in the resting state and open only in response to an external signal to allow for the rapid electrical potential changes necessary for cell signaling.

Resting channels help to create the equilibrium potential, the point at which there is no net flow of ions across the cell membrane. The equilibrium potential is created by the concentration gradient of a single ion across the cell membrane as calculated by the **Nernst equation**:

$$V_m = RT/FZ \times \ln [Ion]_{out}/[Ion]_{in}$$

where V_m is the membrane potential in volts, R is the gas constant (8.3143 joules/mole degree), T is the absolute temperature, F is Faraday's constant (96,487 coulombs/mole), Z is the ionic valence, and $[Ion]_{out}/[Ion]_{in}$ is the concentration gradient for an ion across the cell membrane. Using potassium as an example, the equilibrium potential represents the voltage at which the chemical gradient driving K+ ions out of the cell is exactly balanced by the electrical gradient keeping K+ ions inside the cell, resulting in no net movement of K+ ions.

While the Nernst equation accounts for a single ion, the **Goldman-Hodgkin-Katz equation** accounts for the major ions that contribute to the resting membrane potential and each ion's permeability coefficient, which is the ease with which the ions pass across the cell membrane:

$$V_m = RT/F \times \ln \left(pK[K^+]_{out} + pNa[Na^+]_{out} + pCl[Cl^-]_{out} \right)/(pK[K^+]_{in} + pNa[Na^+]_{in} + pCl[Cl^-]_{in})$$

where the permeability coefficient for a specific ion is represented by pK, pNa, and pCl. If we are to take a hypothetical cell membrane with only resting potassium channels open, we would find that K^+ ions leave the cell because of its concentration gradient. However, the resulting net negative charge inside the cell (from the remaining organic anions) would limit the K^+ ions efflux by balancing the concentration gradient with the increasingly negative electrical gradient inside the cell. Adding sodium channels into the equation, Na^+ ions enter the cell not only down its concentration gradient but also down its electrical gradient. This continues until the system reaches equilibrium with a balance of concentration and electrical gradients for Na^+ and K^+ ions until there is no net movement of ions across the membrane. It is this steady state that creates the resting membrane potential for neurons (−65 mV). Because the permeability of K^+ ions is greater than Na^+ ions, the resting membrane potential is closer to the equilibrium potential of K^+ than it is to Na^+. Since the resting membrane potential for Cl^- ions is similar to the resting membrane potential for the neuron as a whole, Cl^- ions do not contribute much to the overall membrane potential.

If resting channels were unopposed, the resting membrane potential would continue to decrease. Thus, there must be some mechanism to offset the continuous, slow leakage of Na^+ and K^+ ions across the membrane. The −65 mV resting membrane potential is maintained by the **sodium-potassium pump**, a large membrane-spanning protein with binding sites for Na^+, K^+, and ATP (Figure 3-1). The intracellular portion of the pump has binding sites for three Na^+ ions. The extracellular portion contains two binding sites for K^+ ions along with a portion for ATPase activity. As one molecule of ATP is hydrolyzed to ADP and inorganic phosphate, the resulting energy is used to move Na^+ and K^+ against their net electrochemical gradients. It is this unequal movement of ions which maintains the negative resting membrane potential of the neuron. But at rest, the concentration of Na^+, K^+, and Cl^- ions inside and outside of the cell are balanced and constant owing to the previously mentioned forces.

Signaling through the nervous system requires large changes in electrical potential to propagate signals through and between neurons. These electrical potentials are created by substantial changes in permeability to Na^+, K^+, and Cl^- ions. The large changes in electrical potential, however, are created by only a very small net movement of ions. During an action potential, there is very little change in the concentration gradients of the ions.

CASE CORRELATES

- See Cases 1-10 (cellular and molecular neuroscience cases).

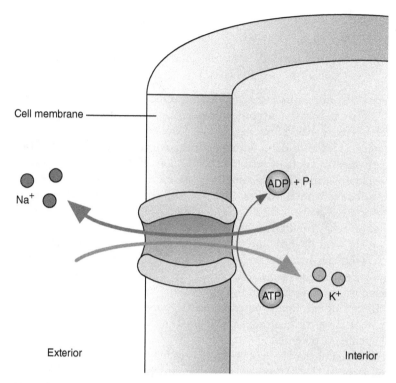

Figure 3-1. Sodium-potassium pump. Na⁺ and K⁺ flux through the resting nerve cell membrane.

COMPREHENSION QUESTIONS

3.1 A 25-year-old woman presents to the ER because of protracted vomiting and diarrhea from what she believed to be food poisoning. Although afebrile, she is pale and obviously volume depleted. She also complains of generalized weakness and fatigue. Initial laboratory tests are remarkable for a K⁺ of 2.8 mEq/L (normal value 4.0 mEq/L). How would the physician expect the marked hypokalemia to affect the resting membrane potential of the nerves?

A. Hypopolarization of the nerve

B. No effect on the resting membrane potential

C. Hyperpolarization of the nerve

3.2 A 47-year-old woman presents to your office complaining of paresthesias. After a rather extensive workup, you are unable to discover the source of her problem and you decide to check the resting membrane potential of her sensory nerves. The microelectrode is inserted, and the intracellular potential is measured as -65 mV (which is normal). What relative ionic concentrations are responsible for maintaining this membrane potential?

A. $[Na^+]_{out} > [Na^+]_{in}, [K^+]_{out} > [K^+]_{in}$
B. $[Na^+]_{out} > [Na^+]_{in}, [K^+]_{out} < [K^+]_{in}$
C. $[Na^+]_{out} < [Na^+]_{in}, [K^+]_{out} > [K^+]_{in}$
D. $[Na^+]_{out} < [Na^+]_{in}, [K^+]_{out} < [K^+]_{in}$

3.3 A researcher in a neuroscience laboratory is investigating the behavior of neuronal membrane potentials in the immediate postmortem period in rats. She notes that immediately following death, the resting membrane potential remains the same as when the animal was alive but that it slowly decreases toward zero over the following hours. What cellular mechanism is most responsible for maintaining the resting potential?

A. Na^+/K^+ ATPase
B. Na^+, K^+, Cl^- cotransporter (NKCC)
C. Ca^{2+} ATPase
D. Na^+/glucose symporter

ANSWERS

3.1 **C.** A decreased extracellular K^+ concentration will result in **hyperpolarization** of the nerve. Because potassium is the major intracellular cation in nerves, major alterations in the body's store of potassium can have a significant effect on the resting membrane potential of the nerve and thus its ability to propagate electrical signals. From the Nernst equation, we can easily see that decreasing the extracellular K^+ concentration will result in a larger negative value for the resting membrane potential for potassium. The nerve is said to be hyperpolarized or more negative, thus making it more difficult for the nerve to depolarize to propagate an electrical signal. Likewise, elevated potassium levels or hyperkalemia affect the resting membrane potential of the neuron in the opposite manner, resulting in a depolarization of the membrane potential.

3.2 **B.** The negative resting membrane potential (Vm) of sensory neurons is maintained by the relative concentrations of ions across the membrane, as well as the permeability of the membrane to these ions. The high relative intracellular concentration of K^+ coupled with the membrane's relatively high resting permeability to K^+ results in a negative Vm. The high relative extracellular concentration of Na^+ and the membrane permeability to Na^+ actually result in a positive Vm. However, since the permeability of the membrane to Na^+ is considerably less than that of K^+ (roughly 100 times less), the primary driving force of the membrane potential is K^+.

3.3 **A.** The **Na⁺/K⁺ATPase** pump maintains the resting potential. Because the neuronal membrane is permeable to both Na^+ and K^+, the ions slowly diffuse down their electrochemical gradients at rest, and without compensatory mechanism the membrane potential would eventually reach zero. However, there is a mechanism that counteracts this diffusion: the Na^+/K^+ ATPase pump, which transports Na^+ out of cells and K^+ into cells, against their concentration gradients. This pump is driven by ATP and therefore no longer functions after cellular metabolism has ceased, as occurs following death. NKCC (Na^+, K^+, Cl^- cotransporter) is an ion-transporting ATPase involved in renal function, the Ca^{2+} ATPase is a membrane enzyme important in muscle cells, and the Na^+/glucose symporter allows absorption of glucose in the intestines. These transporters are not involved in neuronal membrane electrochemistry.

NEUROSCIENCE PEARLS

- ▶ Concentrations of Na^+ and Cl^- are greater on the outside of the cell, while concentrations of K^+ and organic anions (ie, charged amino acids and proteins) are greater on the inside of the cell.

- ▶ The neuronal membrane contains two types of channels: resting channels and gated channels.

- ▶ Resting channels are normally open in the resting state of the neuron and are important for the establishment of the **resting membrane potential.**

- ▶ Gated channels open in response to an external signal and allow for the rapid electrical potential changes necessary for **cell signaling.**

- ▶ The equilibrium potential is created by the concentration gradient of a single ion across the cell membrane and is calculated by the **Nernst equation:** $V_m = RT/FZ \times \ln [Ion]_{out}/[Ion]_{in}$

- ▶ The resting membrane potential occurs when there is a balance of concentration and electrical gradients for Na^+ and K^+ ions.

- ▶ Resting membrane potential for neurons is **−65 mV** and is maintained by the **sodium-potassium pump.**

REFERENCES

Bear MF, Connors B, Paradiso M, eds. *Neuroscience: Exploring the Brain.* 3rd ed. Baltimore, MD: Lippincott Williams & Wilkins; 2006.

Kandel E, Schwartz J, Jessell T. *Principles of Neural Science.* 5th ed. New York, NY: McGraw-Hill; 2012.

Squire LR, Berg D, Bloom FE, du Lac, S, eds. *Fundamental Neuroscience.* 4th ed. San Diego, CA: Academic Press; 2012.

A 37-year-old man has had several bouts of neurological deterioration, which have each improved with time. His initial complaint was of several bouts of blurry vision, which spontaneously resolved. Following these complaints, he developed temporary weakness of his right leg and difficulty with walking. Again, these symptoms resolved in time. This time he presents to your office in Minneapolis, MN, with double vision when he attempts to look either to his left or right side. Following several further tests, you diagnose him presumptively based on the multiple attacks and remissions of weakness and eye problems as having multiple sclerosis (MS), which is a disease of the white matter of the central nervous system.

- ► Describe the characteristics of white matter of the brain.
- ► What is the mechanism of the weakness caused by white matter degeneration?
- ► What are the treatment options available to him?

ANSWERS TO CASE 4:
The Myelin Sheath and Action Potentials

Summary: A 37-year-old man with bouts of various waxing and waning neurological deficits is diagnosed with multiple sclerosis.

- **White matter characteristics:** White matter within the brain consists of myelinated axons that connect various gray matter areas where neuron cell bodies are located to each other by carrying nerve impulses between neurons.

- **Mechanism of weakness:** MS causes the gradual deterioration of the protective myelin sheaths around axons (demyelination) that facilitate signal conduction. Without myelin, the neurons cease to effectively conduct their electrical signals, resulting in a myriad of clinical symptoms (some presented above). MS is categorized as an autoimmune disease in which lymphocytes attack the myelin surrounding axons as if it were a foreign agent.

- **Treatment options:** Beta interferon can help to reduce the number of exacerbations. High-dose intravenous corticosteroids or intravenous immunoglobulin can be administered to reduce the severity and duration of active attacks. Several other medications have been approved by the Food and Drug Administration to treat symptoms of MS.

CLINICAL CORRELATION

MS is a demyelinating disease affecting only the white matter of the cerebrum, brain stem, and the spinal cord. For an unknown reason, MS is more common in northern latitudes, with an incidence of 30–80 per 100,000 in northern United States and Canada compared to 1 per 100,000 near the equator. MS is generally believed to be an autoimmune disease with the myelin sheaths in the CNS coming under attack from the immune system. The **plaques** start as an inflammatory response with monocyte and **lymphocytic perivascular cuffing**, followed by the formation of glial scars. Imaging and pathological studies demonstrate white matter plaques of various ages distributed throughout the CNS. The typical course of MS is of exacerbations and remissions over time. The most common symptoms include visual disturbances, spastic paraparesis, and bladder dysfunction. **Internuclear ophthalmoplegia (INO)** is a visual disturbance commonly associated with MS that results from a lesion in the midbrain affecting the medial longitudinal fasciculus (MLF). This lesion prevents the oculomotor nuclei and the abducens nuclei from coordinating their movements through the MLF. With lateral gaze, the contralateral medial rectus fails to adduct the eye, while the ipsilateral lateral rectus abducts the eye. The dyscoordinated movements of the eye result in diplopia. **MRI scans are the preferred imaging modality** for diagnosis of this disease. Approximately 80% of individuals with a clinical diagnosis of MS will exhibit white matter abnormalities on MRI scans. Analysis of the CSF from a lumbar puncture generally demonstrates normal opening pressures and increased CSF protein levels. CSF IgG is increased relative to other CSF proteins in approximately 90% of patients with confirmed MS. Gel electrophoresis of the CSF reveals oligoclonal bands. Diagnosis relies on a combination of patient history and physical examination supported by imaging and CSF study results. While there is no cure for MS, there are several treatment strategies available to slow the progression of the disease, reduce the frequency of attacks, and shorten the duration of attacks.

APPROACH TO:
The Myelin Sheath and Action Potentials

OBJECTIVES

1. Describe what constitutes the myelin sheath, and understand its role.

2. Describe how the action potential is propagated through the axon.

3. Describe the variables that affect the conduction velocity of the axon.

DEFINITIONS

NODES OF RANVIER: The region of the axon between consecutive myelin sheaths, where high concentrations of ion channels allow for current flow.

SALTATORY CONDUCTION: A type of rapid conduction of an electrical potential at the nodes of Ranvier skipping myelinated segments along the axon. Because the cytoplasm of the axon is electrically conductive and because the myelin inhibits charge leakage through the membrane, depolarization at one node of Ranvier is sufficient to elevate the voltage at a neighboring node to the threshold for action potential initiation, allowing action potentials to hop along an axon instead of propagating in waves.

THRESHOLD: An irreversible point at which the initial depolarization of the membrane results in the rapid opening of all of the voltage-gated ion channels; usually action threshold potentials are around -45 mV.

AXON HILLOCK: The region of the neuron between the cell body and the axon which generates action potentials.

REFRACTORY PERIODS: The period of time in which it is either impossible or difficult for additional stimuli to generate another action potential.

DISCUSSION

Myelin sheaths are produced by the oligodendrocytes in the central nervous system and by Schwann cells in the peripheral nervous system. It is important to note, however, that not all axons are encased in myelin sheaths. Myelin consists of lipids and membrane proteins, which are wrapped in circumferential layers around segments of axons. The intervals between adjacent myelin sheaths are called the **nodes of Ranvier** and are important for conduction of the action potential.

During the development of the peripheral nervous system, Schwann cells become closely associated with developing bundles of axons within the nerve. A single

Schwann cell provides a single segment of the myelin sheath for the developing axon. As axons grow and elongate, the Schwann cells divide by mitosis to ensure complete coverage of the selected axon. This is in contrast to the central nervous system, where oligodendrocytes extend processes to multiple axons to provide myelin sheaths. One oligodendrocyte can provide a segment of myelin for up to 60 different axons. Myelin sheaths are laminated when examined by electron microscopy; this is caused by the sequential wrapping of the myelin around the axon. During this wrapping process, the cytoplasm of the Schwann cells and oligodendrocytes are squeezed out of the developing myelin sheath. In the end, the layers of the cell membrane are opposed and are secured in place by proteins, such as **myelin basic protein, proteolipid protein,** and **protein zero,** that are embedded in the lipid membrane.

At the nodes of Ranvier, the myelin sheaths meet but do not join. In this region, the axon is exposed to the extracellular environment. The voltage-gated ion channels necessary for the **saltatory conduction** of the action potential are concentrated in this region of the axon.

A typical action potential lasts for less than 1 ms and is elicited in an all-or-nothing fashion. Neurons code the intensity of a stimulus by the frequency and not by the amplitude of action potentials. The action potential is crucial for the passage of information via electrical impulses over long distances and is generated by voltage-gated ion channels. These channels alter their selective permeability to a specific ion based upon changes in the transmembrane potential. Specific voltage-gated channels include Na^+, K^+, and Ca^{2+} channels. The action potential is a regenerative signal that does not lose amplitude as it travels down the axon and rely on voltage-gated Na^+ and K^+ channels.

The tertiary structure of the voltage-gated ion channel is determined by the transmembrane potential in the local environment. With changes to the transmembrane potential, the channel either opens or closes to modulate the flow of ions. These channels are typically open for a very short period of time, 10 µs or less, and function as an "all-or-nothing" type response. At the resting membrane potential of neurons, the voltage-gated Na^+ and K^+ channels are generally in their closed configuration.

The action potential is generated at the **axon hillock**, the region between the soma of the neuron and the axon, which contains a higher number of voltage-gated Na^+ channels than anywhere else in the neuron. As the neuron receives signals from the dendrites and soma, they converge at the axon hillock. As the stimulus is received, some of the voltage-gated Na^+ channels open, resulting in the influx of positive Na^+ ions into the cell and a small depolarization of the axon hillock. If the stimulus is strong enough, it will cause more and more of the voltage-gated Na^+ channels to open through positive feedback by the Na^+ ions until a point is reached where the depolarization becomes irreversible. This point is called the **threshold** (Figure 4-1). The Na^+ channels remain open for a brief period of time before they are inactivated by a conformational change that results in channel closing.

The depolarization also affects the voltage-gated K^+ channels. However, these channels open more slowly and for a longer period of time in response to the initial

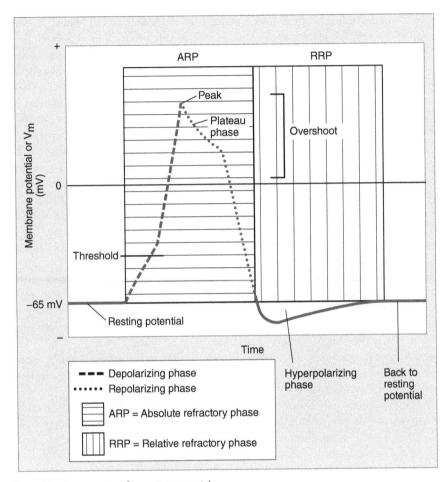

Figure 4-1. Components of an action potential.

depolarization and allow K^+ ions to flow out of the cell. The peak flow of K^+ ions occurs as the sodium current is decreasing. This results in repolarization. In the later stages of the action potential, the K^+ efflux is unopposed and drives the transmembrane potential past the resting potential and closer to the equilibrium potential for K^+ ions. As the voltage-gated K^+ channels close, the resting channels allow for the reestablishment of the resting membrane potential.

Shortly after threshold is reached, there is a period of time where additional stimuli will not result in any action potential. This is termed the **absolute refractory period** and is owing to the already maximal opening of the Na^+ channels. The Na^+ channels begin to close even in the presence of continued presence of the depolarization; the channels become inactivated, and it takes time for them to recover from the inactivation before they are able to open again. Na^+ is critical for the action potential: if the extracellular concentration of Na^+ falls the amplitude of the action potential decreases. Following the peak of the action potential, additional supernormal stimuli may result in the generation of another action potential. This period is the **relative refractory period**. As the K^+ channels close, the membrane slowly approaches its resting membrane potential. Typically, repolarization of the cell by closing K^+ channels results in an overshoot of resting membrane potential, creating a refractory period where the membrane potential equalizes to resting state. During this period, the nerve can be restimulated to fire an action potential by a supernormal stimulus. A larger than normal depolarizing stimulus is required to overcome the overshoot during repolarization by the voltage-gated K^+ channels.

For a neuron to signal other cells over distances, the action potential must be propagated down the length of the axon. The initial depolarization and resulting action potential occurs in only a small segment of the axon, creating a local current. This depolarizing current travels distally down the axon and results in the next under-threshold segment reaching threshold. This segment then generates another action potential, ensuring that the amplitude of the signal is not attenuated as it travels. The signal can only travel in one direction because of the refractory period of the channels.

The **conduction velocity** of the fastest axons in the human body transmits action potentials at a rate of 120 m/s through large, myelinated axons. Smaller, unmyelinated axons conduct action potentials at approximately 0.5 m/s. In the axons of invertebrates, such as the well-studied giant squid axon, conduction velocities are increased by increasing the diameter of the axon up to 1 mm. This strategy decreases the resistance through the axon. Mammals, which tend to have much smaller axons, employ myelin sheaths to increase the membrane resistance of the axon, resulting in faster conduction velocities.

Because of the myelin sheaths, voltage-gated channels are clustered at the nodes of Ranvier and are the only points along a myelinated axon where currents can be generated. Furthermore, the myelin sheaths prevent any significant loss in amplitude of the action potential at the **internodal segments**, the region of the axon covered by myelin. This allows the signal to jump quickly from one node of Ranvier to another while skipping the internodal segments, a process termed saltatory conduction (Figure 4-2). In unmyelinated axons, the current spreads more slowly because of the decreased membrane resistance and lack of saltatory conduction.

Although the two-channel system examined in the Nobel Prize–winning work by Alan Hodgkin and Andrew Huxley in the squid giant axon is found in almost every type of neuron, there are several other types of voltage-gated ion channels. This diversity allows for a much more complex information processing system.

One interesting clinical observation is that if chronic hyponatremia is corrected too quickly, patients may develop osmotic demyelination of the axons of the central part of the pons. Central pontine myelinolysis presents as quadriplegia due to demyelination of the corticospinal tracts, and pseudobulbar palsy due to demyelination of the corticobulbar tracts (cranial nerves X, XI, and XII), resulting in dysphagia, dysarthria, and neck muscle weakness.

CASE CORRELATES

- See Cases 1-10 (cellular and molecular neuroscience cases).

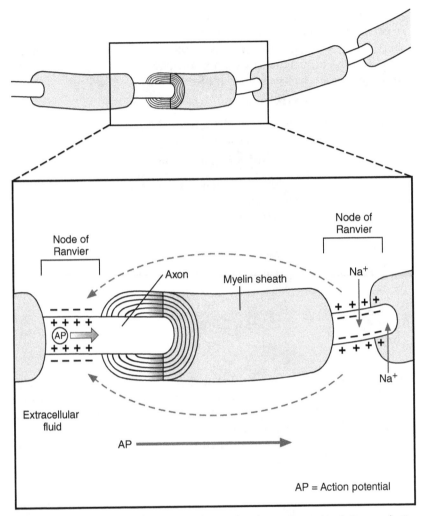

Figure 4-2. Schematic diagram of action potential traveling down an axon via saltatory conduction.

COMPREHENSION QUESTIONS

4.1 In preparation for the excision of a mole from a patient's back, the physician injects lidocaine into the surrounding skin to anesthetize the area. Lidocaine acts by binding to and preventing opening of the voltage-gated sodium channels, thus preventing transmission of impulses down the axon. What segment of the action potential is blocked by the action of this drug?

A. Rapid depolarization

B. Delayed rectifier current

C. Repolarization

D. Resting potential

4.2 During a neuroscience laboratory exercise, you are experimenting with the response of a neuron to various electrical stimuli. You notice that there is a short period following the action potential when no matter how much you depolarize the membrane, you cannot stimulate an action potential. What is the molecular mechanism responsible for this phenomenon?

   A. Undershoot hyperpolarization

   B. Inactivation of sodium channels

   C. Closing of voltage-gated sodium channels

   D. Opening of voltage-gated potassium channels

4.3 Immediately following the time frame in the previous question is a period when an action potential can be generated, but only when a larger than normal stimulus is applied. What is the molecular mechanism responsible for this phenomenon?

   A. Opening of voltage-gated sodium channels

   B. Opening of voltage-gated potassium channels

   C. Slow closing of voltage-gated potassium channels

   D. Slow closing of voltage-gated sodium channels

ANSWERS

4.1 **A.** Lidocaine binds to inactivated sodium channels, preventing the **rapid depolarization** needed for the initiation of the action potential. The opening of the voltage-gated sodium channels is responsible for the rapid upstroke of the action potential. Slower acting voltage-sensitive potassium channels are responsible for the delayed rectifier current, which, in combination with the closing of voltage-sensitive sodium channels, is responsible for repolarizing the axon.

4.2 **B.** The period referred to in the question is caused by **inactivation of voltage-gated sodium channels** and is known as the absolute refractory period. During this time period, which immediately follows repolarization, the voltage-gated sodium channels assume a configuration in which they will not open, regardless of membrane potential. Undershoot hyperpolarization, the time following the action potential when the membrane is more negative than the resting potential, is responsible for the relative refractory period. During this time, an action potential can be elicited, but it requires a larger stimulus, as the membrane must be depolarized further to reach threshold. Closing of voltage-gated sodium channels and opening of voltage-gated potassium channels both contribute to neuronal repolarization.

4.3 **C.** The question refers to the relative refractory period, which is caused by the **slow closing of voltage-gated potassium channels.** During this period an action potential can be triggered, but because the membrane is hyperpolarized, a larger stimulus is required to reach threshold. The reason that the membrane becomes hyperpolarized following an action potential is because of the slowness of response of the voltage-gated potassium channels. This slowness of response results in the potassium channels staying open longer than is necessary to repolarize the membrane, resulting in a transient hyperpolarization or undershoot. As the potassium channels close, the hyperpolarization resolves and the membrane returns to its resting potential

NEUROSCIENCE PEARLS

- ► MS is a demyelinating disease of the white matter of the brain with symptoms of visual disturbances, spastic paraparesis, and bladder dysfunction.

- ► The nodes of Ranvier, the intervals between adjacent myelin sheaths, are important for characteristic saltatory conduction of the action potential down an axon.

- ► Whereas **Schwann** cells each only myelinate **one axon**, **oligodendrocytes** extend processes to myelinate **multiple** axons.

- ► Action potentials are elicited in an all-or-nothing fashion. Intensity of the stimulus is encoded in **frequency**, *not* **amplitude** of the action potential.

- ► Signals from the neuron's dendrites and soma converge at the axon hillock, resulting in a small depolarization of the axon hillock. A sufficiently strong stimulus will cause more voltage-gated sodium channels to open through positive feedback by the Na^+ ions until depolarization becomes irreversible (threshold).

- ► The absolute refractory period occurs when the sodium channels, after having been maximally open, assume an inactivated configuration in which they will no longer open, regardless of membrane potential.

- ► The relative refractory period occurs after the absolute refractory period and is owing to the slow closing of the potassium channels.

- ► Chronic hyponatremia that is corrected too quickly can result in central pontine myolysis (quadriplegia, and pseudobulbar palsy).

REFERENCES

Bear MF, Connors B, Paradiso M, eds. *Neuroscience: Exploring the Brain.* 3rd ed. Baltimore, MD: Lippincott Williams & Wilkins; 2006.

Purves D, Augustine GJ, Fitzpatrick D, Hall WC, eds. *Neuroscience.* 5th ed. Sunderland, MA: Sinauer Associates Inc; 2011.

Squire LR, Berg D, Bloom FE, du Lac, S, eds. *Fundamental Neuroscience.* 4th ed. San Diego, CA: Academic Press; 2012.

A woman brings her 68-year-old father to your office for an examination. She states that over the past several years, "he has become increasingly forgetful." Initially, her father could not remember where he had left his wallet and keys from earlier in the day. Now she has to constantly remind him of what they have just discussed. While he used to be an avid storyteller, he has become much less gregarious of late. Most recently, she states that the neighbors found her father wandering around the block without an explanation of where he was going or from where he had just come. You diagnose her father with Alzheimer disease (AD).

▶ What is the most likely finding on a postmortem brain biopsy?
▶ Which neurotransmitter would most likely be deficient?
▶ What treatments are available for this condition?

ANSWERS TO CASE 5:
Neural Synapses

Summary: A 68-year-old man with a several-year history of progressive cognitive decline is diagnosed with Alzheimer disease.

- **Biopsy findings:** Amyloid plaques and neurofibrillary tangles are clearly visible by microscopy. Amyloid plaques are hard, insoluble plaques of protein fragments that form between neurons. Neurofibrillary tangles consist of insoluble microtubules that accumulate because of abnormal tau proteins.

- **Neurotransmitter:** The oldest of the three major competing theories on the cause of the disease is the "cholinergic hypothesis," which posits that AD is caused by restricted biosynthesis of acetylcholine; thus **acetylcholine** would be the most deficient.

- **Treatments:** There is no definitive treatment for AD. Antipsychotic medications or benzodiazepines may be administered to help treat certain behavioral disturbances common with AD.

CLINICAL CORRELATION

Alzheimer disease is the most common degenerative disorder of the brain, accounting for 50% of all diagnosed dementias. It results in a relentlessly progressive cognitive decline. Initially, the individual exhibits forgetfulness with day-to-day occurrences. Items may be frequently misplaced and appointments may go unattended. Following the establishment of memory loss, further cognitive decline becomes more evident. The patient may develop a halting manner to his or her speech because of the failure of recall of certain words. As the speech deficits increase, characteristics of expressive and/or receptive aphasias become more pronounced. Other cognitive difficulties include dyscalculia, disruption of visuospatial orientation, and ideational and ideomotor apraxias. The patient may eventually develop difficulty with locomotion and may become confined to bed. The incidence and prevalence of AD increase with age. The majority of patients are over the age of 60, but AD has also been diagnosed in much younger patients. Certain forms of AD have a familial occurrence, but these account for less than 1 percent of all cases. The most well-described familial trends follow an autosomal dominant inheritance. The early onset of AD has also been linked with Down syndrome. The most important aspect of the diagnosis of AD is the exclusion of treatable forms of dementia. While imaging of the brain may demonstrate diffuse atrophy with thinning of the cerebral gyri and enlargement of the sulci and ventricles in the advanced stages, it is more important to identify mass lesions, such as chronic subdural hematomas, which may account for the symptoms. On histological specimens, AD is associated with the diffuse loss of neurons in the cerebral cortex. In particular, neuronal loss in the **nucleus basalis of Meynert** is associated with decreased levels of the neurotransmitter **acetylcholine**. There are intracytoplasmic deposition of **neurofibrillary tangles** in neurons composed of paired, helical filaments, immunoreactive for tau protein and **neuritic plaques** composed of paired, helical filaments of fibrillar beta amyloid. In the **amyloid hypothesis** of AD, it was the deposition of fibrillar beta amyloid in the form of neuritic plaques which was believed to be the main cause of the progressive cognitive dysfunction. However, the spatial and temporal patterns of plaque formation did not correlate well with the level of cognitive decline. Newer research into the pathophysiology of AD points to the role of nonfibrillar amyloid beta peptide (A-β), the precursor of neuritic plaques, accumulating in synapses. This has been termed the **synaptic amyloid beta hypothesis**. Amyloid precursor protein (APP), a protein, is embedded in the membrane of the presynaptic terminal of the neuron. Cleavage of APP results in the release of A-β into the synapse, which acts as a synaptotoxin impairing glutaminergic transmission and compromising synaptic function. The elucidation of this new model of AD may point to the development of new therapeutic treatments in the future.

APPROACH TO:
Neural Synapses

OBJECTIVES

1. Identify the parts of the neural synapse.

2. Describe the two types of neural synapses.

3. Know how synapses function.

DEFINITIONS

SYNAPTIC BOUTON: The terminal end of an axon.

SYNAPTIC CLEFT: The space between the synaptic bouton and the postsynaptic cell.

ELECTRICAL SYNAPSES: A type of synapse connected by gap junctions that allows for the direct propagation of an electrical signal.

CHEMICAL SYNAPSES: A type of synapse utilizing neurotransmitters from the synaptic bouton which diffuse across the synaptic cleft and bind to receptors on the postsynaptic cell.

DISCUSSION

Synapses are the means by which neurons communicate with one another. On an average, a single neuron will receive approximately 1000 different synapses to process. In its simplest form, a synapse consists of the **synaptic bouton** or **terminal**, the terminal end of the axon; the **synaptic cleft**, the space between the presynaptic and postsynaptic cell; and the postsynaptic terminal. The postsynaptic terminal can be

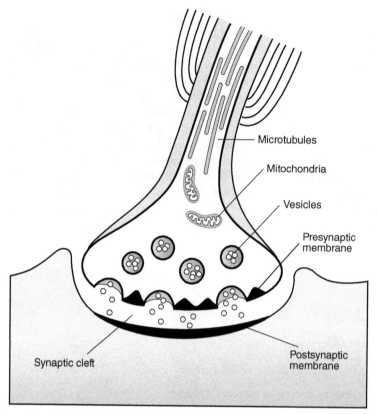

Figure 5-1. Schematic drawing of a synaptic terminal.

the dendrite or soma of another neuron or even a muscle fiber. The synaptic cleft is not simply an open space but instead is spanned by membrane proteins from both terminals, which form connections to maintain the distance between the two cells (Figure 5-1).

There are two types of synapses in the nervous system, **electrical** and **chemical synapses**. An electrical synapse is only 3-4 nm wide and the signal can travel bidirectionally, whereas the chemical synaptic cleft is much larger, measuring approximately 20-40 nm, with the signal traveling more slowly and in one direction, either to another neuron or a muscle cell. The majority of synapses in the brain are chemical.

In an **electrical synapse**, the pre- and postsynaptic cells are connected via **gap junctions** that allow for communication of the cytoplasm of both cells, accounting for the short distance between the two cells. This provides a low-resistance pathway for the direct flow of electrical potential from one cell to the other. Action potentials that reach the presynaptic terminal directly depolarize the postsynaptic membrane. If the depolarization reaches the threshold for the postsynaptic membrane, an action potential can be generated and propagated through that cell.

The transmission of the electrical signal occurs through specialized proteins, which physically connect the two cells. Gap junctions consist of two sets of hemichannels or **connexons**, one each on the pre- and postsynaptic cell membranes. Connexons are in turn composed of six identical, membrane-spanning protein subunits called **connexins**. The connexins arrange in a radial array with an open central channel. Proteins on the extracellular side of the connexins identify and link with proteins on the postsynaptic connexins to form the complete conducting channel. Like voltage-gated ion channels, gap junctions can undergo conformational changes to either open or close the channel depending upon the local milieu. These gap junctions allow for the nearly instantaneous transmission of electrical signal. Because of the nature of the synapse, the pre- and postsynaptic membranes lie in very close apposition to one another.

Chemical synapses must convert an electrical signal, the action potential, into a chemical signal, **neurotransmitter** release, which diffuses across the synaptic cleft to affect the postsynaptic cell. The presynaptic terminal contains large numbers of membrane-bound vesicles. These **synaptic vesicles** are lined by specialized membrane proteins and are formed from the invagination of the membrane of the synaptic terminal. The vesicles contain proteins and amino acids manufactured from the endoplasmic reticulum and Golgi apparatus in the soma and transported down the axon to the terminal via fast anterograde axonal transport by the action of kinesin association to microtubules. These neurotransmitters comprise the chemical messenger system of the nervous system and are concentrated in vesicles clustered around a region of the presynaptic terminal called the **active zone**. These vesicles remain in reserve until needed for secretion.

When an action potential reaches the terminal, it simply escapes through the ion channels in the presynaptic cell. However, it causes a series of reactions resulting in synaptic vesicles binding to the membrane of the presynaptic cell. Depolarization opens Ca^{2+} channels, allowing Ca^{2+} to enter the axon terminal and phosphorylate the vesicle-binding protein synapsin, which frees the vesicle from the actin microfilaments and binds to the active zone. Vesicle and plasma membrane proteins undergo Ca^{2+}-dependent association and the vesicle fuses to the plasma membrane and releases neurotransmitters by **exocytosis** into the synaptic cleft. The vesicle membrane is recaptured from the synaptic terminal plasma membrane to prevent enlargement of the nerve ending and provide a constant supply of vesicles. The neurotransmitters then diffuse across the synaptic cleft and bind to receptors on the postsynaptic membrane.

Based upon the type of neurotransmitter and receptor, the chemical synapse can result in either an excitatory or inhibitory postsynaptic potential. Excitatory neurotransmitters result in a depolarization of the postsynaptic cell. Inhibitory neurotransmitters hyperpolarize the postsynaptic cell. Proteins such as dynein (which associates vesicle with microtubule) used for synaptic transmission are targeted and returned to cell body to be recycled through a process called slow retrograde axonal transport.

Based on their structural and functional characteristics in the nervous system, neurotransmitter receptors can be classified into two broad categories: **metabotropic** and **inotropic** receptors. Inotropic receptors form transmembrane ion channels that open and allow transcellular flow of ions upon the binding of a chemical messenger (ligand), as opposed to an electrical potential changes as with voltage-gated ion channels or mechanical changes as with stretch-activated ion channels. Metabotropic receptors, on the other hand, are indirectly linked with ion channels rather than forming them. Upon binding of a ligand, the metabotropic receptor leads to a cascade of intracellular signals, which then results in opening of the transmembrane ion channel. Because the transmission of the signal from the presynaptic cell to the postsynaptic cell relies on the diffusion of neurotransmitters, it is significantly slower than an electrical synapse. However, unlike an electrical synapse, the chemical synapse is capable of amplification of the initial signal. Many vesicles, each containing thousands of neurotransmitters, release their contents into the synaptic cleft. This process can lead to the activation of thousands of postsynaptic receptors following binding of the neurotransmitter. Transmission is terminated in the synapse by diffusion of the transmitter from the active zone by chemical degradation by enzymes in the synaptic cleft (eg, acetylcholinesterase) and reuptake of the neurotransmitter into pre- and postsynaptic neurons.

The development of the neural synapse relies partially on several specialized **synaptic laminin glycoproteins.** Laminin-11, concentrated in the synaptic cleft, acts as an inhibitor to Schwann cells in the neuromuscular junction to prevent them from entering into the cleft. This allows the pre- and postsynaptic cells to directly oppose one another and helps maintain the long-term stability of synapses and allows for the rapid transmission of information across the synaptic cleft. Laminin-β2 binds directly to calcium channels, which are critical for the release of neurotransmitters in the presynaptic membrane at the neuromuscular synapse. This leads to clustering of calcium channels and the recruitment of other presynaptic components to form the active zone of the presynaptic membrane. Perturbations of either of these synaptic laminin glycoproteins may lead to the pathogenesis of synaptic disease.

CASE CORRELATES
- See Cases 1-10 (cellular and molecular neuroscience cases).

COMPREHENSION QUESTIONS

5.1 A patient presents to your office complaining of weakness and diplopia. After a thorough workup, she is diagnosed with myasthenia gravis, a disease caused by autoimmune destruction of acetylcholine neurotransmitter receptors. In what part of the synapse are these receptors located?

A. Presynaptic membrane

B. Postsynaptic membrane

C. Synaptic cleft

D. Synaptic vesicle

5.2 On electron microscopic examination of the neuromuscular junction of the patient from the previous question, you note that on one side of the synaptic cleft there are numerous round structures clustered together. A sample of one of these structures shows it to contain acetylcholine. What is the specific name of the area in question?

A. Presynaptic terminal

B. Active zone

C. Postsynaptic membrane

D. Terminal bouton

5.3 When stimulated by an action potential, the vesicles seen under electron microscope in the previous question will fuse with the presynaptic membrane. Where will the contents of the vesicles be released?

A. Presynaptic cytoplasm

B. Postsynaptic cytoplasm

C. Synaptic cleft

D. Golgi apparatus

ANSWERS

5.1 **B.** Acetylcholine receptors are located on the **postsynaptic membrane**. There are a variety of different types of acetylcholine receptors, but those affected in myasthenia gravis are inotropic and located on the postsynaptic membrane in the neuromuscular junction. In patients with myasthenia gravis, acetylcholine is released appropriately into the synaptic cleft when stimulated, but the target for the neurotransmitter is reduced, so muscle cells are insufficiently stimulated, resulting in weakness.

5.2 **B.** The portion of the presynaptic neuron containing synaptic vesicles is called the **active zone**. The round structures in the question are synaptic vesicles, as evidenced by the fact that they contain acetylcholine, a neurotransmitter. Presynaptic terminal and terminal bouton are synonyms and refer generally to the end of the axon from which neurotransmission occurs, but neither is as specific for the location of the synaptic vesicles as the active zone. The postsynaptic membrane contains neurotransmitter receptors but not synaptic vesicles.

5.3 **C.** On stimulation, synaptic vesicles fuse with the presynaptic membrane and release their contents (neurotransmitters) into the **synaptic cleft**, where they diffuse across and interact with receptors on the postsynaptic membrane. Neurotransmitters do not actually enter the postsynaptic cells; they interact with membrane proteins and cause their effects through the receptors. Likewise, neurotransmitters are typically not found in large quantities in the presynaptic neuron cytoplasm; they are contained in the synaptic vesicles. Synaptic vesicles do not fuse with or release their contents into the Golgi apparatus; however, depending on the type of neurotransmitter, they may originate there.

NEUROSCIENCE PEARLS

▶ Most AD cases are sporadic and not genetic.

▶ Genetic AD is usually autosomal dominant and related to mutations of one of the three genes that encode amyloid precursor protein, presenilin 1 or presenilin 2.

▶ The most common genetic factor in sporadic AD is the e4 allele of apoliporotein E.

▶ There are two types of synapses in the nervous system: electrical synapses and chemical synapses.

▶ Electrical synapses contain gap junctions, which physically connect adjacent neurons and allow for faster conduction of the signal.

▶ Chemical synapses comprise the majority of synapses and convert the electrical signal (the action potential) into a chemical signal (the release of a neurotransmitter).

▶ Presynaptic terminals contain larger numbers of preformed membrane-bound vesicles full of neurotransmitters, which are released via **calcium-dependent** exocytosis once the action potential reaches the terminal.

▶ **Inotropic receptors** directly form an ion channel pore that allows movement of ions once the receptor is activated.

▶ Activation of **metabotropic receptors** (ie, via binding of a ligand) leads to a cascade of intracellular signals that then results in the opening of an adjacent ion channel.

REFERENCES

Bear MF, Connors B, Paradiso M, eds. *Neuroscience: Exploring the Brain*. 3rd ed. Baltimore, MD: Lippincott Williams & Wilkins; 2006.

Squire LR, Berg D, Bloom FE, du Lac, S, eds. *Fundamental Neuroscience*. 4th ed. San Diego, CA: Academic Press; 2012.

Purves D, Augustine GJ, Fitzpatrick D, Hall WC, eds. *Neuroscience*. 5th ed. Sunderland, MA: Sinauer Associates Inc; 2011.

An 18-year-old man was involved in a motor vehicle accident 2 years ago. At that time, he sustained a Chance fracture (a complete bony injury from anterior to posterior of the spine) at T11 and a complete spinal cord injury at that level. The fracture was repaired surgically, and the patient was eventually discharged from the hospital.

He presents to the physical medicine and rehabilitation clinic for routine assessment. The patient is sitting in his wheelchair with obvious scissoring of his legs. There is normal muscle bulk but complete loss of voluntary movements. There is a marked increase in tone to his legs with passive movement, increased patellar reflexes, and upgoing toes bilaterally. He has a no sensation below his umbilicus. There is also evidence of developing contractures in his distal lower extremities.

- ▶ Which descending tracts account for the loss of voluntary motor movement?
- ▶ Which tracts account for the loss of sensation at the umbilicus?
- ▶ The loss of what type of postsynaptic potential on lower motor neurons might account for the increased tone in the lower extremities?

ANSWERS TO CASE 6:
Synaptic Integration

Summary: An 18-year-old paraplegic man presents for routine assessment following a severe spinal cord injury at T11.

- **Descending tracts:** Corticospinal, rubrospinal, and reticulospinal tracts.

- **Sensory ascending tracts:** Spinothalamic tract and dorsal columns.

- **Loss of postsynaptic potential:** The loss of **inhibitory postsynaptic potential (IPSP)** on lower motor neurons might account for the increased tone in the lower extremities. In the absence of IPSPs, the lower motor neurons will only be innervated by excitatory postsynaptic potentials (EPSPs), making muscle contraction more likely.

CLINICAL CORRELATION

Based on CDC 2010 reports, there are approximately 16,000 new spinal cord injuries per year in the United States. The cervical spine is the most common location, followed by the thoracolumbar spine. The thoracolumbar spine is particularly susceptible to injury because of its location at the transition zone from the rigid kyphosis of the thoracic spine to the mobile lordosis of the lumbar spine. This region serves as a fulcrum between the thoracic and lumbar spine and places it in a relatively neutral position, resulting in the application of maximal stress to this region during any traumatic event. Up to half of all thoracolumbar spine fractures result in some type of neurological injury. These can be divided into **complete** and **incomplete spinal cord injuries** (SCIs). In a complete SCI, there is complete loss of all motor, sensory, and autonomic function below the level of the injury. Incomplete SCIs have preservation of some neurological function, however minimal it may be. Initially, a complete thoracolumbar SCI presents as a flaccid paralysis with loss of all sensation. As time progresses, the flaccid paralysis is replaced by the characteristic **upper motor neuron** findings in the lower extremities because of loss of the descending motor tracts. **Lower motor neuron** findings are generally because of injury to the peripheral nerves (Table 6-1). The spastic weakness and increased muscle stretch reflexes associated with upper motor neuron lesions result from the loss of the **descending motor tracts** to the alpha motor neurons in the ventral horn of the spinal cord.

Table 6-1 • FINDINGS ASSOCIATED WITH LOWER MOTOR NEURON AND UPPER MOTOR NEURON INJURIES	
Lower Motor Neuron	Upper Motor Neuron
Decreased tone or flaccid	Increased tone or spastic
Decreased muscle bulk	Normal muscle bulk
Decreased muscle stretch reflexes	Increased muscle stretch reflexes
Fasciculations, fibrillations, or signs of atrophy on EMG	No signs of denervation on EMG

This leads to a condition known as **hyperreflexia**. The 1a axons, found in a muscle fiber's muscle spindle, keep track of how fast a muscle stretch changes. Additionally, they innervate interneurons within the spinal cord, which modulate the activity of the alpha motor neurons. Some of these interneurons result in inhibitory synapses on the alpha motor neurons of antagonist muscles. Supraspinal tracts also connect with interneurons to help modulate the activity of the alpha motor neurons. With loss of the supraspinal influences, the system becomes underdamped, and the response to the muscle stretch reflex is out of proportion to the stimulus. The alpha motor neurons are in a more hyperpolarized state, resulting in the exaggerated responses seen during testing of the monosynaptic reflex response (MSR).

APPROACH TO:
Synaptic Integration

OBJECTIVES

1. Describe the excitatory synaptic pathway.

2. Describe the inhibitory synaptic pathway.

3. Know how the length constant and the time constant affect postsynaptic potentials (PSPs).

4. Describe how a neuron integrates excitatory and inhibitory synapses to form a single response.

DEFINITIONS

POSTSYNAPTIC POTENTIAL (PSP): The local potential developing on the postsynaptic membrane following binding of a neurotransmitter to its specific receptor.

EXCITATORY POSTSYNAPTIC POTENTIAL (EPSP): Potential leading to the depolarization of the postsynaptic membrane.

INHIBITORY POSTSYNAPTIC POTENTIAL (IPSP): Potential leading to the hyperpolarization of the postsynaptic membrane.

SPATIAL SUMMATION: The total potential resulting from multiple simultaneous PSPs.

TEMPORAL SUMMATION: The total potential resulting from consecutive PSPs at one site.

DISCUSSION

Neurons in the central nervous system receive synapses from multiple other neurons on their dendrites and soma, which form local currents that spread throughout the neuron. These **PSPs** can be either excitatory or inhibitory, depending on the type of neurotransmitter and receptor. Whether a synapse is excitatory or inhibitory depends less on the type of neurotransmitter released by the presynaptic terminal and more on the type of ligand-gated ion channel that opens as a result of binding by the neurotransmitter. The PSP from each of the various synapses travels to the axon hillock. If the sum of all of the EPSPs and IPSPs is sufficient to depolarize the voltage-gated ion channels to threshold at the axon hillock, an action potential is generated and propagated down the axon.

Glutamate is the predominant excitatory neurotransmitter in the brain and spinal cord. There are two types of glutamate-gated channels: inotropic glutamate receptors and metabotropic glutamate receptors. Once opened, the glutamate-gated ion channels are permeable to both Na^+ and K^+ ions. Since the electrochemical gradient for K^+ is close to the resting membrane potential (RMP), there is less flow of K^+ ions through the channel. There is a significant difference between the electrochemical gradient for Na^+ and the RMP, however, resulting in a larger Na^+ conductance. As the ion currents reach equilibrium, the membrane potential approaches 0 mV, resulting in an **EPSP**. When glutamate binds to the metabotropic receptors, a second messenger system is activated in the neuron, which indirectly gates an ion channel.

The predominant inhibitory neurotransmitters in the central nervous system are the amino acids glycine and gamma aminobutyric acid (GABA). GABA receptors are more prevalent than glycine receptors. Binding of inhibitory neurotransmitters to their respective receptors leads to the generation of an **IPSP** in the neuron.

A single neuron can receive upward of 10,000 different synapses. Depending on the size and location of these synapses, they can have a strong or weak influence on the neuron. The amplitude and shape of the PSP diminishes as the electrical distance from the synapse to the axon hillock increases. The amount a PSP decreases as it spreads passively is determined by the **length constant** of the neuron and can be determined mathematically by the following formula:

$$\lambda = \sqrt{[(d \times R_m)/(4R_i)]}$$

where λ is the length constant, d is the diameter of the neuron, R_m is the membrane resistance, and R_i is the internal resistance of the axon. The greater the length constant, the better the signal is retained as it travels through the neuron (less signal degradation). In a process called **spatial summation**, multiple PSPs originating from different presynaptic neurons are summed to determine the cumulative effect on the postsynaptic neuron.

Like the length constant, the **time constant** is an intrinsic property of a neuron. It is determined by the relationship between the capacitance and resistance of the membrane and is determined by the following formula:

$$t = R_m \times C_m$$

where t is the time constant, R_m is the membrane resistance, and C_m is the membrane capacitance. The time constant helps determine the cumulative effects of consecutive PSPs from the same site on the postsynaptic neuron in a process called **temporal summation**. Conduction velocity is proportional to $[1/C_m] \times \sqrt{[d/(4R_m R_i)]}$ so that increasing the diameter or reducing the capacitance increases velocity of transmission. Myelin decreases the membrane capacitance, thus increasing conduction velocity.

Whether a synapse is excitatory or inhibitory is related to its location. Postsynaptic sites on the dendrites are often excitatory, while sites on the cell body tend to be inhibitory. This localization is owing to the greater effect inhibitory Cl^- channels can have at the base of the dendrite and on the cell body than on the distal dendrite. The open Cl^- channels act as a sink for positive current flowing from the dendrites to the axon hillock via the cell body. Presynaptic sites on the axon terminal often modulate the amount of neurotransmitter released by the presynaptic neuron.

As a neuron receives innervation from multiple presynaptic sites, the EPSPs and IPSPs from these synapses travel through the neuron to the axon hillock and are combined in a process termed **neuronal integration**. As we learned previously, this region of the neuron has the highest concentration of voltage-gated Na^+ channels. If the cumulative effect of all of the signals depolarizes the neuron to threshold, an action potential is generated. This represents the most fundamental decision-making process of the central nervous system.

CASE CORRELATES

- See Cases 1-10 (cellular and molecular neuroscience cases).

COMPREHENSION QUESTIONS

6.1 A patient comes to your office with a diagnosis of multiple sclerosis (MS), a disease that results in demyelination of central nervous system (CNS) neurons. Since the function of the myelin sheath is to decrease the capacitance of the neuronal membrane (C_m), how would you expect this demyelination to alter the time constant of the neuron?

A. Increase the time constant

B. Decrease the time constant

C. No change in the time constant

6.2 In studying the response of a CNS neuron to various neurotransmitters, you note that the application of neurotransmitter X to the area around the soma virtually eliminates all action potentials generated by the neuron, no matter how much excitatory input is given to the neuron. Which neurotransmitter is X most likely to be?

A. Acetylcholine (ACh)

B. GABA

C. Glutamate

D. Glycine

6.3 Still studying the response of a CNS neuron to various stimuli, you note that when stimulated by a single excitatory input, there is no action potential generated. However, when stimulated by three pulses from the same excitatory input in rapid succession, an action potential is generated. This is an example of what principle?

A. Spatial summation

B. Temporal summation

C. Excitatory neurotransmission

D. Inhibitory neurotransmission

ANSWERS

6.1 **A.** Increasing the capacitance of a membrane increases its ability to store charge, which **increases its time constant** and therefore increases temporal summation. In vivo, however, the removal of myelin does very little to affect temporal summation because for the most part dendrites are not myelinated. Demyelination primarily affects axonal propagation in the CNS, where an increase in the time constant results in slower action potential propagation, accounting for the pathology of MS.

6.2 **B.** X is most likely to be **GABA.** Since the application of X around the soma eliminates most action potentials, we can conclude that X induces IPSPs in the neuron. Since this is a CNS neuron, the two most common inhibitory neurotransmitters are GABA and glycine. GABA is a more common inhibitory neurotransmitter than glycine, so it is more likely that the neuron in question has GABA receptors than glycine receptors. Glutamate is an excitatory neurotransmitter, so it cannot account for the changes seen. ACh can have various postsynaptic responses, depending on the specific receptor, but in the CNS, it primarily acts as an excitatory neurotransmitter.

6.3 **B.** The situation described is an example of **temporal summation** of PSPs. The excitatory input is not strong enough to cause the postsynaptic neuron to reach threshold, but if subsequent inputs reach the axon hillock before the subthreshold depolarization has decayed to the resting potential, they build on the prior partial depolarization, further depolarizing the postsynaptic cell until threshold is reached. Spatial summation is the combination of different inputs received simultaneously at different sites on the postsynaptic neuron. Excitatory and inhibitory neurotransmission are combined by spatial and temporal summation to result in the net effect on the neuron.

NEUROSCIENCE PEARLS

▶ The sum of all the excitatory and inhibitory potentials converging at the axon hillock must exceed the threshold in order to generate an action potential.

▶ Glutamate is the predominant excitatory neurotransmitter, while glycine and GABA are the predominant inhibitory neurotransmitters in the central nervous system.

▶ Postsynaptic sites on the dendrites are often excitatory, while sites on the cell body tend to be inhibitory.

REFERENCES

Bear MF, Connors B, Paradiso M, eds. *Neuroscience: Exploring the Brain.* 3rd ed. Baltimore, MD: Lippincott Williams & Wilkins; 2006.

Squire LR, Berg D, Bloom FE, du Lac, S, eds. *Fundamental Neuroscience.* 4th ed. San Diego, CA: Academic Press; 2012.

Purves D, Augustine GJ, Fitzpatrick D, Hall WC, eds. *Neuroscience.* 5th ed. Sunderland, MA: Sinauer Associates Inc; 2011.

A 53-year-old man presents to his family doctor complaining of tremors in his hands that he first noticed about a year ago. Since then, the tremors have also involved his legs. He complains of feeling stiffness throughout his body. As he speaks, his expression does not change much, and he does not blink very often. He does not have any significant history of head trauma. He denies any recreational drug use or metal exposures. His only medications are a daily aspirin, an antihypertensive medication, and a cholesterol-lowering drug.

On physical examination, he is alert and attentive. There is an obvious 3-4 Hz resting tremor in his hands, which disappears with movement. He has a slow, shuffling gait with minimal arm swing. He has cogwheel rigidity to passive movement of his extremities. The muscle stretch reflexes are within normal limits, and there is no clonus. After careful consideration, the patient is diagnosed with Parkinsonism, a disease of dopaminergic transmission.

▶ To what class of neurotransmitters does dopamine (DA) belong?
▶ From what precursor is DA synthesized?
▶ By what two enzymes is DA degraded?

ANSWERS TO CASE 7:
Neurotransmitter Types

Summary: A 53-year-old man with a resting tremor, shuffling gait, and cogwheel rigidity to passive movement of his extremities is diagnosed with Parkinsonism.

- **Class of neurotransmitter:** DA belongs to the **catecholamine** class of neurotransmitters.

- **Precursor of DA:** DA is synthesized from the precursor L-**dopa.**

- **Degradation enzymes:** DA is degraded by the enzymes catechol-O-methyl transferase (**COMT**) and monoamine oxidase (**MAO**).

CLINICAL CORRELATION

Parkinsonism is relatively common, with approximately 1% of the population greater than 65 years of age in North America afflicted by the disease. It may be idiopathic, or secondary to other known conditions: viral infections, such as encephalitis lethargica (von Economo encephalitis); repeated head trauma; drug use, such as antipsychotics, phenothiazines, or MPTP (1-methyl-4-phenyl-1, 2, 3, 6-tetrahydropyridine); or poisonings, such as carbon monoxide or manganese toxicities. The idiopathic form is known as **Parkinson disease (PD)**, while the others are known as **secondary Parkinsonism.**

The classic triad of symptoms consists of the pill-rolling tremor of 3-5 Hz, cogwheel rigidity, and bradykinesia. Other symptoms include masked facies, postural instability, and a festinating shuffle. Diagnosis is based upon the history and clinical features. If in the early stages of the disease the diagnosis is in question, repeated evaluations at a later stage are warranted. The most difficult aspect of the diagnosis is to distinguish the idiopathic form from the secondary form. A positive response to the administration of levodopa helps to confirm the diagnosis. The most consistent finding on postmortem examination in both PD and secondary Parkinsonism is the loss of pigmented cells in the substantia nigra and in other pigmented nuclei. This corresponds to the loss of the **dopamine**-producing cells in the **substantia nigra pars compacta** (SNpc) and subsequent gliosis. Eosinophilic intraneuronal hyaline inclusions called **Lewy bodies** are found in the remaining cells of the SNpc. The constellation of clinical findings is caused by the loss of DA in the **neostriatum**, which includes the caudate nucleus, putamen, and globus pallidus. The net result of the striatal pathway is the increased inhibition of the thalamus by the globus pallidus interna (GPi), thus preventing the thalamus from exciting regions of the supplementary motor cortex.

Currently, there is no known treatment to stop the progression of this disease. Medical management is aimed at augmenting or replacing the effects of DA in the striatal pathway. Medications include levodopa-carbidopa, which replaces the loss of DA in the central nervous system, and Selegiline, a monoamine oxidase inhibitor which prevents the metabolic degradation of DA.

> # APPROACH TO:
> ## Neurotransmitter Types

OBJECTIVES

1. Know the characteristics of a neurotransmitter.

2. Identify the small-molecule neurotransmitters and the neuroactive peptides.

3. Know how and where neurotransmitters are produced.

4. Know how neurotransmitters are inactivated and cleared from the synapse.

DEFINITIONS

ACETYLCHOLINE (ACh): A small-molecule neurotransmitter synthesized from choline and acetyl coenzyme A (acetyl CoA).

GLUTAMATE: The most common excitatory neurotransmitter in the central nervous system.

GAMMA AMINOBUTYRIC ACID (GABA): An inhibitory small-molecule neurotransmitter synthesized from glutamate.

GLYCINE: The major inhibitory neurotransmitter in the spinal cord.

CATECHOLAMINES: A group of small-molecule neurotransmitters consisting of DA, norepinephrine (NE), and epinephrine (Epi) formed from the metabolism of tyrosine.

DISCUSSION

Neurotransmitters are the means by which a neuron signals other neurons and cells. Some are produced in the soma of the neuron by the free ribosomes and the rough endoplasmic reticulum, packaged in vesicles, modified by the Golgi apparatus, and transported down the axon to the presynaptic terminal. Other neurotransmitters are produced by enzymes in the cytoplasm and concentrated in synaptic vesicles. The vesicles are stored in the terminal and await the signal for release into the synaptic cleft where the neurotransmitter can diffuse across to the postsynaptic membrane, bind to receptors, and effect a change in the cell. There are specific mechanisms in place to remove neurotransmitters from the synaptic cleft.

In the nervous system, there are two main types of neurotransmitters, **small-molecule neurotransmitters**, such as acetylcholine, glutamate, and GABA, and **neuroactive peptides**, such as enkephalin and substance P.

Small-molecule transmitters are charged molecules that are derived from the metabolism of carbohydrates. Most of these neurotransmitters are amino acids or their derivatives. The precursors to the neurotransmitters are enzymatically altered in the cytosol and concentrated into synaptic vesicles for storage. As in all biosynthetic pathways, there is generally one enzyme that regulates the production of the neurotransmitter and functions as the rate-limiting step for its production.

ACh is the one small-molecule neurotransmitter that is not a derivative of an amino acid. It is synthesized from dietary choline and endogenous acetyl CoA by the enzyme choline acetyltransferase. ACh is packaged into vesicles via a transporter protein, which exchanges H$^+$ ions for ACh. Once released into the synaptic cleft, ACh is hydrolyzed by acetylcholinesterase into acetate and choline. ACh is found in the motor neurons of the spinal cord, where it is released at the neuromuscular junction, and in all preganglionic terminals in the autonomic nervous system and in the post-ganglionic terminals of the parasympathetic nervous system. It is also found widely in many synapses of the brain (see Figure 7-1).

Glutamate is the main excitatory neurotransmitter in the central nervous system and is synthesized from α-ketoglutarate, an intermediary of the tricarboxylic acid cycle. It binds to several different receptor types and acts on both inotropic and metabotropic receptors. Glutamate is cleared from the synaptic cleft by glial cells, which then convert it to glutamine by glutamine synthase. Glutamine diffuses across the plasma membrane, is synthesized back into glutamate in the presynaptic terminal, and is then repackaged into vesicles.

GABA and **glycine** are important inhibitory neurotransmitters in the central nervous system. GABA is synthesized from glutamate by glutamic acid decarboxylase with the help of the cofactor pyridoxal phosphate. Glycine is likely synthesized from serine and is the major inhibitory neurotransmitter in the spinal cord. Both neurotransmitters bind to receptors that lead to the opening Cl$^-$ channels in the postsynaptic neuron. GABA activity in the synapse is terminated by reuptake into presynaptic nerve terminals and surrounding glial cells. The energy needed to drive GABA reuptake is provided by the movement of Na$^+$ down its concentration gradient. While GABA taken back into nerve terminals is available for reutilization, GABA in glia is converted to glutamine, which is then eventually used to reform GABA via a number of metabolic steps. Glycine activity in the synaptic cleft is terminated by reabsorption into the presynaptic cleft via active transport.

Another group of small-molecule neurotransmitters are the **catecholamines**, consisting of **DA**, **NE**, and **Epi**. They are all synthesized from the amino acid tyrosine from a common pathway. Tyrosine is first converted into L-dihydroxyphenylalanine (L-dopa) by tyrosine hydroxylase. This is the rate-limiting step for the production of both DA and NE. L-dopa is then decarboxylated to form DA, which is important in the nigrostriatal pathway in the brain. DA can be converted to NE in the synaptic vesicle by dopamine hydroxylase. NE acts as an important neurotransmitter in the autonomic nervous system and is also found in large concentrations in brain in the locus ceruleus. NE can be methylated by phenylethanolamine-N-transferase to form Epi in the adrenal medulla.

The **neuroactive peptides** are produced by ribosomes on the endoplasmic reticulum of the cell body and, following modifications, are transported down the axon to the terminal. Several different peptides can be encoded by a single mRNA molecule. They are produced as a large precursor protein called a polyprotein in the membrane-limited organelles of the neuron. These larger proteins are cleaved to form the neuroactive peptides, which are removed by both diffusion and breakdown by extracellular proteases. Since polyproteins can encode different neurotransmitters, this processing of the polyprotein is crucial in determining which of the

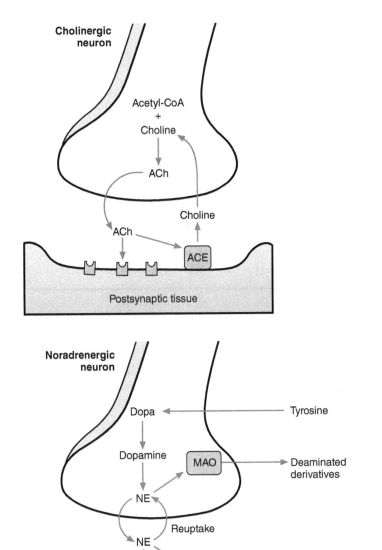

Figure 7-1. Comparison of the biochemical events at cholinergic endings with those at noradrenergic endings.

neuroactive peptides are ultimately released by the neuron. Neuroactive peptides have long-lasting effects because they all work through G protein–coupled receptors.

Neuroactive peptides differ from the **small-molecule neurotransmitters** in several ways. First, because they rely on protein synthesis and modification, neuroactive peptides can only be produced in the cell body. Small-molecule neurotransmitters rely on enzymes found throughout the cytosol and are predominantly manufactured at the presynaptic terminal. They are also taken up and concentrated within the synaptic vesicles, unlike the neuroactive peptides, which are packaged into vesicles by the Golgi apparatus. Because of the different processing steps, the type of synaptic vesicles also differs between the two classes. The vesicles for small-molecule neurotransmitters can be recycled quickly at the nerve terminal following exocytosis to produce more synaptic vesicles. The membrane that constitutes the vesicles for neuroactive peptides come from the Golgi apparatus and is transported from the cell body in a more time-consuming fashion.

Despite these differences, neuroactive peptides and small-molecule neurotransmitters often coexist within the same neuron. They can be released together to function synergistically on the postsynaptic cell. Additionally, several different neuroactive peptides processed from a single polyprotein can be released into the synaptic cleft.

Following release, the neurotransmitters must be removed from the synaptic cleft to prevent desensitization of the postsynaptic receptors and to allow future transmissions to occur. As learned previously, enzymes in the synaptic cleft degrade and inactivate certain neurotransmitters, such as acetylcholine. Neuroactive peptides are cleared more slowly from the synapse by simple diffusion. Most neurotransmitters, however, are taken up by the neuron to terminate their action. Transporter proteins in the neuron often rely on the electrochemical gradient for the active reuptake of the neurotransmitter.

CASE CORRELATES

- See Cases 1-10 (cellular and molecular neuroscience cases).

COMPREHENSION QUESTIONS

7.1 A 62-year-old man presents to your office complaining of tremor and difficulty with movements. He is noted to have masked facies and a shuffling gate. Based on clinical presentation and additional studies, you diagnose him with Parkinson disease. Which molecule is the immediate precursor in the synthetic pathway leading to the neurotransmitter involved in this disease?

A. L-dopa

B. NE

C. Tryptophan

D. Tyrosine

7.2 A 41-year-old woman presents to your office complaining of chronic widespread pain, particularly in "trigger points" throughout her body. She is ultimately diagnosed with fibromyalgia, a disorder that is associated with elevated levels of the neurotransmitter "Substance P." In which of the following locations is substance P synthesized?

A. Cell body

B. Presynaptic terminal

C. Synaptic vesicles

D. Synaptic cleft

7.3 You are following a patient who has been previously diagnosed with Alzheimer disease and is undergoing pharmacological treatment. One of the proposed pathologic mechanisms of this disease is a lack of cholinergic neurotransmission in certain areas of the brain. Some treatments are therefore aimed at increasing cholinergic neurotransmission. By which of the following mechanisms is a drug that increases acetylcholine in the synaptic cleft *most likely* to act?

A. Inhibition of diffusion out of the synaptic cleft

B. Inhibition of reuptake of acetylcholine into the neuron and surrounding glia

C. Inhibition of acetylcholinesterase-mediated degradation of acetylcholine

D. Increased synthesis and release of acetylcholine

ANSWERS

7.1 **A.** The neurotransmitter implicated in the pathogenesis of Parkinson disease is DA, the immediate precursor of which is L-dopa. In the synthetic pathway leading to DA, tyrosine is converted to L-dopa by tyrosine hydroxylase; L-dopa is then converted to DA by dopa decarboxylase. DA can be further processed by dopamine β-hydroxylase to yield NE. Tryptophan is the first step in the synthetic pathway leading to serotonin.

7.2 **A.** Substance P is a neurotransmitter that belongs to the neuroactive peptide class and, like all peptides, is synthesized in the **cell body** on the rough endoplasmic reticulum. After synthesis, these peptides are further processed by the Golgi apparatus, which also packages them into vesicles. These vesicles are transported down the axon via fast anterograde axonal transport to the presynaptic terminal, where they are secreted into the synaptic cleft when properly stimulated. Many of the small-molecule neurotransmitters (ACh, DA, GABA, etc) are synthesized in the cytoplasm of the presynaptic terminal and subsequently packaged into secretory vesicles. DA is converted to NE inside the synaptic vesicles. Neurotransmitters are degraded, not synthesized, in the synaptic cleft.

7.3 **C.** While all of the above are potentially mechanisms by which the acetylcholine levels in the synaptic cleft could be increased, the most likely candidate is **inhibition of acetylcholinesterase-mediated degradation**. The primary method of removal of acetylcholine from the cleft is enzymatic degradation by acetylcholinesterase. Inhibition of this enzyme, therefore, increases the level of ACh in the cleft. There are numerous drugs that do just this, and they are in fact used for treatment of AD as well as other disorders. Diffusion from the cleft is the primary means of removal of neuropeptides, and reuptake is the primary means of removal of amino acid neurotransmitters like GABA and glycine. While increasing the synthesis and release of ACh would also increase its concentration in the synaptic cleft, this is a considerably more complicated process than inhibiting acetylcholinesterase and is therefore a less prominent drug target.

NEUROSCIENCE PEARLS

▶ The typical presentation of Parkinson disease is for gait issues (shuffling) and tremor, followed by problems with memory and language.

▶ Parkinson disease is associated with dopamine depletion of the nigrostriatal tracts.

▶ Acetylcholine is synthesized from choline and acetyl coenzyme A (acetyl CoA) by the enzyme choline acetyltransferase.

▶ In the synaptic cleft, acetylcholine is hydrolyzed by acetylcholinesterase into acetate and choline.

▶ In the synaptic cleft, glial cells facilitate the clearance of glutamate by converting it to glutamine, which then diffuses across the plasma membrane to be converted back to glutamate and then repackaged into vesicles for future release.

▶ Glycine is the major inhibitory neurotransmitter of the spinal cord.

▶ Conversion of tyrosine to L-dopa by tyrosine hydroxylase is the rate-limiting step for DA and NE production.

REFERENCES

Bear MF, Connors B, Paradiso M, eds. *Neuroscience: Exploring the Brain.* 3rd ed. Baltimore, MD: Lippincott Williams & Wilkins; 2006.

Purves D, Augustine GJ, Fitzpatrick D, Hall WC, eds. *Neuroscience.* 5th ed. Sunderland, MA: Sinauer Associates Inc; 2011.

Squire LR, Berg D, Bloom FE, du Lac, S, eds. *Fundamental Neuroscience.* 4th ed. San Diego, CA: Academic Press; 2012.

A very distraught 22-year-old mother brings her 9-month-old infant boy to the emergency room in the early evening. She states that he started to have difficulty eating this afternoon and that he has been "drooling" more. He has also been much less active, and his crying is weaker today. She did not have to change his diaper since morning. He had his normal diet since last night but was given a special treat for his good behavior. He has been meeting all of his developmental milestones previously. On physical examination, the baby is diffusely weak with poor head control. He has pooling of saliva in his mouth. He appears lethargic with bilateral ptosis. During examination he moves only minimally and gives a quiet cry. Based on the clinical presentation, the baby is diagnosed with botulism, a disease that interferes with neurotransmission and the neuromuscular junction.

▶ What ion is involved in the normal release of neurotransmitter?
▶ What neurotransmitter is involved at neuromuscular junctions?
▶ How would the child have most likely acquired botulism?

ANSWERS TO CASE 8:

Neurotransmitter Release

Summary: A 9-month-old infant boy presents to the emergency room with poor feeding, drooling, weakness, and constipation.

- **Ions involved in neurotransmitter release:** Ca^{2+} ions are usually required for the release of neurotransmitters.
- **Neurotransmitter at the neuromuscular junction:** Acetylcholine.
- **Source of botulism toxin:** Eating **honey**.

CLINICAL CORRELATION

Infant botulism is a rare disease resulting from the ingestion of the spores of the bacterium *Clostridium botulinum* in contaminated soil, home-canned goods, and honey. Based on the 2014 CDC report, there are approximately 150 cases per year in the United States, with more than half of all cases occurring in California. Most of the cases involve infants between the ages of 6 weeks and 9 months. Symptoms, including constipation, cranial nerve abnormalities, hypotonia, and respiratory difficulties, classically appear 12-36 hours following ingestion of the contaminated food. Parents will complain that breast-feeding babies will have poor suction and weak cries. The flaccid paralysis usually descends as the disease progresses. In severe cases, the respiratory muscles may be affected. Following ingestion of contaminated food or soil, the spores germinate and colonize in the infant's gastrointestinal tract. Once the bacteria are established, they begin to produce the **botulinum exotoxin**, which is absorbed throughout the intestinal tract. The exotoxin makes its way to the presynaptic neurons of the neuromuscular junction, where they bind irreversibly to the **presynaptic cholinergic receptors** and enter the cell by endocytosis. Once inside the cell, the toxin functions as a protease and cleaves integral membrane proteins of the acetylcholine-containing synaptic vesicles. This prevents fusion of the vesicles to the presynaptic membrane and, ultimately, exocytosis of the neurotransmitter. The decreased levels of acetylcholine at the neuromuscular junction produce the weakness that is the hallmark of botulism poisoning.

APPROACH TO:

Neurotransmitter Release

OBJECTIVES

1. Describe the role of Ca^{2+} ions in the release of neurotransmitter.
2. Know what constitutes a quantum of neurotransmitter.

3. Identify the proteins involved in the fusion of synaptic vesicle to the presynaptic membrane.

4. Describe how vesicles are recovered following exocytosis.

DEFINITIONS

PRESYNAPTIC CHOLINERGIC RECEPTORS: Receptors for the neurotransmitter acetylcholine found on the presynaptic terminal of the nerve.

ACTIVE ZONE: The region of the axon terminal with high concentrations of synaptic vesicles.

QUANTUM: A term for the amount of neurotransmitters contained in one synaptic vesicle.

QUANTAL SYNAPTIC POTENTIAL: The postsynaptic potential generated by the release of neurotransmitters from one synaptic vesicle.

DISCUSSION

Stated simply, the release of neurotransmitter at the neural synapse requires the depolarization of the presynaptic terminal by the action potential. This in turn leads to the binding of vesicles to the membrane at the **active zone** and the release of neurotransmitter into the synaptic cleft.

The voltage-gated Na^+ and K^+ channels in the presynaptic terminal depolarize the surrounding membrane in response to the action potential. In addition to the Na^+ and K^+ channels, the presynaptic terminal has a large concentration of **voltage-gated Ca^{2+} channels** clustered around the active zone of the presynaptic terminal. Although they are slower to open than voltage-gated Na^+ and K^+ channels, their relative proximity to the site of neurotransmitter release allows for rapid opening of the channels. It is the influx of Ca^{2+} (and not Na^+ or K^+) as a result of membrane depolarization that is responsible for vesicle release of neurotransmitter into the synaptic cleft. The amount of neurotransmitter released is directly proportional to the influx of Ca^{2+} ions.

Experiments have shown that neurotransmitters are released in a discrete package called a **quantum**. Each synaptic vesicle carries one quantum of neurotransmitter. A single quantum results in a fixed **quantal synaptic potential** in the postsynaptic cell. The total postsynaptic potential is comprised of multiple quantal synaptic potentials. In the central nervous system, one action potential can result in the release of 1-10 synaptic vesicles. At the neuromuscular junction, a single action potential can result in the release of up to 150 vesicles. Furthermore, the release of neurotransmitters from the vesicles is an all-or-none phenomenon. If the vesicle binds to the presynaptic membrane, the entire quantum of neurotransmitter is released.

The synaptic vesicles are found clustered in the active zone of the presynaptic terminal. Synaptic vesicles in the cytosol are anchored in place to cytoskeletal filaments by proteins called **synapsins**. These vesicles cannot release their contents until they move adjacent to the presynaptic membrane. With influx of Ca^{2+} ions, the synapsins are phosphorylated, and the vesicles are freed from their anchors and move toward the active zone. Membrane-bound proteins, **Rab3A** and **Rab3B**, are thought to be important for the targeting of the vesicles to the target zone. These proteins bind and hydrolyze GTP into GDP and inorganic phosphate.

One hypothesis for vesicle docking relies on the interaction of a group of integral **SNARE proteins**. Integral SNARE proteins in the vesicle bind to their counterpart on the presynaptic membrane and hold the vesicle directly adjacent to the membrane. Neurotransmitter release relies on specialized transmembrane proteins that serve as **fusion pores**. These pores are likely preassembled hemichannels and may resemble gap junction channels. As Ca^{2+} ions enter the terminal, the fusion pore opens and allows for the exocytosis of neurotransmitter into the synaptic cleft in a mechanism that is not completely understood at this time. The membrane of the vesicle is incorporated into the presynaptic membrane following release of the neurotransmitter.

Calcium ions must be cleared from the terminal to prevent the exhaustion in supply of the synaptic vesicles. Cytosolic proteins rapidly bind and sequester the Ca^{2+} ions and prevent further vesicle release. The Ca^{2+} ions are also actively transported into storage cisterns in the terminal. Finally, Na^+/Ca^{2+} exchange transporters use the concentration gradient of Na^+ ions to pump Ca^{2+} out of the terminal and into the extracellular space.

In order to replenish the supply of vesicles, the vesicle membrane must be recovered from the presynaptic membrane. The recovered vesicles are then transported to the membrane-bound organelles in the presynaptic terminal and readied for future synaptic vesicle formation. There are several methods described for the retrieval of the synaptic membrane. The **classical pathway** relies on endocytosis of the synaptic membrane by means of **clathrin-coated pits**. Clathrin is a cytosolic protein that coats the invaginating membrane and helps to form the new synaptic vesicle. In the **kiss-and-run pathway**, the vesicle does not completely fuse with the presynaptic membrane. Once the fusion pore opens, neurotransmitter can be released from the vesicle. As soon as all of the neurotransmitter is released, the pore closes and the vesicle can be recycled. Finally, in the **bulk endocytosis pathway,** recovery of excess membrane occurs without the use of clathrin-coated pits.

CASE CORRELATES

- See Cases 1-10 (cellular and molecular neuroscience cases).

COMPREHENSION QUESTIONS

8.1 A 63-year-old man with a known history of lung cancer complains of progressive weakness. You perform a thorough history and physical examination of the patient and order additional ancillary studies. Based upon your findings, you diagnose him with Eaton-Lambert syndrome, an autoimmune disease which causes destruction of the presynaptic voltage-gated Ca^{2+} channels. Which of the following events involved in normal neurotransmission would *most likely* be disrupted by the destruction of these channels?

A. Propagation of an action potential to the presynaptic terminal

B. Release of neurotransmitter in response to nerve terminal depolarization

C. Response of the postsynaptic cell to neurotransmitter

D. Removal of neurotransmitter from the synaptic cleft

8.2 A 27-year-old man is brought into the emergency room with complaints of severe nausea and vomiting as well as diplopia and weakness. His friend reports he cans his own vegetables and had eaten some the previous evening. Your initial suspicion is that this man has botulism, and you immediately administer botulinum antitoxin. Which of the following processes involved in neurotransmission is inhibited by botulinum toxin?

A. Anterograde axonal transport which delivers neurotransmitter precursors to the synaptic terminal

B. Synthesis of neurotransmitter in the presynaptic terminal

C. Packaging of neurotransmitter into synaptic vesicles

D. Fusion of the synaptic vesicles with the neuronal membrane, resulting in neurotransmitter release

8.3 Which of following best describes the kiss-and-run model of synaptic vesicle recovery?

A. Synaptic vesicles completely fuse with the presynaptic membrane, releasing their contents into the synaptic cleft. The vesicle components are then recovered by clathrin-mediated endocytosis.

B. Synaptic vesicles completely fuse with the presynaptic membrane, releasing their contents into the synaptic cleft. The vesicle components are then recovered by endocytosis without clathrin-coated pits.

C. Synaptic vesicles approach the presynaptic membrane but do not completely fuse with it. They open pores with the membrane, releasing their contents into the synaptic cleft, and then separate from the membrane intact.

D. Synaptic vesicles fuse with the presynaptic membrane, releasing their contents into the synaptic cleft. The vesicle components are not recovered, but rather new components are synthesized in the soma and transported to the presynaptic terminal.

ANSWERS

8.1 **B.** Because Eaton-Lambert syndrome is a disorder that stems from destruction of the presynaptic calcium channels, it follows that those steps in neurotransmission that directly involve calcium channels, specifically **release of neurotransmitter in response to nerve terminal depolarization,** would be disrupted. Recall that opening of voltage-gated calcium channels results in calcium influx, which results in fusion of secretory vesicles with the presynaptic membrane and release of neurotransmitter into the cleft. Action potentials depend on sodium and potassium channels, so destruction of calcium channels will not prevent them. In Eaton-Lambert syndrome, there is a deficient response of the postsynaptic cell, but this is not caused by a disruption in response to neurotransmitter. Rather, it is because there is insufficient neurotransmitter released. The ability of the postsynaptic cell to respond remains intact; there is simply nothing to respond to. Removal of neurotransmitter from the synaptic cleft is unaffected by the loss of calcium channels.

8.2 **D.** Botulinum toxin binds to and inactivates the acetylcholine synaptic vesicle docking complex, which normally facilitates the **fusion of the synaptic vesicles with the neuronal membrane, allowing neurotransmitter release.** This prevents fusion of the vesicle with the presynaptic membrane and release into the synaptic cleft. The other steps listed in the question are not inhibited by botulinum toxin. Packaging of norepinephrine into vesicles is inhibited by the drug reserpine.

8.3 **C.** Answer C accurately describes the kiss-and-run pathway. Answers A and B describe the classical pathway for vesicle recycling and the bulk endocytosis pathways, respectively. Answer D is not a described method for vesicle recycling.

NEUROSCIENCE PEARLS

► Botulinum toxin is an exotoxin secreted by *Clostridium botulinum*, and has a protease effect, cleaving integral membrane proteins of the acetylcholine-containing presynaptic vesicles.

► The decreased levels of acetylcholine at the neuromuscular junction produce the weakness that is the hallmark of botulism poisoning.

► Ca^{2+} (the influx of which is triggered by membrane depolarization) is necessary for the fusion of synaptic vesicles with the neuronal membrane and the release of neurotransmitters into the synaptic cleft.

► The amount of neurotransmitter released is directly proportional to the influx of Ca^{2+} ions.

► Synaptic vesicles in the cytosol are anchored in place to cytoskeletal filaments by proteins called **synapsins**.

► The three known pathways by which synaptic membranes are retrieved are the classical pathway, the kiss-and-run pathway, and the bulk endocytosis pathway.

► The classical pathway describes retrieval by endocytosis of the synaptic membrane by means of clathrin-coated pits.

► The kiss-and-run pathway states that instead of the vesicle completely fusing with the presynaptic membrane, the vesicle's fusion pore opens to release the neurotransmitter and then closes so the vesicle can be recycled.

► The bulk endocytosis pathway describes retrieval of excess membrane without the use of clathrin-coated pits.

REFERENCES

Bear MF, Connors B, Paradiso M. *Neuroscience: Exploring the Brain*. 3rd ed. Baltimore, MD: Lippincott Williams & Wilkins; 2006.

Purves D, Augustine GJ, Fitzpatrick D, Hall WC, eds. *Neuroscience*. 5th ed. Sunderland, MA: Sinauer Associates Inc; 2011.

Squire LR, Berg D, Bloom FE, du Lac, S, eds. *Fundamental Neuroscience*. 4th ed. San Diego, CA: Academic Press; 2012.

A 65-year-old man presents to your office in the morning complaining of double vision and weakness with exertion. This has been ongoing for several months. He states that he feels best first thing in the morning. As the day progresses, however, he becomes more fatigued and will occasionally experience diplopia, or double vision. He has even felt too tired to make it through dinner. On physical examination, his extraocular movements are intact, and he currently denies any double vision. He appears to have a mild ptosis. Muscle strength testing and muscle stretch reflexes are within normal limits. He has a narrow-based gait with appropriate arm swing. You notice that he is speaking progressively more softly and less clearly during the course of the examination. After some thought, you conclude that he has myasthenia gravis, a disorder of the postsynaptic acetylcholine receptor (AChR).

► What are the two major types of neurotransmitter receptors?
► What type of receptor is involved at the neuromuscular junction in this case?
► What are the possible treatments for this condition?

ANSWERS TO CASE 9:

Neurotransmitter Receptors

Summary: A 65-year-old man presents with double vision and progressive fatigue over the course of the day.

- **Receptors:** The two types of neurotransmitters receptors are **inotropic** and **metabotropic.**

- **Receptors involved in this case:** The inotropic receptors at the neuromuscular junction.

- **Treatments:** Anticholinesterase medications, thymectomy, corticosteroids, immunosuppressive agents, plasma exchange, and intravenous immunoglobulin.

CLINICAL CORRELATION

Myasthenia gravis (MG) is predominantly an autoimmune disorder affecting the **nicotinic AChRs** at the neuromuscular junction (NMJ), although a rarer, heritable form has also been described. MG is characterized by a fluctuating weakness which may vary over several minutes or days. The most commonly affected muscle groups include the levator palpebrae and the extraocular muscles, resulting in ptosis and diplopia. The muscles of mastication, facial expression, and speech can also be affected, resulting in dysphagia, expressionless faces, and dysarthria. It can also affect limb, abdominal, and respiratory muscles to varying degrees. MG has a prevalence estimated at around 40-80 per million of the population and an annual incidence of approximately 1 in 300,000 individuals. Tumors of the thymus gland are found in approximately 10%-15% of myasthenic patients, while fully 65% have evidence of lymphoid hyperplasia in the medulla of the thymus. Other autoimmune disorders, such as lupus erythematosus, rheumatoid arthritis, Sjögren syndrome, and polymyositis, have also been associated with MG. In the autoimmune form, antibodies to the AChRs are produced that interfere with synaptic transmission at the neuromuscular junction by several mechanisms. First, the antibodies act as a competitive antagonist of acetylcholine (ACh) at the NMJ. Second, there are fewer AChRs at the NMJ in myasthenic patients. The anti-AChR antibodies also form cross-links with multiple AChRs, which results in the clustering of AChRs, the internalization of the receptors by endocytosis, and, ultimately, the degradation of the AChRs. Because of the reduced numbers of AChRs and the competitive antagonism of ACh, the endplate potentials are not sufficient to generate action potentials in the muscle. This leads to the recruitment of fewer muscle fibers and the loss of overall contractile power in the muscle.

The diagnosis of MG is facilitated by several ancillary studies, including electromyography, a serum assay for anti-AChR antibodies, imaging of the chest to rule out the presence of a thymoma, and edrophonium and neostigmine tests, which test motor strength before and after injection with either of these anticholinesterase medications. These two drugs decrease the clearance of ACh at the NMJ and, in myasthenic patients, result in a marked improvement in strength following administration.

There are several possible treatments for MG. Longer acting anticholinesterase drugs are often beneficial for patients afflicted with purely ocular myasthenia. If an enlarged thymus gland is detected, a thymectomy should be performed. Up to 80% of individuals less than 55 years of age without a thymoma and who have had poor response to anticholinesterase medications also receive some benefit from thymectomy. Corticosteroids and immunosuppressive agents can also ameliorate the symptoms of MG.

APPROACH TO:
Neurotransmitter Receptors

OBJECTIVES

1. Know that the two types of neurotransmitter receptors are inotropic and metabotropic.

2. Describe how inotropic and metabotropic receptors function.

3. Describe the differences between G protein–coupled receptors and tyrosine kinase receptors.

4. Describe the differences between inotropic and metabotropic receptors.

DEFINITIONS

INOTROPIC RECEPTORS: Receptors that directly gate ion channels.

METABOTROPIC RECEPTORS: Receptors that rely on a variety of second messenger systems to indirectly gate ion channels.

G PROTEIN–COUPLED RECEPTORS: A type of metabotropic receptor that uses G proteins to activate various second messenger cascades.

TYROSINE KINASE RECEPTORS: A type of metabotropic receptor that uses tyrosine kinases to phosphorylate proteins to initiate a second messenger pathway.

DISCUSSION

Neurotransmitter receptors are generally located on the postsynaptic membrane and have two important functions: to recognize and bind specific neurotransmitters and to alter the membrane potential of the postsynaptic cell. A single neurotransmitter can bind to several types of receptors, resulting in different effects at different synapses. There are two general classes of receptors: inotropic and metabotropic receptors. **Inotropic receptors** have one or more binding sites for neurotransmitters, which are directly coupled to ion-gated membrane channels. With binding of a specific neurotransmitter, the channel opens to allow passage of specific ions and alters the membrane potential. **Metabotropic receptors** are coupled to second messenger systems within the postsynaptic cell and gate ion channels indirectly.

Inotropic receptors open with binding of the neurotransmitter and close with dissociation. The open channel allows for the passage of ions through the postsynaptic membrane and results in a brief, local **postsynaptic potential** (PSP). Unlike the action potential, the amplitude of the PSP varies depending on the amount of channels opened by the release of neurotransmitter.

One of the most prevalent neurotransmitter receptors is the ACh receptor. It is found throughout the autonomic nervous system and at the NMJ. Two types of AChR, **nicotinic** and **muscarinic**, have been identified. Nicotinic receptors are found in the NMJ and the preganglionic endings of both the sympathetic and parasympathetic nervous system. Muscarinic receptors are found in the postganglionic endings of all parasympathetic endings and in certain sympathetic endings.

The nicotinic receptor is made up of five subunits: two alpha subunits and one beta, one gamma, and one delta subunit. The alpha subunits function as the extracellular binding site for the ACh molecules. The subunits form a channel which remains closed without ligand binding. The pore of the channel contains a ring of negatively charged molecules, which select for positively charged ions. When two molecules of ACh bind to the alpha subunits, the channel undergoes a conformational change and opens, allowing the passage of Na^+ and K^+ ions. Sodium ions flow into the cell and K^+ ions flow out when the postsynaptic cell is at its resting membrane potential. This results in the depolarization of the postsynaptic cell. Certain GABA and glycine receptors also gate ion channels but are selective for anions.

Glutamate has two different inotropic receptors: non-NMDA receptors, which are permeable to Na^+ and K^+, and NMDA receptors, which are permeable to K^+, Na^+, and Ca^{2+}. NMDA receptors are normally blocked by a Mg^{2+} plug in the channel and require both ligand binding of glutamate and depolarization to open. NMDA glutamate receptors are very important in long-term potentiation.

Metabotropic receptors gate ion channels through a different mechanism. The receptor is coupled to one of two second messenger systems: **G protein–coupled receptors** and **tyrosine kinase receptors**. The G protein–coupled receptor consists of a single subunit with seven transmembrane spanning regions. Binding of the neurotransmitter activates a GTP-binding protein, which in turn activates one of several enzymes: adenylyl cyclase in the cyclic AMP pathway, phospholipase C in the IP3-DAG pathway, or phospholipase A2 in the arachidonic acid pathway. These enzymes trigger the second messenger cascade, which ultimately results in the gating of an ion channel. Muscarinic AChRs found in the CNS and autonomic parasympathetic system and adrenergic receptors found in the CNS and peripheral autonomic sympathetic system function through G protein–coupled systems.

Tyrosine kinase receptors consist of a single spanning protein, an extracellular receptor-binding domain, and an intracellular protein kinase domain. Binding of the neurotransmitter to the extracellular site results in the dimerization of two receptors, which activate the intracellular kinases. The kinase phosphorylates itself and other proteins on tyrosine residues. This leads to the activation of a second messenger cascade, which can alter gene transcription within the cell and can also modulate the activity of ion channels. These types of receptors are typically activated by neuropeptides and hormones.

There are several key functional differences between inotropic and metabotropic receptors. The action and duration of inotropic receptors are immediate; binding of a neurotransmitter results in the rapid opening of ion channels. The dissociation of the neurotransmitter closes the ion channel in a process that spans milliseconds. Metabotropic receptors act in a more delayed fashion because of their reliance on a series of reactions. The opening of the ion channel may take from tens of milliseconds to seconds to occur and the duration may last from seconds to minutes.

Inotropic channels function to create either excitatory or inhibitory postsynaptic potentials in a well-localized area, the postsynaptic membrane. These potentials, when summed, can create or inhibit an action potential through their effect on neighboring voltage-gated ion channels. Metabotropic receptors are also excitatory or inhibitory, but they work through freely diffusible second messenger systems, which can interact with channels anywhere on the postsynaptic cell. The second messenger cascade can influence the activity of resting membrane channels, voltage-gated ion channels, or ligand-gated channels. Metabotropic receptors, unlike inotropic receptors, can not only open channels, but they can close them as well.

CASE CORRELATES

- See Cases 1-10 (cellular and molecular neuroscience cases).

COMPREHENSION QUESTIONS

9.1 A 67-year-old man presents to your clinic for management of his long-standing hypertension. He is currently taking atenolol (a norepinephrine β-receptor blocker) for the management of his blood pressure. Which of the following processes is inhibited through the use of this drug?

A. Opening of voltage-gated sodium and potassium channels

B. Opening of voltage-gated chloride channels

C. Stimulation of adenylate cyclase

D. Activation of phospholipase A2

9.2 You are studying the behavior of a neuron to the application of neurotransmitter. Immediately following the application of neurotransmitter to the postsynaptic membrane, you note a local alteration in the membrane potential. Several seconds later, however, the postsynaptic membrane behaves exactly as it did before the application of neurotransmitter. Through what mechanism is this neurotransmitter *most likely* to act?

A. Activation of adenylate cyclase

B. Activation of PLC

C. Activation of PLA2

D. Opening of a voltage-gated ion channel

9.3 As part of a routine examination of a 27-year-old woman, you check her knee-jerk reflex. You tap her patellar tendon with your reflex hammer, and her quadriceps muscle contracts appropriately. You recall that the motor nerve innervating the muscle releases ACh into the neuromuscular junction. What effect does this ACh release have on the postsynaptic muscle membrane?

   A. Opening of ligand-gated sodium and chloride channels

   B. Opening of ligand-gated sodium and potassium channels

   C. Opening of ligand-gated potassium and chloride channels

   D. Opening of ligand-gated sodium, potassium, and chloride channels

ANSWERS

9.1 **C.** Norepinephrine β-receptors **activate adenylate cyclase** via G proteins. Norepinephrine acts in the CNS and the PNS through activation of receptors that are G protein linked. There are a number of different receptors used by this neurotransmitter, and they fall into two broad categories, alpha and beta. Alpha receptors in general are excitatory and are linked via the Gq second messenger to phospholipase C. Beta receptors tend to be inhibitory (although not always) and are linked via Gs or Gi to adenylate cyclase. Both answers A and B refer to inotropic receptors, which are involved in ACh neurotransmission and with amino acid neurotransmitters like glutamate and glycine. Phospholipase A2 is involved as a second messenger system in some neuropeptide neurotransmitters.

9.2 **D.** The neuron described in this question appears to respond to neurotransmitter application via a **voltage-gated ion channel.** The effect of the neurotransmitter is immediate and local, and there appears to be no lasting effect on the neuron from its application. The other mechanisms listed are G protein–coupled responses, which would take longer to have an effect, may alter cell physiology throughout the neuron, and could have a lasting effect.

9.3 **B.** The application of ACh to the neuromuscular junction results in the **opening of ligand-gated sodium and potassium channels.** The opening of these channels alters the permeability of the membrane to ions such that the membrane potential approaches zero, which is sufficiently depolarized to result in the opening of voltage-gated sodium channels. The opening of voltage-gated sodium channels triggers an action potential, resulting in muscular contraction.

NEUROSCIENCE PEARLS

▶ Myasthenia gravis is a rare autoimmune disorder in which antibodies are formed against postsynaptic nicotinic acetylcholine receptors at the neuromuscular junction of skeletal muscle.

▶ Neurotransmitters can bind two general classes of receptors: inotropic and metabotropic receptors.

▶ Inotropic receptors, when activated by a ligand, directly open a transmembrane ion channel.

▶ Metabotropic receptors, when activated by a ligand, set off a cascade of intracellular molecular signals, which leads to the indirect opening of a transmembrane ion channel.

▶ Metabotropic receptors are coupled to either G protein–coupled receptors or tyrosine kinase receptors.

▶ ACh can bind to two types of AChR: nicotinic receptors and muscarinic receptors.

▶ Nicotinic receptors are found at the neuromuscular junction and in the preganglionic endings of both the sympathetic and parasympathetic nervous system.

▶ Muscarinic receptors are found in the postganglionic endings of the parasympathetic nervous system and in certain sympathetic endings.

▶ NMDA receptors require both ligand binding of glutamate and depolarization to open, and they are very important in long-term potentiation.

REFERENCES

Bear MF, Connors B, Paradiso M, eds. *Neuroscience: Exploring the Brain.* 3rd ed. Baltimore, MD: Lippincott Williams & Wilkins; 2006.

Purves D, Augustine GJ, Fitzpatrick D, Hall WC, eds. *Neuroscience.* 5th ed. Sunderland, MA: Sinauer Associates Inc; 2011.

Squire LR, Berg D, Bloom FE, du Lac, S, eds. *Fundamental Neuroscience.* 4th ed. San Diego, CA: Academic Press; 2012.

A 51-year-old man presents to the neurosurgery clinic with a 4-month history of sharp, stabbing pain from his low back to the front of his thigh. He has difficulty walking because of the pain which is exacerbated by physical activity. However, he finds relief when leaning forward or sitting, but minimal improvement with rest. He complains of mild weakness when he kicks with his left leg but denies any difficulty with bowel or bladder control. There is no history of recent trauma. On physical examination, there is mild weakness in his left quadriceps femoris muscle. Muscle stretch reflex in the left patellar tendon is decreased. There is no clonus. Babinski response is negative bilaterally. He has a normally based gait with appropriate arm swing. It is determined that the patient has a lumbar disk herniation at L3-L4.

▶ What imaging study would the physician obtain to further confirm the diagnosis?
▶ What treatments are available?

ANSWERS TO CASE 10:

The Neuromuscular Junction

Summary: A 51-year-old man presents with pain in his low back and left anterior thigh with mild weakness in the left quadriceps muscle and decreased reflex in his left patellar tendon.

- **Further imaging studies:** MRI of the lumbar spine

- **Available treatments:** Conservative management versus surgical excision of the herniated disk

CLINICAL CORRELATION

Only 1%-3% of low back pain, one of the most common ailments for which patients seek medical treatment, is caused by **lumbar disk herniations**. The intervertebral disk is composed of a central **nucleus pulposus** surrounded by a fibrous **annulus** which acts to provide support to the spine and allow for stable motion. With aging, the proteoglycans within the **nucleus pulposus** desiccate, which results in a loss of disk space height and a greater susceptibility to injury. Tears in the annulus allow the nucleus pulposus to protrude or herniate out of the disk space and impinge on the passing nerve root. Impingement of the nerve root can result in symptoms such as pain radiating down an extremity, motor weakness in the distribution of a nerve root, dermatomal sensory changes, and/or decreased muscle stretch reflexes.

The most common location for lumbar disk herniations is L5-S1, with the second most common being L4-L5. The L3-L4 level is a less common site of pathology. The herniated disk will most likely impinge on the nerve root from the lower lumbar level; that is, an L5-S1 disk herniation will affect the S1 nerve root.

The clinician can use the straight leg raising test or **Lasegue's sign** to distinguish possible radicular pain from pain secondary to hip pathology. In this maneuver, the patient is supine and the clinician raises each leg independently while keeping the knee extended. Pain from a herniated disk will generally be reproduced before the leg is raised more than 60 degrees. The **FABER** test stresses the hip joint and does not elicit radicular pain.

The initial treatment of an acute disk herniation and radiculopathy is conservative management since upward of 85% of patients will improve on their own within 5-8 weeks.

If the patient develops **cauda equina syndrome** or a progressive motor deficit, then emergency surgery should be considered. Cauda equina syndrome may develop from a very large midline disk herniation, most commonly at the L4-L5 levels. The patient may experience low back pain or radicular pain. There can be significant motor weakness, which can progress to paraplegia if not treated. The most common sensory deficit is a **saddle anesthesia** involving the anus, lower genitalia, perineum, inner thighs, and buttocks. Also, the patient may have difficulty with bladder and bowel control to varying degrees. The surgical treatment of lumbar disk herniations consists of removal of the offending disk material and decompression of the nerve root through a posterior approach called a **lumbar diskectomy**.

APPROACH TO:
The Neuromuscular Junction

OBJECTIVES

1. Identify the components of the neuromuscular junction.

2. Describe the morphology of the endplate potential waveform.

3. Describe how the endplate potential is converted into an action potential (AP).

4. Describe changes that occur with denervation of a muscle.

DEFINITIONS

MOTOR ENDPLATE: The region of the muscle fiber innervated by a motor nerve.

ENDPLATE POTENTIAL: The postsynaptic potential in the muscle fiber that occurs following the release of acetylcholine (ACh) from the nerve terminal.

DISCUSSION

The **neuromuscular junction** (NMJ) is the interface between a motor neuron and a skeletal muscle fiber at a region called the **motor endplate** (Figure 10-1). The axon of the motor neuron loses its myelin sheath as it approaches the muscle fiber and splits into multiple fine **synaptic boutons**. The **synaptic cleft** is approximately 100 nm wide from the bouton to the surface of the muscle fiber, which contains multiple deep invaginations called **junctional folds**. The cleft contains the basement membrane composed of collagen and extracellular matrix proteins that anchor **acetylcholinesterases** to hydrolyze and inactivate the neurotransmitter ACh. The surface

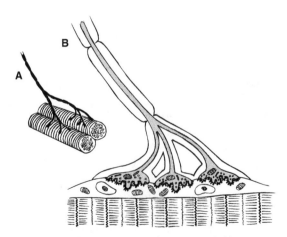

Figure 10-1. Schematic illustrations of a neuromuscular junction. **A:** Motor neuron supplying several muscle fibers. **B:** Cross-section as seen in an electron micrograph. (*With permission from* Waxman's Clinical Neuroanatomy. *25th ed. New York, NY: McGraw-Hill; 2002:3, Fig. 3-11.*)

of the postsynaptic membrane contains **nicotinic acetylcholine receptors** (AChR) at the surface of the muscle fiber and voltage-gated Na^+ channels deep within the junctional folds.

As the AP travels to the bouton, the synaptic vesicles release ACh into the synaptic cleft in a process described in the previous chapters. ACh rapidly diffuses across the synaptic cleft and binds to the AChRs. It is cleared from the cleft by diffusion out of the synaptic cleft and hydrolysis by the acetylcholinesterases.

As previously discussed, the neurotransmitter binds to the inotropic receptors and rapidly depolarizes the membrane at the motor endplate, resulting in a postsynaptic potential called the **endplate potential**. The amplitude of the postsynaptic potential is greatest at the endplate and diminishes as it passively spreads away because of the leakage of current along the muscle fiber. This is in contrast to the AP, which is capable of regeneration along its course.

Recordings of the endplate current demonstrate a rapid depolarization followed by a more gradual repolarization. This wavelike morphology is due to several factors. Stimulation of a motor neuron releases large numbers of ACh molecules, which bind to and rapidly open more than 200,000 AChRs. This results in a rapid and large depolarization in the postsynaptic membrane. However, ACh is rapidly cleared from the synaptic cleft and receptors, and the channels begin to close in a random manner, producing small decreases in the endplate potential in a step-like fashion. However, owing to the sheer number of channels involved, the step-like decrease in the magnitude of the current appears more smooth and gradual.

The depolarization that results from the stimulation of a single motor neuron is up to 70 mV at the neuromuscular junction. This is in contrast to the postsynaptic potentials produced in the central nervous system, which reach an amplitude of approximately 1 mV. The motor endplate potential is usually sufficient to activate the voltage-gated Na^+ channels in the junctional folds. The endplate potential is converted into an AP and is propagated throughout the muscle fiber, which results in an increase in the intracellular Ca^{2+} ion concentration and the contraction of the muscle fiber.

Injury to a nerve supplying muscle fibers leads to denervation changes to the muscle, which occurs in several stages. The distal segment of the axon produces spontaneous injury potentials from hypopolarization of the nerve membrane. These injury potentials travel to and stimulate the muscle fiber, resulting in coordinated contractions called **fasciculations,** which are visible to the eye and are one of the earliest indications of denervation. As the distal segment of the injured nerve continues to degenerate, the multiple terminals of the axon are separated. They continue to produce injury potentials and isolated muscle fiber contractions, but in an uncoordinated fashion. These **fibrillations** are not visible to the eye but can be detected by an electromyogram. Finally, following complete degeneration of the nerve, the muscle no longer receives any type of potential and is electrically silent. **Degeneration atrophy** occurs, resulting in significant loss of bulk and tone. Denervated muscles will initially upregulate their AChRs to serve as targets for the regenerating nerve. However, if reinnervation has not occurred within 2 years, the receptors are lost.

> **CASE CORRELATES**
>
> - See Case 1 (cell types), Case 2 (the neuron), Case 3 (electrical properties and resting membrane potential), Case 4 (myelin sheath and action potentials), Case 5 (neural synapses), Case 6 (synaptic integration), Case 7 (neurotransmitter types), Case 8 (neurotransmitter release), and Case 9 (neurotransmitter receptors).

COMPREHENSION QUESTIONS

10.1 A 27-year-old man is brought into the emergency room immediately after having injected himself with a rather large dose of atracurium (a nondepolarizing skeletal muscle relaxant). He has flaccid paralysis throughout his body and is being mechanically ventilated by the paramedics who brought him in. Which of the following events at the neuromuscular junction is inhibited by this drug?

   A. Influx of calcium ions at the presynaptic terminal as a result of depolarization

   B. Release of synaptic vesicles as a result of Ca^{2+} influx

   C. Depolarization of the postsynaptic membrane by AChR activation

   D. Removal of ACh from the synaptic cleft by acetylcholinesterase

10.2 A 35-year-old woman presents to your office with complaints of weakness and diplopia, both of which are worse at the end of the day. She is concerned that she may have myasthenia gravis (MG), and you order a Tensilon test. In this test a short-acting acetylcholinesterase inhibitor is administered to see if it results in improvement of symptoms. In what location in the neuromuscular junction does this drug act?

   A. Presynaptic membrane

   B. Inside the synaptic vesicle

   C. Synaptic cleft

   D. Postsynaptic membrane

10.3 In what way is the motor end plate potential (EPP) different from the AP?

   A. Involves the opening of sodium and potassium channels

   B. Results in membrane depolarization

   C. Decreases as it travels down the length of the cell membrane

   D. Occurs as a result of normal depolarization of the motor nerve

10.4   A 59-year-old man complains of generalized weakness, including the inability to get out of a sitting position. He was diagnosed with lung cancer a month ago. The physical examination shows weakness of both lower extremities and upper arms. Electromyographic studies show poor signal transmission at the neuromuscular junction. This patient most likely has:

A.   Antibodies to the acetylcholine receptor

B.   Antibodies to presynaptic calcium channels

C.   Long-standing diabetes mellitus

D.   Adult polymyositis

ANSWERS

10.1   **C.** Atracurium (and all nondepolarizing skeletal muscle relaxants) work by binding to AChRs in the neuromuscular junction and preventing their activation by ACh, thus preventing **AChR-mediated depolarization of the postsynaptic membrane.** This results in total flaccid paralysis, which can result in death secondary to paralysis of the respiratory muscles unless mechanical ventilation is continued until the patient has recovered from the effects of the drug. Calcium influx into the presynaptic terminal is impaired in Eaton-Lambert myasthenia, synaptic vesicle release is impaired in botulism, and removal of ACh by acetylcholinesterase is blocked by a number of therapeutic drugs and also by nerve agents such as sarin gas.

10.2   **C.** Acetylcholinesterase is the enzyme primarily responsible for the degradation of ACh and its removal from the **synaptic cleft.** The enzyme is located within the synaptic cleft. The defect in MG is a paucity of AChRs on the postsynaptic membrane, resulting in impaired neurotransmission at the neuromuscular junction. By inhibiting acetylcholinesterase within the synaptic cleft, the concentration of ACh is increased, which increases the amount of neurotransmission. This results in symptomatic improvement in a large number of cases.

10.3   **C.** The EPP only depolarizes in the vicinity of AChRs, so it **decreases in amplitude as it travels along the length of the cell membrane.** The motor EPP triggered by the release of ACh from the presynaptic motor nerve is very similar to an AP but has some important differences. Both potentials are the result of opening of both sodium and potassium channels in the membrane, and both serve to depolarize the membrane from its normal negative resting potential. The ion channels involved, however, are different. The EPP arises from opening of ligand-gated channels, while the AP arises from the opening of voltage-gated channels. An important consequence of this is that the AP is self-propagating. The depolarization of the membrane causes the opening of more voltage-gated channels, so it spreads rapidly throughout the postsynaptic cell. The EPP only depolarizes in the vicinity of AChRs, so it decreases along the length of the cell membrane. In a normal cell, however, the depolarization from the EPP is sufficient to trigger an AP, which then propagates throughout the muscle cell, so both types of potential normally occur following the depolarization of a motor nerve.

10.4   **B.** This patient with lung cancer and weakness due to problems at the neuro-muscular junction has Eaton-Lambert myasthenic syndrome. It is due to **antibodies to presynpatic calcium channels** and can mimic myasthenia gravis. The Tensilon test does not improve strength.

NEUROSCIENCE PEARLS

▶ The cauda equina is the bundle of intradural nerve roots at the end of the spinal cord. Compression of the cauda equine leads to back pain and bowel and bladder dysfunction, and is a surgical emergency.

▶ Injury to a nerve causes distal segments of the axon to produce spontaneous injury potentials, manifesting as fasciculations.

▶ As the distal segment of the injured nerve continues to degenerate, the multiple terminals of the axon separate and produce uncoordinated muscle fiber contractions. These are called fibrillations and can only be detected by an electromyogram.

▶ If reinnervation does not occur within 2 years, the postsynaptic AChRs will be lost.

REFERENCES

Bear MF, Connors B, Paradiso M, eds. *Neuroscience: Exploring the Brain.* 3rd ed. Baltimore, MD: Lippincott Williams & Wilkins; 2006.

Purves D, Augustine GJ, Fitzpatrick D, Hall WC, eds. *Neuroscience.* 5th ed. Sunderland, MA: Sinauer Associates Inc; 2011.

Squire LR, Berg D, Bloom FE, du Lac, S, eds. *Fundamental Neuroscience.* 4th ed. San Diego, CA: Academic Press; 2012.

A 23-year-old graduate student brings her 20-month-old boy to the student health clinic. She is concerned that he doesn't talk yet and still looks "small" for his age with a head that looks "tiny." She noticed that he was a lot smaller than his peers in day care and has been small since birth. Despite his size, he has gotten into trouble for his hyperactive behavior. On examination, the boy's head measures 46 cm (5th percentile for age). Though both she and her boyfriend are both over 6 ft tall, their son is less than the 10th percentile in both height and weight. Examination of his face revealed small eyes and thin, smooth lips. The results of her prenatal tests from her college health clinic were normal. She notes that the pregnancy was unplanned and occurred during her senior year of college. The patient's constellation of symptoms points to an early developmental insult, either from a genetic or congenital factor. The physician elicits a maternal history of alcohol consumption during the pregnancy. She states she has stopped the alcohol use.

▶ What is the most likely cause of the boy's symptoms?
▶ What is the underlying mechanism of this disorder?
▶ Are there any preventative measures available?

ANSWERS TO CASE 11:
Central Nervous System Development

Summary: This young mother gave birth to a baby boy who has many developmental issues that affect his appearance (morphology), size (growth), and behavior (neurological development). She admitted to binge drinking and having "partied" too much during her senior year, resulting in a conviction for intoxicated driving, but has since sought help and is now sober. Examination of the toddler revealed poor motor dexterity and coordination for his age as well as speech delay.

- **Cause of the patient's symptoms:** Fetal alcohol syndrome (FAS) is a likely diagnosis. CNS manifestations of alcohol exposure result in structural deficits like small head circumference and facial abnormalities. The neurological sequelae can cause difficulty with motor and coordination tasks. The functional deficits can be the most detrimental: developmental delay, mental retardation, hyperactive behavior, problems with daily living, and poor reasoning and judgment.

- **Underlying mechanism:** Dysmorphia occurs when normal development of the organism is altered, resulting in features that are altered in shape, size, or positioning. Alcohol is a known teratogen that affects the development of the CNS, through the disruption of biochemical signals, altered gene expression, and changes in cell growth and survival. Despite the early developmental nature of the insult, the CNS deficits generally persist for life.

- **Preventive measures:** FAS is completely preventable—if a woman abstains from drinking alcohol while pregnant or when she could become pregnant. There is no level of alcohol consumption in pregnancy that has been determined to be safe.

CLINICAL CORRELATION

FAS is one of the most common causes of mental retardation in the United States and is the most severe form of fetal alcohol spectrum disorders (FASDs). Data from the US Centers for Disease Control and Prevention show FAS to have an incidence of 0.2-1.5 cases per 1000 live births. Alcohol usage is widespread; based on the 2012 CDC report, more than half of women of childbearing age reported alcohol consumption during the prior month. Most drank only occasionally, but 15% could be classified as moderate or heavy drinkers. Some 13% of women had consumed five or more drinks on one occasion (binge drinking) in the last month. Nearly half of all US pregnancies are unplanned, so the risk of alcohol exposure in pregnancy is significant.

Diagnosis of FAS requires the presence of three hallmark characteristics: facial abnormalities, growth deficits, and CNS abnormalities. The CNS abnormalities can be subdivided into structural, neurological, and functional domains. FASDs represent a continuum of symptoms that range from the mildest functional perturbations to full-blown FAS, with all of them linked by early maternal alcohol consumption.

APPROACH TO:

Central Nervous System Development

OBJECTIVES

1. Relate the various CNS developmental structures.

2. Understand the sequence of steps involved in CNS formation.

3. Understand the tissue-specific signals involved in CNS development.

DEFINITIONS

ECTODERM: The outermost of the three main embryonic germ cell layers, which gives rise to the tissues of the central and peripheral nervous system as well as the skin. The **endoderm**, or innermost layer, generates the gut, lungs, and liver. The **mesoderm**, or middle layer, develops into muscle, the vascular system, and connective tissue.

NEURAL PLATE: The earliest structure formed in the CNS, which appears as a shoe-shaped thickening of the ectoderm beginning around the third week of development.

NEURAL GROOVE: A longitudinal invagination of the neural plate that occurs soon after the plate is formed.

NEURAL TUBE: The lateral edges of the groove form a neural fold that becomes elevated, approaches the midline, and fuses together to form a tubelike structure. This fusion first occurs in the cervical region and proceeds in both cranial and caudal directions, finishing around the 27th day of gestation.

FOREBRAIN (PROSENCEPHALON): The cranial end of the neural tube forms three distinct dilations, with the forebrain being the most cranial. The prosencephalon later forms the **telencephalon**, which ultimately develops into the cerebral hemispheres (most rostral) and the **diencephalon**, which becomes the thalamus, hypothalamus, epithalamus, and posterior pituitary.

MIDBRAIN (MESENCEPHALON): This is the second dilation of the neural tube.

HINDBRAIN (RHOMBENCEPHALON): The third and most caudal of the initial dilations and consists of two parts, the **metencephalon**, which later forms the pons and cerebellum, and the more caudal **myelencephalon**, which becomes the medulla. The spinal cord develops later from the most caudal portion of the neural tube.

NOTOCHORD: A mesodermal structure that lies ventral to the neural tube and induces the formation of ventral neural structures.

DIFFERENTIATION: The process of progressive changes that a cell undergoes to mature from a cell that has the ability to become multiple cell types into a cell with a more restricted fate.

INDUCING FACTORS: Inducing factors are signaling molecules that are provided by other cells that can influence the differentiation of a particular cell.

COMPETENCE: The ability of a cell to respond to developmental signals like inducing factors.

DISCUSSION

The central nervous system (CNS) forms through progressive rounds of differentiation, which are orchestrated by a series of exposures to signals mediated by inducing factors.

One of the earliest stages is the **segregation** of a population of ectodermal cells, which will become neural tissue from others that will become epidermis. This differentiation of ectoderm is mediated by exposure to inductive signals from underlying mesoderm that block the activity of a family of growth factors called bone morphogenetic proteins (BMPs), causing the ectoderm to form neural tissue. Examples of these protein signals include **chordin, noggin, and follistatin**. Cells that do not receive these signals develop into epidermis and can no longer form neural structures.

Once these cells become committed to become neural tissue, they coalesce to form the **neural plate** (Figure 11-1). Within the neural plate, further signaling occurs that organizes the cell fates of these neural cells based upon their location within the plate. Through the process of **neurulation**, the neural plate invaginates to form the neural groove, which subsequently fuses into a complete neural tube. The elevation of the lateral edges of the groove is the first sign of brain development. Two major axes become important for the specification of neural identity. The mediolateral axis in the flat neural plate develops into the dorsoventral axis in the closed neural tube. The rostrocaudal axis running the length of the organism determines the four major subdivisions within the CNS (from rostral to caudal): forebrain, midbrain, hindbrain, and spinal cord. The rostral two-thirds of the neural tube develop into the brain, while the caudal one-third develops into the spinal cord. Failure of the cranial closure results in a condition called anencephaly.

The dorsoventral axis in the developing CNS relies upon further inductive factors produced by nonneural mesodermal cells. The **notochord**, a ventral midline structure composed of mesoderm produces the inducing factor, sonic hedgehog, which then signals the neurons in the ventral neural tube to form motor neurons and ventral interneurons. In a complementary fashion, cells in the dorsal epidermal ectoderm produce several BMPs that pattern the dorsal neural tube to form dorsal sensory interneurons. In a piece of developmental alchemy, the amount or concentration of these inducing signals sets up gradients along the dorsoventral axis, which are critical in specifying the different cell types along the axis.

The rostrocaudal axis relies upon several layers of patterning and signals. The original inductive signals that produce the neural plate, mediated by BMP and its inhibitors chordin, noggin, and follistatin, appear sufficient to specify the rostral CNS and forebrain. The more posterior (caudal) portion of the neuraxis uses the fibroblast growth factor (FGF) family of proteins and the steroid-like retinoic acid to pattern the midbrain, hindbrain, and spinal cord. There are eight segments of the hindbrain called rhombomeres. Rhombomeres one through five comprise the metencephalon, while rhombomeres six through eight comprise the myelencephalon. *Hox* genes are expressed in rhombomeres three through eight and are regulated by retinoic acid.

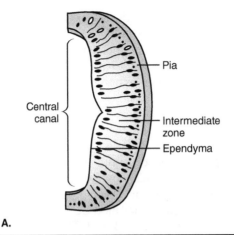

A.

Coronal view
of embryo brain

Lateral
ventricles

Cortical plate

Ventricular layer

Choroid plexus

III Ventricle

Cortical plate

Ventricular layer

Subventricular layer

Intermediate zone

Marginal layer

Ventricle

Dividing   Migrating   Developing
neuroblasts   neuroblasts   cortex

B.

Figure 11-1. Two stages in the development of the neural tube (only half of each cross-section is shown). **A.** Early **B.** Later.

To better understand the finer patterning of the neuraxis, the organization of the hindbrain serves as a model system. Within the hindbrain, discrete swellings called **rhombomeres** occur in a distinct pattern. The rhombomeres contain sensory and motor neurons that form the cranial nerves that leave the CNS to innervate specific target tissues that develop from the branchial arches, an evolutionary remnant from the gills of aquatic vertebrate ancestors. For example, rhombomeres 4 and 5 contain the neurons/nuclei that give rise to the facial nerve (cranial nerve VII), which goes on to innervate the muscles of facial expression that develop from the second branchial arch.

The *Hox* family of transcription factors demonstrates a very specific pattern of expression within the hindbrain rhombomeres. This family of genes is clustered on the chromosomal DNA, with the genes in the 5′ end of the cluster preferentially expressed in the more caudal portions of the neural tube. Conversely, the rostral rhombomeres contain the *Hox* gene products from the 3′ end. Thus, each rhombomere is defined by a certain combination of *Hox* gene products. These transcription factors delineate the identity of the neurons within each rhombomere—for example, *Hoxb-1* is highly expressed in rhombomere 4, which gives rise to facial neurons. Elimination of this gene results in the development of trigeminal neurons instead of facial neurons in rhombomere 4. Other genes influence the expression of the *Hox* genes, like *Krox-20*, which are themselves subject to other inducing factors. In this way, CNS development is built upon layers of specific, regulated signaling between cells that orchestrate a symphony of gene activation and inhibition resulting in precise cell type identity.

This precise system is reiterated throughout embryonic development in many organ systems and tissues. Because the nervous system is the most structurally and functionally complex organ system, it is also the most susceptible to perturbation and damage. A number of internal (genetic mutations) and external (teratogens) can upset the development of this system, with maternal alcohol exposure being a prime example.

COMPREHENSION QUESTIONS

11.1 A 54-year-old man presents to your clinic with a 1-year history of progressively worsening gait disturbance. Additionally, he has recently begun to have some slurred speech. After extensive workup and testing, you diagnose him with a sporadic form of cerebellar ataxia. From what embryologic structure does the cerebellum arise?

    A. Telecephalon

    B. Mesencephalon

    C. Metencephalon

    D. Myelencephalon

11.2 You are following a 44-year-old man who was recently diagnosed with amyotrophic lateral sclerosis (ALS) after presenting with difficulty writing secondary to hand and finger weakness. ALS is a disease of the anterior horn cells (motor neurons of the spinal cord). From what part of the neural tube did the cells affecting this man's writing arise?

A. Ventral aspect of the cranial two-thirds of the neural tube

B. Dorsal aspect of the cranial two-thirds of the neural tube

C. Ventral aspect of the caudal one-third of the neural tube

D. Dorsal aspect of the caudal one-third of the neural tube

11.3 A 24-year-old woman with severe nodulocystic acne presents to your clinic and is interested in beginning therapy with isotretinoin (Accutane), which is a derivative of retinoic acid. You agree to prescribe the medication after administering a pregnancy test, which is negative. In order for her to get the medication, however, you require that she also takes birth control and uses at least one other form of contraception because Accutane is a known teratogenic agent, interfering with the action of endogenous retinoic acid. Endogenous retinoic acid is important for what stage of differentiation of the developing CNS?

A. Formation of the neural plate

B. Fusion of the neural groove to form the neural tube

C. Craniocaudal axis differentiation of the neural tube

D. Dorsoventral axis differentiation of the neural tube

ANSWERS

11.1 **C.** The cerebellum arises from the **metencephalon.** The cranial two-thirds of the neural tube (which develops into the brain) is divided into five dilations along its craniocaudal axis. From cranial to caudal, they are the telencephalon, which becomes the cerebral hemispheres; the diencephalon, which becomes the thalamus, hypothalamus, epithalamus, and subthalamus; the mesencephalon, which becomes the mid-brain; the metencephalon, which becomes the pons and cerebellum; and the myelencephalon, which becomes the medulla. The one-third of the neural tube caudal to these dilatations becomes the spinal cord.

11.2 **C.** ALS is a disease that affects the motor neurons in the spinal cord, which arise from the **ventral aspect of the caudal one-third of the neural tube.** Recall that the cranial two-thirds of the neural tube differentiate into the brain and brainstem, and the caudal one-third becomes the spinal cord. Also recall that the dorsal aspect of the neural tube becomes sensory neurons, while the ventral aspect becomes motor neurons. Therefore, motor neurons in the spinal cord (those controlling the hands and fingers) would have arisen from the ventral aspect of the caudal one-third of the neural tube.

11.3 **C.** Retinoic acid is an important signaling molecule involved in the **craniocaudal differentiation of the neural tube.** A gradient of retinoic acid is established, with higher concentrations at the cranial end of the neural tube. Maternal ingestion of Accutane results in higher than normal concentrations of retinoic acid and disrupts this gradient, resulting in birth defects and spontaneous abortions. Accutane is FDA pregnancy category X.

NEUROSCIENCE PEARLS

▶ Fetal alcohol syndrome is caused by ethanol consumption during pregnancy, leading to growth problems, mental retardation, and craniofacial abnormalities.

▶ The CNS develops through a series of regulated spatial and temporal signals that pattern and differentiate primitive ectoderm into the CNS.

▶ Structurally, the CNS begins as a flat plate that progressively folds into a tube and elongates, with dilated regions corresponding to specialized structures (for example, cerebral cortices).

▶ CNS development is very sensitive to perturbation and is susceptible to numerous exogenous toxins and teratogens.

REFERENCES

Calhoun F, Warren K. Fetal alcohol syndrome: historical perspectives. *Neurosci Biobehav Rev.* 2007;31(2):168–171. Review.

Kandel ER, Schwarz JH, Jessell TM, Siegelbaum SA, Hudspeth AJ, eds. *Principles of Neural Science.* 5th ed. New York, NY: McGraw-Hill; 2012.

Sadler TW, ed. *Langman's Medical Embryology.* 7th ed. Baltimore, MD: Williams and Wilkins; 1995.

A 32-year-old mother brings in her 5-day-old newborn boy for his first postnatal checkup. The pregnancy was uneventful—all the screening tests were normal; he was born after a full-term gestation via vaginal delivery without complication; and height, weight, and head circumference were within normal ranges. The mother remarked that her son has been increasingly difficult over the last 2 days. He had been breast-feeding well but has started to spit up more of his feedings. She is also concerned about his lack of bowel movements. He averaged 10 wet diapers a day, but has only had one small watery soiled diaper since he left the hospital. Initial physical examination revealed a crying but normal-appearing infant with a moderately distended abdomen. Rectal examination revealed an empty rectal vault and abnormal muscle tone. An abdominal x-ray was obtained that demonstrated a distended colon up to the transverse colon with gas and feces present. The diagnosis is determined to be Hirschsprung disease, which results from the absence of parasympathetic ganglion neurons in the myenteric and submucosal plexus of the rectum and/or distal colon. Without these neurons, the gut musculature remains tonically contracted, unable to distend.

▶ What is the cause of the patient's symptoms?
▶ What is the underlying mechanism causing the symptoms?

ANSWERS TO CASE 12:
Peripheral Nervous System Development

Summary: A 5-day-old newborn boy presents with a progressive history of constipation, vomiting, and abdominal distention. Initial plain radiograph imaging confirms large bowel obstruction with proximal dilation with gas and feces. Further radiographic images taken during a barium enema study confirm obstruction at the mid-transverse colon. A diagnosis of congenital aganglionic megacolon or Hirschsprung disease is made. A rectal biopsy is performed that demonstrates a lack of ganglion neurons in the myenteric plexus and increased acetylcholinesterase staining—pathognomonic for Hirschsprung disease.

- **Cause of the patient's symptoms:** The abdominal obstruction is caused by a lack of nervous system control of the intestinal musculature. Normal gut function requires motility that is mediated by the peripheral nervous system (PNS).

- **Underlying mechanism:** The neurons that are fated to innervate the distal gut travel along the vagus nerve (cranial nerve X) to populate the enteric gut plexus. It is failure of neurons to successfully navigate this distant trip that results in Hirschsprung disease.

CLINICAL CORRELATIONS

Congenital aganglionic megacolon, or Hirschsprung disease, was first recognized in 1886 as a cause of constipation in early infancy. The ganglion cells that innervate the enteric nervous system are derived from neural crest cells. These cells are of ectodermal origin and form along the lateral "crest" as the neural plate undergoes neurulation. They have multiple fates, with some forming the autonomic ganglia in the PNS. A subset of these cells travels with the vagus nerve along the intestinal tract to populate the enteric plexus. These ganglion cells arrive in the proximal colon by 8 weeks of gestational age and in the rectum by 12 weeks of gestational age. Arrest in this migration leads to an aganglionic segment and Hirschsprung disease. Hirschsprung disease occurs in around 1 in 5000 live births in the United States. It is associated with enterocolitis in about one-third of cases, which represents the majority of the mortality. More severe cases are diagnosed in neonates with signs and symptoms of bowel obstruction and inability to pass meconium or stool. Milder cases are diagnosed at later ages with chronic constipation, abdominal swelling, and malnutrition (decreased growth). Most cases are sporadic in nature, though a family history of a similar condition is present in up to 30% of cases. Male cases outnumber female cases four to one. A strong association exists with Down syndrome—5%-15% of patients with Hirschsprung disease also have trisomy 21. Mutations in a variety of genes have been associated with Hirschsprung disease: the *RET* proto-oncogene, glial cell–derived neurotropic factor (GDNF), the endothelin-signaling system, and *SOX10* (sex-determining region Y-box).

APPROACH TO:
Peripheral Nervous System Development

OBJECTIVES

1. Relate the various PNS developmental structures.

2. Understand the sequence of steps involved in PNS formation.

3. Understand the tissue-specific signals involved in PNS development.

DEFINITIONS

PNS: The part of the nervous system that is not a part of the CNS, that is, everything except the brain and spinal cord. It can be subdivided into somatic and autonomic components.

SOMATIC NERVOUS SYSTEM: Composed of efferent nerves that control skeletal muscle and external sensory receptors, as well as afferent nerves that convey sensory information from skin and muscle receptors.

AUTONOMIC NERVOUS SYSTEM (ANS): Composed of efferent and afferent nerves that control/regulate homeostatic processes and internal organ physiology. The ANS can be further subdivided into the **sympathetic** and **parasympathetic** systems.

NEURAL CREST CELL (NCC): An ectodermal-derived cell originating at the lateral edge of the neural plate. As the neural folds rise up and fuse to form the neural tube, the NCCs are brought together along the dorsal "crest" of the tube. These cells then dissociate from the neuroepithelium, allowing them to migrate throughout the organism to form the PNS.

ECTODERMAL PLACODE: These placodes are discrete areas of thickened ectoderm that appear on the heads of all embryos and develop into the peripheral sensory nervous system. Along with NCC, these placode cells form the PNS.

EPITHELIO-MESENCHYMAL TRANSFORMATION: A critical step in the differentiation of neural crest cells as they separate from the neuroepithelium. This delamination process is mediated by changes in the cell–cell and cell–intracellular matrix interactions.

DISCUSSION

The entire PNS of vertebrates is descended from two embryonic cell populations: the neural crest and the cranial ectodermal placodes. Both cell populations form at the lateral border of the neural plate, with the placodes occupying the most rostral end of the neuraxis and the neural crest starting at the level of the diencephalon and descending caudally. They both give rise to a vast assortment of cell types. Neural crest cells form a tremendous array of different cell types: bone and cartilage in the head, teeth, endocrine cells (adrenal medulla), peripheral sensory neurons (including the dorsal root ganglia neurons), all peripheral autonomic neurons (enteric, postganglionic sympathetic, and parasympathetic neurons), all peripheral glial cells, and all melanocytes. Cranial ectodermal placodes are mainly responsible for the numerous peripheral sensory functions in the head, including olfaction, mechanosensory hair cells, trigeminal sensation to the head, taste, as well as all the endocrine cells in the anterior pituitary.

Neural crest development into the PNS is characterized by a complex sequence of events that can be summarized into three stages: induction, migration, and differentiation. The neural plate is defined by early BMP signaling from axial mesoderm (see Case 11), and in much the same way, the neural crest is induced by signals from paraxial mesoderm. These signals appear to involve intermediate levels of the BMP inhibitors, noggin and follistatin. Additional factors involving fibroblast growth factor (FGF) and the WNT gene family continue the process of forming neural crest. The epithelial-mesenchymal transition is the last step in defining a neural crest cell, without which the cell is not a *bona fide* neural crest cell. Multiple signals and genes contribute to this transformation; however, it appears again that BMP signaling is implicated. This delamination process proceeds in a rostral to caudal direction.

Neural crest cells now must migrate to their target tissues. Cranial neural crest cells are observed to travel in characteristic streams associated with the branchial arches as coherent populations. The paraxial mesoderm again plays a vital role in providing the cues (mainly repulsive) to guide the traveling cells. The extracellular matrix plays an important role by providing a permissive pathway for migration. In the rest of the body, the paraxial mesoderm develops into repeating units, called **somites**. Somites are masses of mesoderm distributed along the two sides of the neural tube that will eventually become dermis, skeletal muscle, and vertebrae. These segmented somites define levels along the rostrocaudal axis. Approximately 44 somites form and give rise to the bones of the face, vertebral column, associated muscles, and overlying dermis.

For the final step in development, the neural crest must assume its final identity through differentiation. There are two main theories to explain the lineage segregation of neural crest cells: instruction and selection. The first theory (instruction) posits that the neural crest is a homogenous group of multipotent cells whose differentiation is *instructed* by environmental signals. The second theory (selection) holds that the neural crest is a heterogeneous population of predetermined cells, which are *selected* to survive in permissive environments, and eliminated from inappropriate ones. The experimental evidence suggests that both multipotent and fate-restricted neural crest cells exist. Most of the signals to which the cells respond, not surprisingly, are from the tissues that they migrate past and that surround their postmigration destination.

Initial therapy for Hirschsprung disease consists of gastrointestinal decompression to avoid perforation and enterocolitis. Nasogastric suctioning combined with frequent digital rectal stimulation or irrigation is sufficient. Intravenous antibiotics or hydration is sometimes necessary to treat enterocolitis or dehydration, respectively. Definitive treatment is surgical removal of the dysfunctional bowel and reanastomosis. Other therapeutics, often reserved for recurrent or intractable cases include rectal dilation or myotomy (physical dilation), application of topical nitric oxide (chemical dilation), and injection of botulinum toxin to block muscular contraction (neurotransmitter blockade).

> **CASE CORRELATES**
> • See Cases 11-17 (nervous system development).

COMPREHENSION QUESTIONS

12.1 A 32-year-old woman presents with recurrent, episodic headaches, palpitations, sweating, and hypertension. After appropriate workup, she is diagnosed with a pheochromocytoma, a catecholamine-secreting tumor of the adrenal medulla. From what embryologic structure do the cells that make up this tumor arise?

A. Neural tube

B. Neural crest

C. Epidermal placodes

D. Mesoderm

12.2 A 35-year-old man comes into the office complaining of lower back pain that radiates down the back of his left leg, as well as weakness of foot plantarflexion on the same side. You suspect that this man has a herniated vertebral disk, which is confirmed by MRI which shows herniation of L5-S1 disk and compression of the S1 nerve root. The muscles innervated by this nerve root are derived from a segmental unit of mesoderm known as what?

A. Branchial arch

B. Somite

C. Motor unit

D. Spinal segment

12.3 Which of the following best describes the mesoderm's effect on the migration and differentiation of neural crest cells?

A. Mesoderm guides migration and neural crest cell differentiation.

B. Mesoderm guides migration of neural crest cells but does not affect differentiation.

C. Mesoderm guides neural crest differentiation but does not affect migration.

D. Mesoderm has no effect on either migration or differentiation of neural crest cells.

ANSWERS

12.1 **B.** The chromaffin (catecholamine-secreting) cells of the adrenal medulla arise from the **neural crest,** which also gives rise to ganglion cells of the PNS, and melanocytes, among other structures. Neural tube cells become the cells of the CNS. Epidermal placode cells become PNS structures in the head and neck, and mesoderm gives rise to the cells of the adrenal cortex.

12.2 **B.** In the developing nervous system, each spinal segment is paired with a segment of mesoderm known as a **somite.** Each spinal segment gives rise to paired dorsal sensory and ventral motor nerve roots. These nerve roots innervate the sensory organs and muscles that derive from the corresponding somite. Dermatomes and myotomes are a result of this segmental development and innervation. Branchial arches are related structures that occur in the developing head and neck. A motor unit comprises a motor nerve and the muscles that it innervates; there are multiple motor units in each spinal segment/somite pair.

12.3 **A.** Mesoderm plays a key role in generating signals that affect both the **migration of neural crest cells and their differentiation into their final identities.** Migration of cells is mostly guided by repulsive signals generated by mesoderm not in the proper location for the specific neural crest cells. The final identity of a neural crest cell is determined in large part by signaling molecules secreted by the cells of the surrounding tissue.

NEUROSCIENCE PEARLS

▶ Hirschsprung disease is a developmental disorder in which there is absence of distal colon ganglion cells (which originated from the neural crest), leading to a functional colonic obstruction.

▶ In Hirschsprung disease, both the myenteric (Auerbach's) plexus and the submucosal (Meissner's) plexus is absent, leading to diminished peristalsis.

▶ The PNS develops from two populations of cells formed on the lateral border of the neural plate: neural crest cells and ectodermal placodes.

▶ Neural crest cells undergo a three-stage process to form the PNS: (1) induction; (2) migration; and (3) differentiation.

▶ The paraxial mesoderm is a critical source of signaling in the establishment of the PNS.

REFERENCES

Kandel ER, Schwarz JH, Jessell TM, Siegelbaum SA, Hudspeth AJ eds. *Principles of Neural Science.* 5th ed. New York: McGraw-Hill Publishers, 2012.

Kessmann J. Hirschsprung's disease: diagnosis and management. *Am Fam Physician.* 2006 Oct 15; 74(8):1319-1322.

Paran TS, Rolle U, Puri P. Enteric nervous system and developmental abnormalities in childhood. *Pediatr Surg Int.* 2006;22:945-959.

Rao MS, Jacobson M. *Developmental Neurobiology.* 4th ed. New York, NY: Kluwer Academic/Plenum Publishers; 2005.

Sadler TW, ed. *Langman's Medical Embryology.* 7th ed. Baltimore, MD: Williams and Wilkins; 1995.

CASE 13

A 38-year-old pregnant woman arrives to the emergency room after her "water breaks." On questioning, she admits to missing most of her prenatal care appointments but remembers one of her blood tests was "high" and believes that she is 4 weeks "early." Initial examination shows cervical dilation of 6 cm and increasing strength and frequency of contractions. Fetal heart rate monitoring reveals decelerations indicative of fetal distress, prompting delivery via cesarean section. The baby boy was delivered and had reassuring APGAR scores. He is able to move his head and arms, but his lower legs are motionless and contorted. Further examination reveals a 3-cm clear fluid–filled sac in the mid-lower back.

▶ Describe the anatomic deformity.
▶ What is the underlying etiology of the fluid-filled sac?
▶ Which vitamin may decrease the incidence of this condition?

ANSWERS TO CASE 13:

Neurulation

Summary: A baby boy was delivered with flaccid distal lower extremities. Physical examination demonstrated moderate iliopsoas, hip adductors, quadriceps, and tibialis anterior function with flaccid antagonists resulting in hip adduction-flexion, knee hyperextension, and foot inversion. A thin-walled cystic mass arises from the surrounding skin of the upper lumbar spine. Mild hydrocephalus is noted. The mass is diagnosed as a myelomeningocele, a form of neural tube defect (NTD). The imaging also revealed an associated Chiari malformation and mild hydrocephalus.

- **Anatomic deformity:** Below the L4 level, **the placode of tissue consists of dysplastic neural elements that were unable to form a closed neural tube.** This dysfunction extends into the nerve roots emanating from the placode. The exposed neural elements remain covered in meninges but are not covered by bone, muscle, or skin. Cerebrospinal fluid (CSF) fills the sac as it would normally fill the spinal canal in the nonpathological state, but because of the paucity of overlying tissue, leakage of fluid is common.

- **Underlying etiology:** The etiology of NTDs is multifactorial and not clearly understood. NTDs constitute a variety of congenital malformations, as severe as anencephaly (complete absence of telencephalic structures) or as mild as a tethered spinal cord, which have at its core the failure of proper *neurulation*.

- **Vitamin that decreases incidence of NTD:** Approximately 400-800 μg of folate per day reduces the risk for both the first occurrence of NTDs and recurrent NTDs.

CLINICAL CORRELATION

Congenital malformations of the brain occur in approximately 0.5% of live births. Causes are generally ascribed to exogenous and endogenous sources. Exogenous causes include nutritional factors, radiation, infections, chemicals, ischemia, and medications. Endogenous causes are mainly genetic. Defective neurulation, also called neural tube defects (NTDs), **spinal dysraphisms** or **spina bifida** (split spine), are among the most common of these congenital CNS malformations. Though the primary defect is a failure of neurulation and of the neuroectoderm, subsequent maldevelopment occurs in the adjacent mesoderm, which, in turn, is responsible for forming the appropriate skeletal and muscular structures surrounding the nervous system. Therefore, NTDs have been classically classified by the severity of this secondary mesodermal disruption: the more severe **spina bifida aperta** (neural structures communicate with the atmosphere) or the less involved **spina bifida occulta** (neural elements are skin covered).

Spina bifida aperta (open NTD) is clinically quite obvious, like myelomeningocele; however, in spina bifida occulta (closed NTD) the lesion is covered by skin, therefore rendering the underlying neurological involvement **occult** or hidden. There may be subtle skin changes present: a hairy patch, dermal sinus tract, dimple, hemangioma, or lipoma. Besides cutaneous stigmata, closed NTD can present with: spinal defects (scoliosis or defects in the lamina), orthopedic deformities (clubbed feet or leg length asymmetry), urological problems (neurogenic bladder or incontinence), and/or neurological symptoms (leg pain, weakness, or numbness and sometimes atrophy or hyperreflexia).

Many suspected teratogens have been identified: radiation, infections, hyperthermia, valproic acid, and folate deficiency. **Folate** is vital during periods of rapid cell growth such as infancy and pregnancy, as it is **needed to replicate DNA.** Folate deficiency hinders DNA synthesis and cell division, affecting most clinically the bone marrow, a site of rapid cell turnover. **Valproic acid is a known folate antagonist** and its association with NTD may be through that action. Folic acid is not protective unless ingested during the time surrounding conception. Screening tests have made a significant impact—both serum markers and imaging studies have proven useful. A raised maternal serum alpha-fetoprotein (AFP) between 15 and 20 weeks of gestation may be indicative of an open NTD. In addition, the use of fetal ultrasonography can diagnose an NTD with nearly 98% specificity and 95% sensitivity. The prognosis for many forms of NTD is variable: anencephalic babies rarely survive more than a few hours or days and open NTDs like myelomeningocele have many associated anomalies. Hindbrain malformations (Chiari II), hydrocephalus, and syringomyelia, as well as brain stem and cranial nerve malformations, are all commonly associated with myelomeningoceles. These anomalies conspire to diminish the functional independence of many affected babies that survive to adulthood. Nevertheless, many individuals with mild NTDs can function independently.

APPROACH TO:
Neurulation

OBJECTIVES

1. Understand the timing of events during neurulation.

2. Be able to relate the clinical result of developmental failure at various stages in neurulation.

3. Appreciate the role of teratogens in the etiology of NTDs.

DEFINITIONS

NEURULATION: The developmental process by which the neural plate fuses into the neural tube. The process begins in the cervical region and progresses in both directions, first closing the *rostral (cranial) neuropore,* followed by the *caudal neuropore.*

PRIMARY NEURULATION: Refers to the transformation of the neural structures from a plate into a tube, thereby forming the brain and spinal cord. Failure of primary neurulation results in open NTDs.

SECONDARY NEURULATION: An independent process from primarily neurulation that refers to the formation of the lower spinal cord from cells derived from the embryological tail bud.

SPINA BIFIDA: Latin for "split spine," a term used to describe certain NTDs.

ANENCEPHALY: Anencephaly or *craniorachischisis* is the most severe form of NTD and refers to a severe deformity in which an extensive defect in the craniovertebral bone causes the brain to be exposed to amniotic fluid. The defect normally occurs after neural fold development at day 16 of gestation but before the closure of the cranial neuropore at day 24-26 of gestation.

ENCEPHALOCELE: An encephalocele represents herniation of the brain through a skull defect. This most commonly occurs in the occipital region in the United States, while in Asian countries, the frontal bone is most involved.

MYELOMENINGOCELE: A condition in which the spinal cord and nerve roots herniate into a sac comprised of the meninges. This sac protrudes through a bone and musculocutaneous defect. The splayed open neural structure is called the neural placode.

MENINGOCELE: A meningocele is a herniation of only the meninges through the bony defect (spina bifida). The spinal cord and nerve roots do not herniate into the dural sac, as in a myelomeningocele. These lesions are important to distinguish from myelomeningocele because their treatment and prognosis are different from myelomeningocele. Neonates with a meningocele usually have normal findings upon physical examination and do not have associated neurological malformations such as hydrocephalus or Chiari II malformations.

DIASTEMATOMYELIA: A split cord malformation in which a bony spur splits the spinal cord in two.

TETHERED CORD: Tethering of the spinal cord is caused by an abnormal adhesion or thickened filum that can cause traction on the spinal cord with subsequent neurological deficits as the child grows.

HYDROCEPHALUS: Latin for "water head," it is an abnormal accumulation of CSF within the cerebral ventricles because of physical obstruction (obstructive hydrocephalus) or inability to absorb (nonobstructive hydrocephalus) the circulating fluid.

CHIARI MALFORMATIONS: Formerly called Arnold-Chiari malformations, these are a series (types I-III) of hindbrain defects. Type I is characterized by the downward herniation of the cerebellar tonsils through the foramen magnum. A type II malformation is herniation of the cerebellar vermis and brainstem below the foramen magnum. The type III malformation is essentially a posterior fossa encephalocele with herniation of the cerebellum through the posterior fossa bone and is a more severe NTD.

SYRINGOMYELIA: Syringomyelia is a cavitation of the spinal cord leaving a cystic space within the cord, which can cause progressive neurological dysfunction.

This cyst, called a syrinx, can expand and elongate over time and destroy the spinal cord.

DISCUSSION

NTDs include a wide range of clinical malformations: anencephaly, encephalocele, myelomeningocele, meningocele, diastematomyelia, and tethered cord. The lesions often present with associated failure of overlying bony structures to fuse, hence the common term spina bifida, Latin for "split spine." The multiple descriptive names highlight the fact that these malformations involve not only the ectodermal derivatives (central and peripheral nervous system and skin), but also the other embryonic layers as well, most notably the muscle and bone formed from axial mesoderm.

Timing is critical during development—the earlier the developmental insult, the more devastating the consequences. In relation to NTDs, the first important process is primary neurulation, which refers to the formation of the neural structures into a tube, thereby forming the brain and spinal cord. Secondary neurulation refers to the formation of the lower spinal cord, which gives rise to the lumbar and sacral elements. The neural plate is formed at gestational days 17-19, the neural fold occurs at days 19-21, and the fusion of the neural folds occurs at days 22-23. Any disruption during these early processes (neural plate formation until fusion into a neural tube) can cause craniorachischisis, the most severe form of NTD. Subsequent events include the closure of the rostral neuropore at days 24-26. Failure at this point can result in anencephaly and encephaloceles. A myelomeningocele is a result of disruption occurring around days 26-28, during the closure of the caudal neuropore. After day 28, disruptions are unlikely to be able to cause an open NTD such as myelomeningocele but may cause more subtle defects, like a closed NTD or tethered cord.

The molecular mechanisms that control neurulation are poorly understood. Experimental evidence suggests that the mechanisms involved in cranial neurulation and closure of the cranial neuropore differs from the closure process at subsequent axial levels. As the flat neural plate physically invaginates and cavitates to form a tube, it is not surprising that aspects of cellular polarity (to direct the movement of cells), cellular cytoskeleton (to dynamically change the shape of cells), and intercellular adhesion (to fuse the neural folds together) are all critical for proper neurulation.

Various studies have provided evidence for **abnormal folate metabolism** in cell lines from a subset of fetuses affected by NTDs, explaining why prenatal folate is so crucial for decreased incidence of NTDs; however, the specific abnormalities are yet to be elucidated. The recommended daily allowance of folate for women of reproductive age is 400 μg, while for pregnant women it is 600-800 μg. Foods that contain folate include leafy vegetables such as spinach and turnip greens, dried beans and peas, fortified cereal products, and sunflower seeds. Normal folate metabolism begins with folate being reduced first to dihydrofolate and then tetrahydrofolate by the enzyme dihydrofolate reductase.

TREATMENT OPTIONS

Initial management of an infant with a myelomeningocele with leaking CSF is to avoid trauma to the exposed neural placode and roots. Prolonged exposure increases the risk for CNS infections manifested as meningitis or encephalitis, sometimes leading to systemic sepsis. Surgical closure and repair of the lesion by a pediatric neurosurgeon is required. Additional placement of a ventriculoperitoneal shunt to treat the hydrocephalus and prevent spinal fluid leakage from the repair site can be done contemporaneously. Multiple other developmental anomalies are associated with NTD, which would need to be addressed by a multidisciplinary team to include a geneticist, urologist, orthopedic surgeon, physical medicine specialist, and a pediatric neurosurgeon.

CASE CORRELATES

- See Cases 11-17 (nervous system development).

COMPREHENSION QUESTIONS

13.1 A baby is born to a 17-year-old woman with poor prenatal care who admits to not taking her prenatal vitamin supplements. On examination immediately following birth, the child is noted to have a cystic swelling over the lower back in the midline, but appears to have both motor and sensory modalities intact in both lower extremities. The child is diagnosed with a meningocele. A failure of what neurodevelopmental event results in this defect?

A. Closure of the cranial neuropore

B. Closure of the caudal neuropore

C. Formation of the neural tube

D. Separation of the prosencephalon into paired telencephalons

13.2 Which of the following **decreases** the risk for development of NTDs?

A. Decreased maternal AFP

B. Previous history of NTDs

C. Folate deficiency

D. Radiation exposure

E. Maternal valproic acid exposure

13.3 A myelomeningocele is most likely to be a result of a neurulation defect that occurs during which gestational period?

A. Days 19-21

B. Days 22-23

C. Days 24-26

D. Days 26-28

E. Days 28-30

13.4 A 30-year-old woman who is pregnant at 20 weeks gestation had an elevated maternal serum alpha fetoprotein level. She undergoes an amniocentesis, which reveals an elevated level of acetylcholinesterase in the amniotic fluid. Which of the following fetal embryonic problems is most likely?

   A. Incomplete hemispheric maturation

   B. Abnormality in migration

   C. Abnormality in proliferation

   D. Abnormality in fusion

   E. Abnormality in apoptosis

ANSWERS

13.1 **B.** Failure of the neural tube to close results in an NTD. Failure of closure of the cranial neuropore results in anencephaly, in which the brain is not formed, a condition not compatible with life. Failure of closure of the **caudal neuropore** results in a spectrum of NTDs, ranging from occult spina bifida to meningocele to myelomeningocele. Failure of the entire neural tube to form would result in spontaneous abortion of the developing embryo, and a failure of prosencephalon to separate results in a condition known as holoprosencephaly.

13.2 **A.** Increased, not decreased, AFP has been associated with open NTDs. A previous history of a child with an NTD, folate deficiency, radiation exposure, and maternal use of valproic acid are all additional risk factors.

13.3 **D.** Disruption of the closure of the caudal neuropore around **days 26-28** is responsible for the development of a myelomeningocele.

13.4 **D.** Neural tube defects lead to an increased leakage of alpha fetoprotein and acetylcholinesterase into the amniotic fluid, which then leads to an elevated maternal serum AFP. Neural tube defects are due to failure of fusion of the neural tube during the fourth week of embryonic development.

NEUROSCIENCE PEARLS

▶ Folate supplementation can reduce the risk of neural tube defects.

▶ The process of neurulation involves multiple sequential steps.

▶ NTDs result from failure of proper neurulation.

▶ Central nervous system neurulation is linked with the development of vertebral bone and skeletal muscle, derived from the axial mesoderm.

REFERENCES

Czeizel AE, Dudas I. Prevention of the first occurrence of neural-tube defects by periconceptional vitamin supplementation. *NEJM.* 1992;327:1832-1835.

Kandel ER, Schwarz JH, Jessell TM, Siegelbaum SA, Hudspeth AJ, eds. *Principles of Neural Science.* 5th ed. New York, NY: McGraw-Hill; 2012.

MRC Vitamin Study Research Group. Prevention of neural tube defects: results of the Medical Research Council Vitamin Study. *Lancet.* 1991;338:131-137.

Rao MS, Jacobson M, eds. *Developmental Neurobiology.* 4th ed. New York, NY: Kluwer Academic/Plenum Publishers; 2005.

Sadler TW, ed. *Langman's Medical Embryology.* 7th ed. Baltimore, MD: Williams and Wilkins; 1995.

Wyszynski DF, ed. *Neural Tube Defects: From Origin to Treatment.* New York, NY: Oxford University Press; 2006.

Police officers bring in a 72-year-old unidentified woman whom they found walking aimlessly around the grocery store parking lot. On questioning, she remembers her name and that she is 72, but she is not sure where she lives or how to contact her family. Cursory examination reveals that she is wearing clean clothes that are slightly wrinkled and with misaligned buttons. She denies any pain or injury. She cannot recall ever seeing a doctor, commenting that she is "okay" and has always been healthy. On neurological examination, she is unable to recall the month/date/year, mistakenly reciting her date of birth several times, and also has difficulty with arithmetic, word recall, and object-naming tasks. The remainder of the physical examination and laboratory results are normal for her age. Her daughter is found and states that her mother has had difficulty with her memory beginning over a year ago, which has slowly progressed over time. She stopped shopping 6 months ago when she could not remember what to buy or to keep her checkbook balanced and organized. Over the last several months, she stopped cooking for herself. Her daughter has had to assist increasingly with her mother's daily activities. Upon careful consideration of the case, you conclude that the patient suffers from dementia, a degenerative brain disorder that seriously affects a person's ability to carry out daily activities.

▶ What is the most common cause of dementia?
▶ What two abnormal intracellular structures are characteristic of the most common cause of dementia?
▶ What are the risk factors?

ANSWERS TO CASE 14:

Neurogenesis

Summary: The most common form of dementia among older people is Alzheimer disease (AD), which initially involves the parts of the brain that control thought, memory, and language.

- **Most common cause of dementia:** Alzheimer disease is the most common of the neurodegenerative disorders, which also includes Parkinson disease (PD), Huntington disease (HD), and amyotrophic lateral sclerosis (ALS). The hallmark of these neurodegenerative disorders is the loss of neuronal function, and eventual neuron death. Often, clinical signs predate any significant neuron loss. In AD, the neuron loss occurs in the hippocampus and cerebral cortices.

- **Abnormal structures associated with AD:** Extensive research has focused on the two well-known abnormalities found in AD, amyloid plaques and neurofibrillary tangles. The amyloid plaques contain small, toxic cleavage products from a larger precursor protein, amyloid precursor protein (APP), while the neurofibrillary tangles consist of abnormally phosphorylated versions of the microtubule-associating protein, tau. There has been controversy surrounding whether these protein accumulations were the causative factors or just the molecular debris left behind after a prior molecular insult.

- **Risk factors:** There are multiple risk factors for AD that cannot be altered: age (incidence of up to 50% in those 85 years and older), family history of AD, sex (females are at higher risk), and certain genetic traits (a form of apolipoprotein E4 [Apo E$_4$] increases the risk). The evidence for lifestyle-modifiable factors is unfortunately less clear and somewhat controversial. Head injury, most notably in boxers and other full-contact athletes, has also been postulated to be a risk factor for dementia.

CLINICAL CORRELATION

Although scientists are learning more every day, right now they still do not know what causes AD, and there is no cure. A Mini Mental State Examination (MMSE) is given to the patient. It is used to screen for the presence of cognitive impairment over a number of domains. Cognition is defined as mental activity such as memory, thinking, attention, reasoning, decision making, and dealing with concepts. This patient scored 18 out of 30 possible points, consistent with a diagnosis of moderate AD.

Neurodegeneration is a general term that covers many neurodegenerative diseases, including AD, PD, HD, and ALS. Neurodegeneration can also result from stroke, heart attack, head and spinal cord trauma, and bleeding in the brain. Neurodegenerative diseases are often characterized by a neurological deficit, a loss of memory, mobility, independence, or well-being. Each disease has its own characteristics, but all progress slowly over time, and all invariably result in premature death. AD is the most common neurodegenerative disorder worldwide, with approximately 4.5 million affected individuals in the United States alone. It is also the most common cause of cognitive decline, or dementia, in the elderly. From the time of diagnosis, life expectancy averages 8 years. Diagnosis hinges upon excluding other conditions that can cause cognitive impairment: small undetected strokes, depression, side effects of medications, and even PD (another neurodegenerative disorder), which has an associated dementia complex. Because AD can only be 100% accurately diagnosed by autopsy, the clinical diagnosis of AD relies upon a clinical picture composed of medical history, blood tests (to rule out other medical conditions), mental status evaluation, neuropsychological testing, and possibly brain imaging (Figure 14-1). The current interest in treating neurodegenerative diseases like AD with cell-based technologies has renewed the importance of understanding the developmental

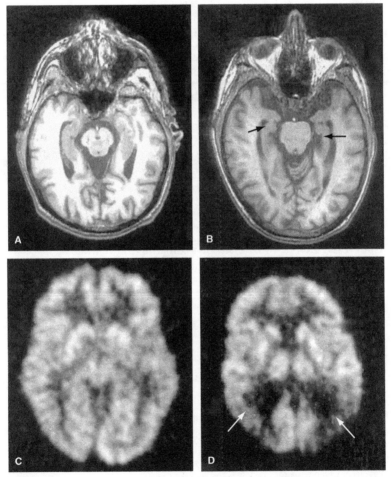

Figure 14-1. Alzheimer disease: Axial T1-weighted MR images through the midbrain of a normal 86-year-old athletic individual (A) and a 77-year-old male (B) with AD. Fluorodexyglucose PET scans of a normal control (C) and a patient with AD (D). Note that the patient with AD has decreased activity in the parietal lobes bilaterally (*arrows*), a typical finding in this condition. AD, Alzheimer disease; PET, positron emission tomography. (*With permission from* Harrison's Principles of Internal Medicine. 15th ed. 2001. Figure 362-1, page 2392.)

biology involved with how neurons are formed in the first place. To recapitulate the developmental process, science must first understand that process. Neurogenesis, or the regulated generation of neurons involves (1) early discrimination of neuron from nonneuron, (2) acquisition of specialized neuronal properties, and (3) determination of which cells will live or die as part of development.

APPROACH TO:
Neurogenesis

OBJECTIVES

1. Understand the cell-cell interactions that progressively restrict the fate of a cell.
2. Recognize the role of spatial patterning and temporal regulation in determining cellular identity.
3. Be able to describe the molecular mechanisms that regulate the early processes in neurogenesis.

DEFINITIONS

NEUROGENESIS: The process by which neurons are generated from neuroepithelial cells.

PROGENITOR CELLS: Neuroepithelial or neural stem cells that have the capability for long-term self-renewal and can generate all the cell types in the nervous system.

VENTRICULAR ZONE (VZ): The area bordering the ventricular space within the neural tube. This is where the neuroepithelial cells proliferate, differentiate into neurons, and migrate away.

SUBVENTRICULAR ZONE (SVZ): A second zone of cells that continues to proliferate and generate neurons and glia at later stages in development, even into adulthood. It supplies neurons to the olfactory region via the rostral migratory stream.

PRONEURAL GENES: A family of basic helix-loop-helix (bHLH) transcription factors that is critical in determining the identity of neural lineages. Neurogenin 1 and 2 are two mammalian examples.

LATERAL INHIBITION: The process of cell-cell interaction whereby one cell, destined to a neural lineage, inhibits its neighbor from acquiring the same neural fate.

DISCUSSION

The early cellular process of neurogenesis is generally viewed as the progression from multipotent stem cells to fate-restricted neuronal precursors, through the gradual reduction of potential fates. Though the multipotent neuroepithelial cells have long processes that span the width of the early neural tube, all cell division occurs at the ventricular surface. A symmetric cell division is important for self-renewal and critical for early expansion of the population. Later, neural progenitors are produced by asymmetric cell division, which results in the loss of the ability to self-renew, and consequently the cells become more fate restricted. Neural progenitors then begin to express genes that promote differentiation. The last step is to leave the cell cycle (mitosis) and become a neuron.

As the nervous system goes through neural induction, spatial patterning is established along the rostrocaudal and dorsoventral axes. This imprints positional identity upon the neuroepithelial cells, which influence the types of neurons that can arise from the precursors, a fate restriction. For example, isolated cells from the spinal region give rise to cells that populate the spine, whereas cells from the forebrain generate cell types appropriate for the cerebral cortex. A similar process occurs along the dorsoventral axis with these progenitors forming dorsal sensory interneurons and ventral motor neurons. Temporal regulation of neurogenesis affects the neuroprogenitor population as well. In the developing cortex, newly born neurons leave the ventricular zone and migrate along specialized glial cells to populate the cortical layers. Neurons migrate from the ventricular zone to the subventricular zone to their destination via radial glial migration. The glial cells have long processes that extend perpendicular to the subventricular zone and form scaffolding for migrating neurons. Interestingly, the time at which the neurons become postmitotic and leave the cell cycle, their "birthdate," correlates precisely with their laminar position. Neurons born earlier reside in the deeper layers, while those that are born later populate the more superficial layers. Transplantation studies demonstrated that the early cortical progenitor cells are multipotent and competent to make both early (deep layer) and late (superficial layer) cell types, whereas later progenitors are restricted in their competence and only make the late cell types.

The molecular mechanisms that regulate neurogenesis, like many developmental programs, constitute a balance of forces: forces that promote neural identity and those that constrain the differentiation process. What molecular events are responsible for driving ectodermal cells to become neurons? The first event is to define a subset of cells that can have the potential or competence to become neural precursors. This cluster of cells within the embryological ectoderm, termed a *proneural cluster*, all express low levels of proneural genes. Through the process of lateral inhibition, neural precursor cells inhibit the expression of proneural genes in neighboring cells, thus preventing them from assuming a neural fate. On a molecular level, this cell-cell interaction is mediated by the Notch-Delta signaling system. The neural precursor cell expresses delta ligand, which activates the notch receptor on adjacent cells. This activity downregulates the proneural genes and delta ligand expression in the neighboring cell. This decreased delta expression in the neighboring cell, by release of inhibition, results in the upregulation of proneural gene expression in the precursor cell, fate-restricting it to becoming a neural precursor cell.

There is no cure for AD. The two drug classes approved by the FDA to help delay the progression of dementia symptoms are cholinesterase inhibitors and memantine. Cholinesterase inhibitors work by blocking the degradation of acetylcholine (ACh), thereby increasing ACh levels in the synaptic cleft. Memantine blocks the N-Methyl-D-aspartate receptor, which is thought to be responsible for glutamate-mediated excitotoxicity.

CASE CORRELATES

- See Cases 11-17 (nervous system development), and Case 5 (Alzheimer disease).

COMPREHENSION QUESTIONS

A 27-year-old man is evaluated in your office for persistent headaches that are unresponsive to medications, and recent onset of instability when walking. An MRI of his head shows an epidermoid cyst (ectopic ectodermal mass) at his cerebellopontine angle. He is referred to neurosurgery for surgical excision.

14.1 The expression of what gene set in embryological ectodermal cells causes them to develop into neural tissue rather than ectoderm?

A. *Hox* genes

B. Proneural genes

C. *BMP* genes

D. *Wnt* genes

14.2 In addition to the developmental signals discussed above, what process aids in the differentiation of certain ectodermal cells into neuroectoderm while adjacent cells continue to differentiate into ectoderm?

A. Lateral inhibition

B. Neurulation

C. Cell fate determination

D. Apoptosis

14.3 What signaling molecule/receptor pair is responsible for the process of lateral inhibition?

A. Slit-Robo

B. Semaphorin-plexin

C. Notch-Delta

D. Laminin-integrin

ANSWERS

14.1 **B. Proneural genes** are necessary for the differentiation of ectodermal cells into neural precursor cells. The differentiation of the embryological ectoderm into neuroectoderm is a very complicated process that depends on a complex interplay of numerous signaling molecules. Certain embryological ectodermal cells express proneural genes, which are necessary for the differentiation of the cells into neural precursor cells. These proneural genes are inhibited by a family of proteins known as bone morphogenic proteins (BMPs). Signals from embryological mesoderm like chordin, noggin, and follistatin antagonize the action of BMPs, allowing the expression of the proneural genes and development of embryonic ectoderm into neuroectoderm. *Hox* genes are involved in a number of embryological developmental stages including differentiation of the rhombencephalon. *Wnt* genes are involved in closing of the neural groove into the neural plate.

14.2 **A.** Through the process of **lateral inhibition,** proneural cells secrete signaling molecules that downregulate proneural gene expression in neighboring cells, thereby decreasing the likelihood that those cells will become neuroectodermal cells. While this process does somewhat restrict the fate of the cells that are prevented from becoming neural cells, lateral inhibition is a more specific answer than cell fate determination and is therefore the *best* answer. Neurulation is the process by which the neural plate becomes the neural tube. Apoptosis (programmed cell death) is a process involved in many other developmental processes.

14.3 **C.** The **Notch-Delta** signaling system is responsible for lateral inhibition. Proneural cells secrete delta ligand, which binds to the notch receptor on neighboring cells, resulting in decreased expression of proneural genes. This also causes a decrease in delta expression in those inhibited cells, thereby relieving inhibition on the initial proneural cell. The other combinations of signaling molecules are involved in axonal guidance.

NEUROSCIENCE PEARLS

▶ During neurogenesis, a multipotent precursor cell goes through a series of developmental steps that sequentially restrict its potential fates.

▶ The timing of a neuron's birth is an important predictor of its cellular fate.

▶ The spatial location of a neuron helps to determine its identity.

▶ Cell-cell interactions during early neurogenesis are critical for specifying neural cell fate.

REFERENCES

Bossy-Wetzel E, Schwarzenbacher R, Lipton SA. Molecular pathways to neurodegeneration. *Nat Med.* 2004 Jul;10(suppl):S2-S9.

Kandel ER, Schwarz JH, Jessell TM, Siegelbaum SA, Hudspeth AJ, eds. *Principles of Neural Science.* 5th ed. New York, NY: McGraw-Hill; 2012.

Lindvall O, Kokaia Z, Martinez-Serrano A. Stem cell therapy for human neurodegenerative disorders—how to make it work. *Nat Med.* 2004 Jul;10(suppl):S42-S50.

Rao MS, Jacobson M, eds. *Developmental Neurobiology.* 4th ed. New York, NY: Kluwer Academic/Plenum Publishers; 2005.

A 66-year-old retired marine sergeant complains of worsening trembling in his hands. He first started noticing the problem a year ago when he was signing checks at work. His wife noticed that he could no longer keep up with her on their daily walks, joking that he now shuffles his feet like a penguin. Apart from a history of tobacco use and mild hypertension, he does not have any significant medical issues and has no other complaints. Family history is negative. His only medication is an inhaler that he uses to help with his breathing. Physical examination reveals a tremor in both hands with a rubbing of the thumb and forefinger—"pill rolling." His muscles are stiff to passive movement, which causes a small amount of discomfort. Examination of his gait shows a wide-based, shuffling walk that has a noticeable loss of normal arm swing. He has a bright intellect and has no difficulties with any cognitive tasks presented to him. He is diagnosed with Parkinson disease (PD). The physician discusses the diagnosis and disease process with the patient, explaining to him that these symptoms usually worsen, and that new symptoms may occur, such as loss of normal facial expressions (smiling, blinking), impaired speech (often very soft and monotonous), difficulty in swallowing, and even cognitive decline or dementia.

▶ What brain structure is affected in this patient?
▶ What is the underlying mechanism of the movement disorder?
▶ Are there any preventative measures available?

ANSWERS TO CASE 15:
Cell Fate Determination

Summary: A 66-year-old man is noted to have progressive tremor, shuffling gait, and muscle rigidity. He is diagnosed with PD.

- **Brain structure affected:** Substantia nigra, a component of the basal ganglia. PD is a neurodegenerative disorder that affects movement. It is one of a group of common neurodegenerative disorders, which includes Alzheimer disease (AD), Huntington disease (HD), and amyotrophic lateral sclerosis (ALS). In PD, the symptoms are caused by the selective death of a specific cell type in the basal ganglia, the pigmented dopaminergic (DA) neurons of the substantia nigra. These diseased neurons often contain Lewy bodies, a specific type of cytoplasmic inclusion.

- **Underlying mechanism:** The basal ganglia are a group of interconnected nuclei consisting of the putamen, caudate, globus pallidus, subthalamic nucleus, and substantia nigra. These nuclei affect multiple cortical functions, including motor control, cognition, emotions, and learning. The loss of dopaminergic neurons leads to reduced levels of the neurotransmitter dopamine, altering the function of the basal ganglia.

- **Prevention:** Herbicide and pesticide exposure has been linked to the development of PD; however, no specific chemicals have been identified. As with all neurodegenerative diseases, the risk increases with age. There is also a genetic contribution—affected first-degree relatives confer a 5% risk. However, neither age nor family genetics are modifiable.

CLINICAL CORRELATION

PD is recognized as one of the most common neurological disorders, affecting approximately 1% of individuals older than 60 years of age, with an incidence of 5-20 cases per 100,000 people a year. Cardinal features include resting tremor, rigidity, bradykinesia, and postural instability. The major neuropathologic findings in PD are a loss of dopaminergic neurons in the substantia nigra pars compacta and the presence of Lewy bodies. The basal ganglia circuit modulates the cortical signals necessary for normal movement. Impulses from the cerebral cortex are processed through the basal ganglia-thalamus circuit and act as a feedback pathway. Output from the basal ganglia motor circuit is directed toward suppressing the thalamocortical pathway and decreasing movement. The loss of dopaminergic signaling in the basal ganglia increases the inhibition of the thalamocortical pathway, leading to further suppression of movement (Figure 15-1).

The molecular mechanisms of PD remain elusive. The protein alpha-synuclein recently was discovered to be a major structural component of Lewy bodies. Improper folding of this protein may result in abnormal aggregation and toxicity to neurons. Lewy bodies also contain the protein ubiquitin, a key player in targeting proteins for degradation. A variety of rare familial forms of PD have shed light

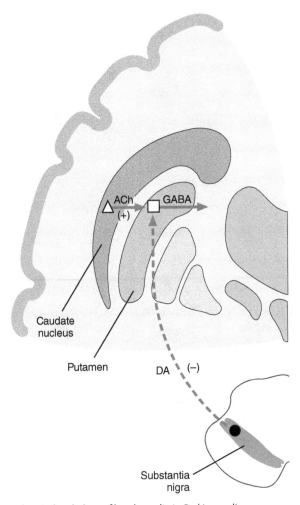

Figure 15-1. Neurochemical pathology of basal ganglia in Parkinson disease.

on several possible molecular mechanisms for the disorder. These mutations have implicated oxidative stress and mitochondrial injury, as well as dysfunction of the ubiquitin system for protein degradation as pathways leading to neuron death in PD. The specific nature of the degenerating system (dopaminergic neurons) and the discrete location (substantia nigra pars compacta) have made PD an attractive target for cell-based therapies. The early successful transplantation of fetal mesencephalic neurons (rich in dopaminergic neurons) into the degenerating Parkinsonian brain demonstrated that transplanted cells could survive and functionally integrate into neuronal circuits. Recent work has been focused on engineering stem cells to assume the phenotype of the degenerating substantia nigra neurons. In order to be successful, the cells must (1) behave like dopaminergic neurons and release dopamine, (2) integrate functionally into the neuronal circuitry, (3) yield sufficient numbers for long-term survival, and (4) reverse the symptoms of PD. Critical to this process is an understanding of the pathway that neurons follow to determine their cellular fate so that it can be replicated to entrain new populations of stem cells.

APPROACH TO:
Cell Fate Determination

OBJECTIVES

1. Understand the stages of cell fate determination in the nervous system.

2. Explain the different types of signals that cells respond to during differentiation.

3. Appreciate the role of neurotrophic factors in cell fate determination.

DEFINITIONS

COMPETENCE: The ability of a cell to respond to development signals.

FATE SPECIFICATION/DETERMINATION: Both terms refer to the mechanism by which precursor cells begin with many possible fates and are progressively limited in their potential, until they finally differentiate into a mature cell.

PHENOTYPE: The characteristics or traits that a cell exhibits. Different cell types display different phenotypes, which are often because of the underlying differences in protein expression, especially transcription factors.

INTRINSIC DETERMINATION: A model that explains progressive cell fate restriction by the influences of the cell's lineage and internal signals. This would be the cell's metaphorical version of the nature versus nurture debate in human development.

EXTRINSIC DETERMINATION: A model that explains progressive fate restriction by the influences of the environment, as experienced by cell-cell interactions and diffusible signaling molecules.

APOPTOSIS: The active process of programmed cell death, typified by cell shrinkage, condensation of chromatin, cellular fragmentation, and phagocytosis of cellular debris. A prominent feature of neuronal development is the death of cells by apoptosis.

TROPHIC FACTOR: A substance that encourages survival and proliferation of a cell or tissue.

DISCUSSION

After a newly born cell exits mitosis and becomes committed to the neuronal lineage, it must continue to respond to signals that will influence the type of neuronal fate it will acquire. There is significant debate as to where neurogenesis ends and neural fate determination begins. As previously discussed, during cortical development the neural precursor cells that are born later have had their fates restricted—they can only populate the superficial cortical layers. In other words, these precursor cells that were born later are no longer competent to form cells in the deep cortical layers. Therefore, during neurogenesis, some aspects of fate determination have already been decided just by the timing of birth. The developmental plan is best viewed as a continuum of overlapping processes.

Many prospective neurons express the same genes early in development, and at some point, as their fates diverge, they begin to express unique genes and proteins required for their eventual fate. In this manner, the cells acquire progressively specialized neuronal phenotypes. The influences which a maturing neuron encounters can be classified as intrinsic or extrinsic signals.

One example of intrinsic determination occurs during early neurogenesis when the neural progenitor pool divides along the ventricular surface in the *ventricular zone* (VZ). The early cell divisions occur with the plane of division perpendicular to the ventricular surface, generating two identical daughter cells, both capable of further proliferation. The later divisions occur with the plane of division parallel to the ventricular surface and are asymmetrical with one daughter cell retaining the ability to proliferate, while the second daughter is postmitotic and migrates out of the VZ. This intriguing process appears to be mediated by two intracellular proteins, Numb and Prospero. These proteins are asymmetrically inherited by the daughter cells during division and thus confer different phenotypes to the progeny.

Neuronal target tissues often influence the fate of neurons via extrinsic determinants, a concept called **target specification.** An example is the specification of neurotransmitter phenotype of the autonomic neurons by sweat glands. Most sympathetic neurons use norepinephrine as their primary neurotransmitter; however, the exocrine sweat glands in the hands and feet use acetylcholine instead. The maturing sympathetic neuron first expresses the norepinephrine phenotype, but when it contacts the sweat glands, the neurons gradually convert their neurotransmitter phenotype to acetylcholine. The signals for this fate specification appear to be a soluble cytokine of the interleukin-6 family secreted by the sweat glands. Target tissues play other critical roles in the fate determination of neurons. Target cells of developing neurons produce a limited amount of a trophic factor that can control the survival of the innervating neurons—the **neurotrophic factor hypothesis**. Nerve growth factor (NGF) was the first isolated neurotrophic factor. Survival of some neurons requires NGF, a target-derived factor. NGF works by binding to tyrosine kinase receptors on the dependent axon and is transported in a vesicle via retrograde fashion to the cell body, where it can affect transcription and cellular differentiation. Many other families of neurotrophic factors have been isolated: the neurotrophins, IL-6 class, transforming growth factor beta (TGF-β) class, as well as fibroblast growth factor (FGF). It now appears that elimination of neurotrophic factors can control survival by activating programmed cell death, or apoptosis.

An important principle in neuron cell fate is the gradual acquisition of progressively more restrictive fates until the final neuron identity is reached. This process is mirrored phenotypically by the expression of various proteins during development. The protein expression patterns can define different developmental stages, resulting in a hierarchy of gene expression. This is best illustrated in the developing spinal cord by the LIM family of transcription factors. This family of proteins is expressed in different overlapping subpopulations of motor neurons. The protein that all the motor neurons express is Islet-1, suggesting that this protein is responsible for characteristics common for all motor neurons. Other LIM proteins are subsequently expressed, which progressively identify specific motor neuron subgroups and even their behaviors like axon projection or target selection.

Treatment Options

Unlike most other neurodegenerative disorders, PD responds well to treatment. The mainstay of treatment is medical treatment with oral levodopa (L-dopa), a precursor to dopamine which can cross from the blood into the brain and be converted by neurons into dopamine. L-dopa is often combined with carbidopa (eg, Sinemet), which helps to decrease the systemic side effects. Other medications help potentiate the effect of L-dopa treatment by inhibiting the degradation of dopamine: Selegiline blocks the enzyme monoamine oxidase, and a class of inhibitors to the enzyme catechol-O-methyltransferase (COMT). Another group of drugs, the anticholinergics, alters the basal ganglia signaling to compensate for the loss of dopamine action. Previously, surgical treatment has been used more frequently, but currently it is reserved for patients with significant symptoms refractory to medical management. Selective placement of destructive lesions to the thalamus (thalamotomy) and globus pallidus (pallidotomy) can be helpful in controlling the symptoms but can carry significant risks. A newer form of surgical treatment called deep brain stimulation (DBS) involves stimulating the subthalamic nucleus with tiny implanted electrodes and a pacemaker-like neurostimulator. Regenerative medical interventions, such as gene therapy and stem cell replacement are areas of considerable current interest and funding.

> ## CASE CORRELATES
> - See Cases 11-17 (nervous system development), and Case 7 (Parkinson disease).

COMPREHENSION QUESTIONS

Refer to the following case scenario to answer questions 15.1-15.2:

A 25-year-old man falls from the roof of his house and injures his spinal cord. Initially, he has flaccid paralysis of his lower extremities, but in the subsequent months his legs become hypertonic, hyperreflexic, and spastic because of destruction of his corticospinal tracts. The axons in these damaged tracts project in large part from pyramidal neurons in the cortical layer V of the primary motor cortex.

15.1 In which location did these neurons originate?

    A. Ventricular zone

    B. Marginal zone

    C. In cortical layer V

    D. In the neural plate

15.2 Which of the following is true of the cells that gave rise to these neurons?

    A. They are capable of giving rise to neurons of all cortical layers.

    B. They are capable of giving rise to only neurons in deep cortical layers.

    C. They are capable of giving rise to only neurons in superficial cortical layers.

    D. They are capable of giving rise to neurons in none of the cortical layers.

15.3 A 22-year old-patient presents to your office complaining of excessive sweating in his axillae, without apparent cause and unresponsive to the application of antiperspirant. You diagnose him with hyperhidrosis and recommend botulinum toxin A (Botox) injections to the sweat glands to help resolve his symptoms. Which of the following best describes the reason that sympathetic nerves innervating sweat glands release acetylcholine, while all other sympathetic neurons release norepinephrine?

    A. Apoptosis

    B. Intrinsic determination

    C. Extrinsic determination

    D. Cell fate restriction

ANSWERS

15.1 **A.** Cortical neurons all originate from neural precursor cells located in the **ventricular zone**. Once generated, these immature neurons migrate (via radial glial migration or tangential migration) to their final location in the cortex. While it is true that the precursor cells that give rise to neurons originally came from the neural plate, it is more accurate to say that the neurons themselves originated in the ventricular zone.

15.2 **A.** One of the interesting features of the neural progenitor cells in the ventricular zone is that as development progresses, their fate becomes more determined. The stem cells that give rise to cortical layer V neurons (relatively deep in the cortex) are **competent to give rise to all neurons** more superficial in the cortex. Conversely, a stem cell that gives rise to a layer I neuron (superficial in the cortex) is not competent to generate a neuron in the deeper layers.

15.3 **C.** These sympathetic neurons release acetylcholine instead of norepinephrine because of **extrinsic determination**. The neurotransmitter ultimately released by the sympathetic neurons innervating the sweat glands is determined by trophic factors released from the sweat glands themselves. These factors induce changes in the neuron that causes it to change from an adrenergic neuron to a cholinergic neuron. This is an example of extrinsic determination: the fate of the neuron is determined by a cell other than itself. While this is a type of cell fate restriction, extrinsic determination is a more specific answer.

NEUROSCIENCE PEARLS

▶ PD is the second most common neurodegenerative disorder worldwide (after AD). It has classic motor features of resting tremor, rigidity, and bradykinesia, and neuropsychiatric manifestations.

▶ PD is associated with a loss of dopaminergic neurons and reduced levels of the neurotransmitter dopamine, altering the function of the basal ganglia.

▶ Cell fate specification relies on both intrinsic and extrinsic cues.

▶ The expression of certain genes dictates the phenotype and eventual fate of developing cells.

▶ Apoptosis, or programmed cell death, is a critical process in the development of the nervous system.

REFERENCES

Bossy-Wetzel E, Schwarzenbacher R, Lipton SA. Molecular pathways to neurodegeneration. *Nat Med.* 2004 Jul;10(suppl):S2-S9.

Kandel ER, Schwarz JH, Jessell TM, Siegelbaum SA, Hudspeth AJ, eds. *Principles of Neural Science.* 5th ed. New York, NY: McGraw-Hill; 2012.

Lindvall O, Kokaia Z, Martinez-Serrano A. Stem cell therapy for human neurodegenerative disorders—how to make it work. *Nat Med.* 2004 Jul;10(suppl):S42-S50.

Rao MS, Jacobson M, eds. *Developmental Neurobiology.* 4th ed. New York, NY: Kluwer Academic/Plenum Publishers; 2005.

Vila M, Przedborski S. Genetic clues to the pathogenesis of Parkinson's disease. *Nature Med.* 2004 Jul;10(suppl):S58-S62.

CASE 16

A 2-week-old baby boy is brought into the emergency room by ambulance after he started shaking uncontrollably during his sleep. When the paramedics and transport team arrived at the home, he was still shaking with his eyes deviated upward. When he stopped shaking, he remained unresponsive with his head next to a puddle of vomit. At the hospital, he is more arousable but is irritable on examination. The mother states that the infant had an uncomplicated gestation and birth history. There haven't been any issues that concerned the parents—no family history of diseases, no recent exposures to sick persons, and no fevers. They did notice that he coughed a lot during and after his feedings. Examination of the baby is normal until a second episode of generalized involuntary stiffness and shaking occurs. The infant is intubated and placed on a mechanical ventilator to help him breathe and stabilize with medications. An emergent head computed tomography (CT) scan is obtained, and the boy is taken to the intensive care unit. The patient is diagnosed with lissencephaly (LIS).

► What would be the gross appearance of the brain tissue in this patient?
► What is the etiology of this disorder?
► Where in the developing brain are neurons formed, and by what process do neurons move from their birthplace to their ultimate location?

ANSWERS TO CASE 16:
Neuronal Migration

Summary: This baby boy was previously healthy before he presented acutely with generalized seizures. There was no evidence of a gross neurological deficit or any significant developmental delay. The head CT demonstrated impaired cerebral cortical structure with loss of the normal cortical involutions and gyrations, giving a "smooth" appearance to the cerebral cortex. A diagnosis of LIS was made. The only other notable abnormality was some mild muscle stiffness in his extremities.

- **Appearance of brain:** Smooth or nearly smooth cerebral surface lacking the normal gyral convolutions. LIS is always associated with an abnormally thick cortex, reduced or abnormal lamination, and diffuse neuronal heterotopia (displaced neurons).

- **Etiology:** LIS occurs because of a disruption of normal embryogenesis in the brain. Migration of postmitotic neurons from the ventricular zone (VZ, ie, area surrounding the ventricles) to the cortical plate during nervous system development is defective, resulting in brain malformations, including LIS.

- **Origins of neurons and neuronal migration:** Though several neurogenic zones have been identified in the brain, most require further study. The VZ is the most well studied neurogenic zone and newly formed neuroblasts in the VZ migrate radially to differentiate into glial cells, and also via the rostral migratory system to form neurons in the olfactory bulb.

CLINICAL CORRELATION

Classical LIS consists of various features: (1) diffuse or widespread agyria-pachygyria, (2) an abnormally thick cortex, (3) enlarged ventricles without hydrocephalus, (4) frequently abnormal corpus callosum, (5) grossly normal brainstem and cerebellum, (6) normal or slightly reduced head circumference, and (7) slow head growth. A study of LIS in the Netherlands demonstrated a prevalence of approximately 12 cases per million births, while data from the United States showed rates of 4-11 per million births. Together, these malformations constitute a continuum, the agyria-pachygyria spectrum of malformations. Like other disorders, LIS can occur in varying degrees; however, the affected children will generally have mental retardation, seizure disorder, difficulty swallowing and eating, difficulty controlling his/her muscles, stiffness or spasticity of arms and legs, and slowed growth and developmental delays. The identification of the human genes responsible for LIS has advanced the knowledge of how neurons migrate. The first human neuronal migration gene to be cloned was *LIS-1* gene. This gene encodes a regulatory subunit of an enzyme, termed platelet-activating factor (PAF) acetylhydrolase. This enzyme catalyzes the release of potent phospholipids that act as signaling proteins within neurons and interact with tubulin to suppress microtubule dynamics. The highest levels of *LIS-1* are detected in the developing cortex, consistent with the protein's putative role in neuronal migration. Mutations in the *LIS-1* gene disrupt the structural cytoskeleton of neurons. This cytoskeleton is made up of proteins (like tubulin) that give cells their distinctive shapes, and normal cellular movement requires regulated rearrangements of this cellular structure. Disruptions of the cytoskeleton and its flexibility affect the ability of neurons to migrate normally and are thought to lead to the disrupted cortical pattern seen in children with LIS. Another gene *XLIS*, which has been identified in X-linked LIS, also interacts with the protein tubulin. Analysis of these human disorders has shown that the cytoskeleton plays a key and critical role in the normal migration of neurons.

APPROACH TO:
Neural Migration

OBJECTIVES

1. Understand the patterns of neural migration in the nervous system.

2. Explain the different types of signals that cells respond to during neural migration.

3. Appreciate the role of the internal and external molecular processes that dictate the process.

DEFINITIONS

RADIAL MIGRATION: A form of neural migration that occurs as nascent neurons travel from the VZ radially out toward the pial surface of the cortex.

RADIAL GLIAL CELLS: A type of glial cell that forms a physical scaffold for migration with its processes that extend from the VZ out to the pial surface.

TANGENTIAL MIGRATION: A form of neural migration, also called nonradial migration, that does not require interaction with radial glial cell processes.

PERMISSIVE FACTOR: A signal that passively allows a process to occur, without actively promoting its occurrence. For example, an abundance of nutrients is a permissive factor in cellular growth.

EXTRACELLULAR MATRIX (ECM): The network of glycoproteins, proteoglycans, and hyaluronic acid that defines the space in a tissue that is not part of a cell.

CELL ADHESION MOLECULES (CAMs): A family of cell membrane–associated proteins that interact and bind with extracellular molecules or **ligands**. They can transmit signals through the cell membrane to the intracellular cytoskeleton.

DISCUSSION

After a newly-born neuron exits mitosis and becomes committed to the neuronal lineage, it must leave its VZ birthplace and travel to the appropriate laminar position. This neural migration can occur by three general mechanisms: two forms of radial migration (somal translocation and locomotion) and a nonradial or tangential migration.

Postmitotic neurons migrate radially away from the VZ toward the outer pial surface to reach the top of the cortical plate. There, they assemble into layers with distinct patterns of cortical connections. **Radial migration** can occur via two methods: somal translocation and glia-guided locomotion. Somal translocation consists of movement of the soma (cell body) and nucleus of the neuron toward the cortical plate by selectively releasing the ventricular attachment site while maintaining the pial attachment site. Glia-guided locomotion occurs by movement along the scaffold formed by the radially oriented processes of the radial glial. The early-generated neurons predominantly use somal translocation when the ventricular-pial distance is shorter. Later-migrating pyramidal neurons first use glia-guided locomotion and then subsequently switch to somal translocation as they move past earlier generated neurons. These neurons use the radial glia processes as their primary migratory guides, forming specialized membrane contacts with the cell processes.

Until recently, the predominant model of neural migration stated that the vast majority of neurons migrate radially along glial processes. However, recent evidence has forced a revision of this model. In the developing neocortex, most excitatory glutamatergic pyramidal neurons do follow radial migration; however, most GABA-ergic, nonpyramidal interneurons travel tangentially. These neurons do not appear to use the radial glial processes or other specific cellular scaffolds.

Movement of neuronal cells from their site of birth in the VZ to their specific cortical destination relies on three cellular events: **initiation, maintenance,** and **termination.** In turn, the molecular mechanisms that regulate all three cellular processes rely upon cell-cell recognition and adhesive interactions between neurons, glia, and the ECM.

The initiation of migration must occur along appropriate pathways. Reorganization of the actin-based cytoskeleton appears responsible for priming the neuron for migration. These instructive factors may be extracellular signals present in the VZ.

Subsequent cell-cell interactions form junctional domains in the cell membrane. In the radial migration stream, the neurons interact with the processes of the radial glia through these domains. The internal microtubule-based cytoskeletal components subsequently associate with these junctional domains to provide the force required for active cellular movement. Mutations in genes responsible for this maintenance phase of migration have been linked to human diseases, like LIS. Many of these mutations occur in proteins responsible for interactions with the microtubule-based cytoskeleton. Extracellular signals from the ECM appear to modulate this motility via interactions with the **integrin** family of CAMs. The many subtypes of integrins have different molecular adhesive characteristics and are associated with varied intracellular signal transduction pathways. Thus, the specific integrin expression on the neuron surface and interactions with the ligands in the ECM and other cells can modulate the pattern of migration from initiation to termination. In fact, the destination tissue itself may relate the signals to stop the migration process by changing the ligand environment—whether from afferent fibers in the target zone, the destination neuronal population, or changes in the terminal radial glia process.

> **CASE CORRELATES**
>
> • See Cases 11-17 (nervous system development).

COMPREHENSION QUESTIONS

Refer to the following case scenario to answer questions 16.1-16.2:

A 14-year-old adolescent girl is brought into your office with the complaint of the recent development of periods of odd behaviors and movements. She is diagnosed with complex partial epilepsy, and a brain MRI is ordered. On the MRI scan, a round nodule of what looks like gray matter is seen adjacent to the lateral ventricle, beneath the normal periventricular white matter.

16.1   An error in what process in responsible for this inappropriately located gray matter?

   A. Neurogenesis

   B. Radial glial migration

   C. Tangential neuronal migration

   D. Termination of migration

16.2   From which area in the developing brain did these improperly located neurons originate?

   A. Ventricular zone

   B. Cortex

   C. Volar plate

   D. Dorsal plate

16.3   A 35-year-old male patient is brought into the emergency department by friends who found him lying unresponsive on the floor of his apartment, with an empty bottle of wine and a bottle of valium on the table near him. On examination you note that the patient is unresponsive, with shallow respirations. You recall that both benzodiazepines (the class of drugs to which valium belongs) and alcohol increase the effects of GABA in the CNS. The neurons that normally release GABA migrate by which mechanism?

   A. Radial glial migration

   B. Somal translocation radial migration

   C. Tangential migration

   D. They do not migrate

ANSWERS

16.1 **B.** The child in this scenario has subcortical heterotopia, a disease which is thought to arise from a **disorder in radial glial migration of neurons**, rather than a defect in tangential migration of neurons. In this condition, neurons in the brain are located inappropriately below the cortex. This is caused by a failure of the neurons to migrate from their originating location in the VZ to their proper place in the cortex. The neurons are generated correctly, so there is no error in neurogenesis, nor is there a failure to terminate migration.

16.2 **A.** Neurons that ultimately reside (or are supposed to reside) in the cortex of the brain **originate from neural precursor cells in the ventricular zone** and then migrate to their final locations. These final locations could include the cortex and the dorsal or volar plate in the developing brainstem and spinal cord.

16.3 **C.** GABAergic neurons migrate via **tangential migration**. Recently it has become known that there are multiple mechanisms by which neurons migrate from the VZ to their final location. It appears that excitatory neurons tend to migrate radially via radial glial migration or somal translocation. Inhibitory GABAergic neurons, however, seem to migrate tangentially, and do not use any sort of glial-mediated guidance.

NEUROSCIENCE PEARLS

▶ Lissencephaly is a rare brain malformation caused by abnormal neuronal migration resulting in lack of gyri (smooth brain).

▶ Specific neuronal connections and networks are a result of appropriate migration and final destination of neurons.

▶ Cell-cell recognition and adhesive interactions between cells and the surrounding environment are critical in regulating migration.

▶ The cellular cytoskeleton undergoes regulated reorganization to make migration possible.

REFERENCES

Couillard-Despres S, Winkler J, Uyanik U, Aigner L. Molecular mechanisms of neuronal migration disorders, quo vadis? *Curr Mol Med.* 2001;1:677-688.

Kandel ER, Schwarz JH, Jessell TM, Siegelbaum SA, Hudspeth AJ, eds. *Principles of Neural Science.* 5th ed. New York, NY: McGraw-Hill; 2012.

Kato M, Dobyns WB. Lissencephaly and the molecular basis of neuronal migration. *Hum Mol Genet.* 2003;12(Review Issue 1):R89-R96.

Rao MS, Jacobson M, eds. *Developmental Neurobiology.* 4th ed. New York, NY: Kluwer Academic/Plenum Publishers; 2005.

A 28-year-old pregnant woman presents to her obstetrician's office for a routine prenatal appointment. She is a healthy person without any medical conditions, and as instructed, has been taking prenatal vitamins and has stopped drinking alcoholic beverages. She and her husband are excited about the pregnancy. The ultrasound examination at 12 weeks confirmed a single fetus with good placental attachment and cardiac function. There was some question regarding the head size, so a second ultrasound was obtained at 18 weeks. This examination was more detailed and revealed incomplete separation of the cerebral ventricles. A perinatologist who specializes in developmental malformations was consulted, and a follow-up magnetic resonance imaging (MRI) scan was obtained. This detailed imaging study confirmed the incomplete division of the embryonic forebrain into distinct lateral cerebral hemispheres. The diagnosis of holoprosencephaly (HPE) was made based on the MRI.

▶ What part of the developing brain is affected in this condition?
▶ What is the mechanism of this developmental abnormality?

ANSWERS TO CASE 17:
Formation of the Cerebral Cortex

Summary: This mid-gestational fetus is diagnosed with holoprosencephaly (HPE). The remainder of the gestation and pregnancy was uneventful. A scheduled elective cesarean section delivery was performed at full-term gestational age.

- **Brain structure affected:** This disorder is characterized by failure of the prosencephalon, or forebrain of the embryo, to develop.

- **Developmental abnormality mechanism:** The prosencephalon develops at the tip of the neural tube at around 3 weeks in human embryos, subdivides into diencephalon and telencephalon, and by the fifth or sixth week it has divided bilaterally to form the cerebral hemispheres. HPE results when this cleavage fails to occur.

CLINICAL CORRELATION

HPE is a complex congenital brain malformation characterized by failure of the forebrain to bifurcate into two hemispheres, a process normally complete by the fifth week of gestation. It is the most common developmental defect of the forebrain and midface in humans, occurring in 1 in 250 pregnancies. Since only 3% of fetuses with HPE survive to delivery, the live birth prevalence is only approximately 1 in 10,000. The majority of cases, up to two-thirds, consist of the most serious form, alobar HPE. The HPE phenotypes are distributed along a continuum:

1. Alobar HPE (most severe)—the brain is not divided and there are severe abnormalities: an absence of the interhemispheric fissure, a single primitive ventricle, fused thalami, and absent midline structures like the third ventricle, olfactory bulbs and tracts, and optic tracts.

2. Semilobar HPE (moderate)—the brain is partially divided and there are some moderate abnormalities: there are two hemispheres in the rear but not in the front of the brain (partially separated cerebral hemispheres and a single ventricular cavity).

3. Lobar HPE (mild)—the brain is divided and there are some mild abnormalities: there is a well-developed interhemispheric fissure but some midline structures remain fused.

4. Middle interhemispheric variant of HPE (MIHV)—the middle of the brain (posterior frontal and parietal lobes) are not distinctly separated.

Prognosis, as would be expected, varies depending on disease severity and associated malformations. Patients with alobar HPE have a survival rate of approximately 50% by 4-5 months of age and approximately 20% at 1 year. Isolated semilobar and lobar HPE have empirical survival rates to 1 year of approximately 50%. Almost all survivors have some degree of developmental delay, presenting with mental retardation which usually correlates to the severity of HPE. Feeding difficulties leading to aspiration pneumonia and/or failure to thrive frequently occur in individuals within all subtypes. The diverse etiologies that lead to HPE highlight the susceptibility of early nervous system development to perturbation, either from genetic alterations, epigenetic factors, or both. **The common denominator in HPE is there is failure of cleavage of the prosencephalon which gives rise to the cerebral hemispheres and diencephalon during early first trimester (5-6 weeks), resulting in persistent fusion of the cerebral cortices.**

The best-known genetic model for the disease is the **sonic hedgehog** (SHH) signaling that occurs during cortical development. The mouse mutants that have disrupted *SHH* gene expression develop the murine equivalent of HPE, complete with cyclopia. In humans, mutations in *SHH* have been found in 17% of familial and 3.7% of sporadic cases of HPE. *SHH* is involved in multiple developmental events at various times during embryogenesis: establishment of the left-right axis; specification of the floor plate and ventral spinal cord; and ventral identity of the brain along the entire rostral-caudal central nervous system. Later during embryogenesis, *SHH* has a crucial role in the development of various structures: the limbs, the pituitary gland, the neural crest cells, the midbrain, the cerebellum, the eyes, and the face. Other genes—such as *PTCH*, *GLI2*, *ZIC2*, and *DHCR7*—that have been linked to the SHH molecular pathway also result in HPE phenotypes when mutated.

APPROACH TO:
Formation of the Cerebral Cortex

OBJECTIVES

1. Understand the multiple events that occur in the development of the cerebral cortex.

2. Be able to relate the types of molecular cues involved.

3. Appreciate the importance of the chemospecificity hypothesis in elaborating the molecular basis of signaling interactions.

DEFINITIONS

NEUROGENESIS: The mechanism for generating a neuron from a population of neuroepithelial sheet of cells.

NEURULATION: The developmental process by which the neural plate fuses into the neural tube. The process begins in the cervical region and progresses in both directions, first closing the **cranial neuropore**, followed by the **caudal neuropore**.

CELL FATE DETERMINATION: The developmental journey during which precursor cells begin with many possible fates and are progressively limited in their potential, until they finally differentiate into a mature cell.

NEURAL MIGRATION: The spatial translocation of a nascent premature neuron from the ventricular zone to its final destination; in the cortex, to a specific cortical layer and position.

AXON GUIDANCE: The mechanism by which a neuron sends a cellular extension or axon over sometimes great distances—bypassing billions of potential but inappropriate targets—before terminating in the correct area.

GROWTH CONE: The specialized terminal apparatus on the leading edge of an axonal process that can recognize (sensory function) and mechanically respond (motor function) to guidance cues.

ATTRACTANT FACTORS: A molecular signal that induces a growth cone to grow *toward* the source.

REPELLANT FACTORS: A molecular signal that induces a growth cone to grow *away* from the source.

SYNAPTOGENESIS: The process of forming a synapse, or functional connection, between two neurons. It consists of the formation of an appropriate connection and the maturation of the pre- and postsynaptic terminals into mature functional units.

DISCUSSION

The central nervous system is a complex system, and the brain as the most specialized area of the CNS is more elaborate yet. The cerebral hemispheres, the latest evolutionary advancement in the nervous system, raise the level of complexity even further. The mathematics are prodigious—the human cerebral cortices contain an estimated 20 billion neurons, and each cortical neuron is connected to upward of 10,000 other neurons. These billions of cells that make trillions of connections all start from a single cell.

The cerebral cortex relies on the same basic principles of development that the rest of the nervous system utilizes. During early **neurogenesis**, neural progenitors must be specified in the neuroectoderm and begin to adopt a fate separate from their ectodermal brethren. These precursor cells must expand and form first a neural plate and then proceed to invaginate and form the neural tube, via continued neurogenesis and **neurulation**. As immature neurons are born in the ventricular zone, they experience various intrinsic and extrinsic signals that begin to establish their identity. This **fate determination** occurs through temporal specification, spatial specification, and other determining factors. As these neurons become increasingly "educated" in their eventual fate, they are forced to migrate to their cortical destination through a complex environment consisting of extracellular matrix (ECM) and numerous cell populations, all sending out an extensive repertoire of attractant and repellant cues in the process of **neuronal migration**. Similar cues guide the leading processes of migrating neurons to form axonal projections toward their destined target (**axon guidance**) and to establish a functional connection or synapse (**synaptogenesis**) with appropriate targets. It is these final steps, the labyrinth of connections and the computational networks they define, that are the distinguishing mark of the cerebral cortex.

The central nervous system begins to adopt a "cortical fate" with the early establishment of spatial patterning. The regional patterning of the forebrain relies on the establishment of an anterior-posterior or rostral-caudal axis. This axis defines the first three initial vesicular enlargements in the neural tube (from anterior to posterior): the prosencephalon (forebrain), mesencephalon (midbrain), and rhomb-encephalon (hindbrain). Further development of the prosencephalon results in an additional enlargement, separating the forebrain into the telencephalon and dien-cephalon. It is this telencephalon that is the embryological precursor to the cerebral cortices. The developmental divergence continues, as the spatial and connectiv-ity patterns of neurons within a functional neural circuit are so complex that they require a special set of mechanisms.

Axon guidance involves the targeting of a specific neuron's axon to the appropri-ate target. Early developmental evidence resulted in the **chemospecificity hypoth-esis**, first proposed by Roger Sperry in 1963, which stated that chemical matching between axons and their target tissues was responsible for appropriate guidance. This model suggested the existence of "recognition molecules" that were respon-sible for matching. The chemospecificity hypothesis posits that signaling molecules at targets provide positional information in the form of gradients that would be detected by complementary gradients of receptors on axons to guide their growth. For this seminal idea in neurobiology, Roger Sperry was awarded the Nobel Prize. The leading process of the axon, the growth cone, is responsible for sensing these recognition cues and responding to them. Axonal growth cone movement toward its final destination occurs through polymerization or depolymerization of actin in the finger-like projections of the growth cone. Molecules bind to receptors on the growth cone and cause a signaling cascade that results in polymerization.

To date, a vast array of molecules that exert either attractant or repellent forces on the growing axons has been discovered. These molecules can be attached to other cells or the extracellular environment to act over short distances, or they can be secreted, diffusible molecules that act over longer distances. The short-range interactions can be attractive: the laminins in the ECM partner with specific inte-grin proteins in the membrane of growth cones or the cadherin family or immu-noglobulin superfamily of cell surface proteins can mediate cell-cell compatibility. They can also be repulsive: the gradient of ephrin protein ligand expressed in the target inhibits growth cones expressing the corresponding ephrin kinases. Similarly, soluble proteins like netrin-1 that act at long distances can have both attractive (if the growth cone expresses the DCC receptor) or repulsive (if it expresses receptors like UNC-40) effects on axonal guidance.

Once axonal guidance leads to the appropriate target tissue, the process of form-ing a functional connection or synapse can occur. This formation of synapses, or **synaptogenesis**, is the key to the final step of cortical and nervous system develop-ment, because it allows the information-processing circuit to function. Once neu-rons are situated, synapse formation occurs when pre- and postsynapse contact each other by molecular interaction between the two cells. Neurologin then allows for the clustering of vesicles within the presynaptic terminal (Figure 17-1).

CASE CORRELATES

- See Case 11 (CNS development), Case 12 (peripheral nervous system), Case 13 (neurulation), Case 14 (neurogenesis), Case 15 (cell fate determination), and Case 16 (neuronal migration).

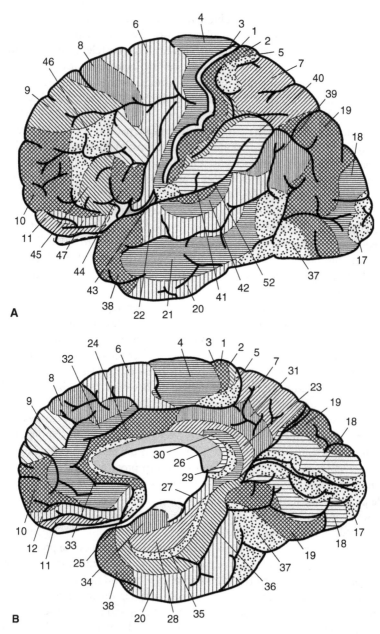

Figure 17-1. Cytoarchitectural zones of the human cerebral cortex according to Brodmann. **A.** Lateral surface, **B.** Medial surface, **C.** Basal inferior surface. (*With permission from* Adam and Victor's Principles of Neurology. *7th ed. New York: McGraw-Hill; 2000. Page 465, fig 22-2.*)

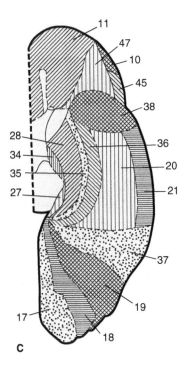

11
47
10
45
38
28
36
34
35
20
27
21
37
19
17
18
C

Figure 17-1. (Continued)

COMPREHENSION QUESTIONS

17.1 An 82-year-old man is being evaluated for memory loss and cognitive difficulties. An MRI shows that he has diffuse cortical atrophy with compensatory enlargement of his ventricles, consistent with Alzheimer disease. From which embryological precursor does the cerebral cortex arise?

A. Diencephalon

B. Mesencephalon

C. Telencephalon

D. Rhombencephalon

17.2 What type of interactions are used to guide the growth of axons so that they reach their appropriate destination?

A. Short-range contact inhibition only

B. Diffusion gradient inhibition and contact attraction only

C. Contact inhibition and diffusion gradient attraction only

D. Contact inhibition and attraction as well as diffusion gradient attraction and inhibition

17.3 Which of the following places the developmental events in the correct chronological order?

    A. Neurogenesis → Neural migration → Synaptogenesis → Axonal pathfinding

    B. Neurogenesis → Neural migration → Axonal pathfinding → Synaptogenesis

    C. Neural migration → Neurogenesis → Synaptogenesis → Axonal pathfinding

    D. Neurogenesis → Axonal pathfinding → Neural migration → Synaptogenesis

ANSWERS

17.1 **C.** The **telencephalon**, which arises from the prosencephalon, is the most rostral swelling of the neural tube, and ultimately develops into the cerebral cortex. The diencephalon becomes the thalamus, epithalamus, subthalamus, and hypothalamus; the mesencephalon becomes the midbrain; and the rhombencephalon becomes the pons, medulla, and cerebellum.

17.2 **D.** In the process of axonal pathfinding, there are many different cues that guide the axon to its ultimate destination such as **attractive/repulsive contact cues** and **attractive/repulsive diffusion gradient cues**. Since the axonal growth cone of each different type of neuron expresses different receptor molecules, they all respond to the cues differently, allowing them all to reach their own appropriate target.

17.3 **B.** The correct order is that first the **neurons are formed** from precursor cells in the ventricular zone, then they **migrate** via one of several methods to their final location in the cortex. Following migration, the neurons send out **axonal growth cones** that respond to attractive and repulsive signals to reach the appropriate target, where they can undergo **synaptogenesis** to mature the connection between presynaptic and postsynaptic cell.

NEUROSCIENCE PEARLS

▶ Holoprosencephaly is a structural brain malformation, arising from partial or total failure of the process of division of the prosencephalon (forebrain).

▶ The cerebral cortices develop through a series of conserved and overlapping mechanisms shared with other parts of the nervous system.

▶ Axonal guidance is reliant on a balance of attractant and repellant factors.

▶ Specific molecules mediate chemical interactions between the growing axon and target tissues.

REFERENCES

Cohen MM, Jr. Holoprosencephaly: clinical, anatomic, and molecular dimensions. *Birth Defects Res.* 2006;76(pt A):658-673.

Hahn JS, Plawner LL. Evaluation and management of children with holoprosencephaly. *Pediatr Neurol.* 2004;31:79-88.

Kandel ER, Schwarz JH, Jessell TM, Siegelbaum SA, Hudspeth AJ, eds. *Principles of Neural Science*. 5th ed. New York, NY: McGraw-Hill; 2012.

Rao MS, Jacobson M, eds. *Developmental Neurobiology*. 4th ed. New York, NY: Kluwer Academic/Plenum Publishers; 2005.

Roessler E, Muenke M. How a hedgehog might see holoprosencephaly. *Hum Mol Genet*. 2003;12(Review Issue 1):R15-R25.

Sadler TW, ed. *Langman's Medical Embryology*. 7th ed. Baltimore, MD: Williams and Wilkins; 1995.

An 18-year-old woman, who is wheelchair-bound, states that since she has been suffering from an irregular, staggering gait with frequent falls since the age of 9. Over the next few years, her upper extremity coordination deteriorated, and she developed weakness during her teenage years. Physical examination reveals abnormal eye movements, slow and poorly coordinated movements of all her extremities, and severe weakness of her legs. She has deficits in position sense, vibratory sense, pain, and tactile discrimination. Her deep tendon reflexes are absent. Hypertrophic cardiomyopathy is present, and she is subsequently diagnosed with Friedreich ataxia.

- ▶ What is the mechanism of this disease pathology?
- ▶ Which pathways are involved?
- ▶ What are the sensory modalities lost with each pathway?

ANSWERS TO CASE 18:

Proprioception

Summary: An 18-year-old woman has progressive ataxia, lower extremity weakness, and loss of coordination of extremities and eye movements. Examination reveals loss of position sense, vibratory sense, pain, and tactile discrimination. She is diagnosed with Friedreich ataxia.

- **Mechanism:** The degeneration of nerve tissue in the spinal cord and of nerves that control muscle movement in the arms and legs. This results in the thinning of the spinal cord and the loss of myelin sheaths in many neurons. The leading theory for the pathogenesis of the disease implicates an overaccumulation of iron in mitochondria, leading to cellular damage and death by the production of free radicals in neurons of the central nervous system (CNS) and peripheral nervous system (PNS).

- **Involved pathways:** This disorder leads to progressive degeneration of the dorsal root ganglia with secondary degeneration of the dorsal and ventral spinocerebellar tracts, medial lemniscal system, spinothalamic tract, and lateral cervical system. The corticospinal pathways are also affected.

- **Sensory modalities lost:** The dorsal and ventral spinocerebellar tracts convey proprioceptive (sense of balance and self) information from the body to the cerebellum. The medial lemniscal system conveys information from specialized touch, pressure, vibration, and joint receptors, and proprioceptive information from the body, to the cerebral cortex. The spinothalamic tract conveys somatosensory information including nociception (pain perception), temperature, itching, and crude touch to the thalamus. The corticospinal tract conveys motor information from the cerebral cortex of the brain to the spinal cord in response to information generated from many different sensory pathways.

CLINICAL CORRELATION

Friedreich ataxia is the most common of the **autosomal recessive** hereditary ataxias. It is caused by a mutation of a gene on chromosome 9 that codes for the protein **frataxin**, which is needed for normal function of the mitochondria. This disorder is characterized by neuron degeneration which begins in the periphery, resulting in loss of neurons and a secondary scarring process known as **gliosis**. Large myelinated axons in peripheral nerves are affected progressively with age and duration of the disease. Unmyelinated fibers in sensory roots and peripheral sensory nerves are spared. **The posterior columns, corticospinal, ventral, and lateral spinocerebellar tracts are all affected by demyelination.** The dorsal root ganglia shrink and eventually show disappearance of neurons. The **posterior column degeneration accounts for the loss of position and vibration senses and the sensory ataxia.** Absent tendon reflexes in the legs occur as a result of loss of afferent fibers from muscle spindle receptors. The progressive degeneration of the corticospinal pathway leads to limb weakness. Kyphosis and foot anomalies are also commonly seen. **Many affected**

individuals also have hypertrophic cardiomyopathy. While research in this area is progressing, there is currently no effective treatment available.

APPROACH TO:
Proprioception

OBJECTIVES

1. Be familiar with the CNS pathways that convey proprioceptive information.

2. Understand the clinical symptoms that result from a lesion to the dorsal column pathways.

DEFINITIONS

KINESTHESIA: The sense of movement.

ASTEREOGNOSIS: Loss or impairment of the ability to recognize common objects by touching or handling them without visual input.

ROMBERG TEST: A test in which the patient stands with feet placed closely together and eyes closed. A positive test results if the body sways abnormally or there is loss of balance, indicating a deficiency in conscious recognition of muscle and joint position.

GOLGI TENDON ORGAN: A proprioceptive sensory receptor located in muscle fibers and tendons.

MEDIAL LEMNISCAL SYSTEM: The pathway carrying proprioceptive information to the cerebral cortex involving position sense, kinesthesia, and tactile discrimination.

DISCUSSION

Two sets of sensory pathways in the spinal cord provide important information to the brain regarding muscle action, joint position, and the objects with which a person is in contact. One of these sets of pathways projects to the cerebellum, where the information is processed for the coordination of movement but not for conscious perception. These pathways include the **dorsal** and **ventral spinocerebellar tracts,** the **cuneocerebellar tract,** and the **rostral spinocerebellar tract.** The second set of pathways consists of three tracts that project to the cerebral cortex by way of the thalamus. This information is perceived consciously. These are the **spinal lemniscus,** the **spinothalamic tract,** and the **lateral cervical system.** Both sets of sensory receptors receive information from mechanoreceptors.

Mechanoreceptors located in the muscles, joints, and skin mediate the various separate and integrated sensations of proprioception, touch, and tactile discrimination. Mechanoreceptors consist of **Golgi tendon organs, pacinian corpuscles, Meissner corpuscles, Ruffini corpuscles,** Merkel cells, and free nerve endings in muscles, tendons, ligaments, joint capsules, and skin. Information about static limb position is carried mainly through **muscle spindle afferents.** Kinesthetic information is mediated by a combination of **joint receptor** afferents and receptors in the skin, muscles, and joints. Pacinian corpuscles, found in the skin and connective tissue, detect vibration. Meissner corpuscles detect superficial phasic touch. The movement of hairs is detected by free nerve endings in hair follicles and also conveys a sense of touch. Contrastingly, Ruffini corpuscles detect skin stretch. Merkel cells are specialized epithelial cells that detect fine touch information. Most of the mechanoreceptors are innervated by large-diameter myelinated fibers. The free nerve endings are exceptions as their cell bodies are located in the dorsal root ganglia, and their central processes enter the medial side of the dorsal root zone. These nerve endings detect pain and temperature. Afferent fibers from mechanoreceptors enter the spinal cord and distribute either to the interneurons and motor neurons in the ventral horn, the neurons in the dorsal and intermediate gray areas where ascending pathways originate, or to the neurons of the dorsal column nuclei in the medulla (Table 18-1).

The ventral and dorsal spinocerebellar tracts carry proprioceptive and other somatosensory stimuli from the lower limbs to the cerebellum. Proprioceptive sensation is the perception of the relative position of the parts of the body. Unlike the other sensory systems that allow us to perceive the world outside our body, proprioception provides feedback information about the body itself. It consists of both static limb position and kinesthesia, or the sensation of motion. Some proprioceptive fiber collaterals from Golgi tendon organs synapse with neurons in the intermediate gray area and the base of the posterior horn of the spinal cord. At the lumbar and sacral levels of the spinal cord, these neurons give rise to the primarily crossed ventral spinocerebellar tract. This is the most peripheral tract in the ventral margin of the lateral funiculus.

The **nucleus dorsalis,** or **Clarke nucleus,** is located at the base of the posterior horn in spinal segments T1 through L2. This column of neurons receives afferents from muscle spindles, cutaneous touch receptors, and joint receptors. The axons of these neurons ascend rostrally and ipsilaterally as the **dorsal spinocerebellar tract** just posterior to the ventral spinocerebellar tract in the lateral funiculus. While the proprioceptive afferents traveling from dorsal roots T1 to L2 synapse in the nucleus dorsalis at the level where they enter the spinal cord, the corresponding afferents from dorsal roots L3 through S5 ascend first in the fasciculus gracilis of the dorsal funiculus to reach the nucleus dorsalis. They synapse there at levels L1 and L2, which becomes the most caudal level of the dorsal spinocerebellar tract.

Table 18-1 • PRESSURE AND TOUCH RECEPTORS AND THEIR FUNCTION	
Mechanoreceptor	Function
Pacinian corpuscle	Sensitivity to deep pressure touch and high-frequency vibration
Meissner corpuscle	Sensitivity to light touch
Merkel cell or nerve endings	Slowly adapting, provide touch information to the brain
Ruffini corpuscle	Slowly adapting, sensitive to skin stretch, and contributes to the kinesthetic sense of and control of finger position and movement

The cuneocerebellar tract and the rostral spinocerebellar tract carry information from mechanoreceptors in the upper extremities to the cerebellum. Afferent fibers from C2 to T5 travel rostrally in the dorsal funiculus in the fasciculus cuneatus before synapsing on neurons in the **accessory cuneate nucleus** in the lower medulla. This is the upper extremity counterpart of the nucleus dorsalis and gives rise to the ipsilateral cuneocerebellar tract or **dorsal arcuate fibers.** This tract also mediated information from muscle spindles, cutaneous touch receptors, and joint receptors.

The **rostral spinocerebellar tract** is the upper-extremity tract that corresponds to the ventral spinocerebellar tract. It originates in the cervical enlargement of the intermediate zone of the spinal cord gray area. After projecting to the cerebellum, the tract synapses with fibers of the ventral spinocerebellar tract.

All of the ascending fibers from the dorsal spinocerebellar, cuneocerebellar, and rostral spinocerebellar pathways enter the cerebellum through the inferior cerebellar peduncle. The ventral spinocerebellar tract, however, travels through the pons prior to entering the cerebellum through the superior cerebellar peduncle. These four tracts terminate primarily in the midline vermis and intermediate zone of the cerebellum ipsilateral to the cells of origin. There is also a significant projection to the anterior lobe and posterior lobe of the cerebellum that contributes to standing and walking.

Proprioceptive information regarding position sense, kinesthesia, and tactile discrimination are carried to the cerebral cortex by afferent fibers from muscle spindles, Golgi tendon organs, and mechanoreceptors in joints and skin by way of the **medial lemniscal system.** This information contributes to conscious position as well as movement sense. The posterior funiculus consists of two large bundles of fibers called fasciculi. Fibers from the leg ascend as the **fasciculus gracilis** adjacent to the dorsal median septum. Fibers from the arm ascend lateral to the leg fibers as the **fasciculus cuneatus.** The fibers of the posterior funiculus maintain somatotopic organization in relation to one another. The fasciculus cuneatus terminates in the lower medulla in the nucleus cuneatus. Similarly, the fasciculus gracilis terminates in the nucleus gracilis, also in the lower medulla. These tracts are frequently referred to as the dorsal column pathways. A portion of the fibers of the medial lemniscal system travel with the lateral cervical system in the dorsal part of the lateral funiculus. The **dorsolateral pathway** can also be used to describe the entire lemniscal pathway of the spinal cord.

The cells of the dorsal column nuclei give rise to the **internal arcuate fibers,** which cross to the contralateral side of the medulla in the **decussation of the medial lemniscus.** From here they ascend as the **medial lemniscus** to the thalamus and terminate in the **ventral posterolateral nucleus (VPL).** The somatotopic organization of the fibers is maintained in both the medial lemniscus and VPL. Thalamocortical fibers from the VPL continue on to the postcentral gyrus of the parietal lobe and terminate in the **primary somatosensory cortex.** This cortex maintains a topographic representation of the body that is similar to that of the parallel motor strip on the opposite side of the central sulcus.

The **spinal lemniscus system** mediates the sense of limb position and movement, the sense of steady joint angles, the sense of motion produced by active muscular contraction or passive movement, the sense of tension exerted by contracting muscles, and the sense of effort. Cortical processing is necessary for the conscious recognition of body and limb posture. The lemniscal pathways also provide important information regarding the place, intensity, and temporal and spatial patterns of neural activity evoked by mechanical stimulation of the skin. This same pathway to the cortex, therefore, is necessary for discriminative tactile sensation and vibration sense.

The **lateral cervical system** responds to light mechanical stimulation of the skin on the ipsilateral side of the body. This system is an alternate pathway to the information conveyed in the posterior column system, and its relevance in humans is not fully understood. The peripheral nerve fibers synapse in the dorsal horn throughout the length of the spinal cord. Heavily myelinated axons from second-order neurons arise in this lamina and ascend ipsilaterally in the dorsal corner of the lateral funiculus to terminate in the **lateral cervical nucleus.** Projections from this nucleus cross the spinal cord in the **ventral white commissure** to join the contralateral medial lemniscus and proceed with it to terminate in the thalamus. Information from here is projected to the somatic sensory areas of the cerebral cortex.

A bilateral interruption of the dorsolateral pathway in the spinal cord will result in complete loss of proprioceptive sensation. A lesion in this location will produce deficits in position sense, vibration sense, and tactile discrimination. A unilateral injury of a dorsolateral pathway will create ipsilateral symptoms. Lesions within the gracile and cuneate nuclei, the medial lemniscus, the thalamus, and the postcentral gyrus will create varying degrees of similar symptoms. A lesion of the lemniscal pathway will preserve simple touch, pain, and temperature, but it will disrupt proprioceptive sensation and result in the following symptoms: inability to recognize limb position, astereognosis or loss of ability to recognize common objects by touching them with the eyes closed, loss of two-point discrimination, loss of vibratory sense, and a positive **Romberg sign.** In Romberg test a patient stands with his or her feet close together. The amount of sway is noted with the eyes open. The patient then closes the eyes, and an abnormal increase in sway or loss of balance without visual input results in a positive sign. The visual system is able partly to compensate for the deficiency in recognition of muscle and joint positions, allowing patients with dorsolateral pathway lesions to maintain balance with their eyes open.

CASE CORRELATES

- See Cases 18-23 (sensory systems).

COMPREHENSION QUESTIONS

18.1 A 63-year-old woman comes into your office for a routine physical examination. She notes that she has some occasional tingling in her feet and toes. You perform a complete neurological examination and find that she has reduced vibration sense in her great toe and at her ankle on both sides. Through which structure does the tract carrying this information to the brain pass?

A. Ventral horn of the spinal cord

B. Lateral funiculus

C. Medial lemniscus

D. Lateral lemniscus

18.2 A 23-year-old man is brought into the emergency department by EMS, restrained and screaming that he feels bugs crawling all over his skin. A friend who is accompanying him reports that he has recently begun using cocaine quite heavily. The physician examines the patient and notes that there are in fact no bugs crawling on him, despite his claims. If there actually were bugs on him, which of the following sensory organs would most likely be responsible for the crawling sensation?

A. Meissner corpuscle

B. Pacinian corpuscle

C. Golgi tendon organ

D. Muscle spindle

18.3 A 16-year-old adolescent girl is brought into the emergency department following a motor vehicle collision in which she was ejected from the vehicle. She has a shard of metal sticking out of her back in the midline at the level of T7. A CT scan of the back is obtained and the tip of the shard appears in the spinal canal, possibly in the area of the dorsal columns, slightly more to the left than the right. If the shard has injured the right dorsal column at T7, which of the following sensory defects would be expected?

A. Loss of proprioception in the left leg

B. Loss of proprioception in the right leg

C. Loss of pain sensation in the left leg

D. Loss of pain sensation in the right leg

ANSWERS

18.1 **C.** The sensation in question here, vibration, is transmitted via the dorsal column–**medial lemniscal** pathway. The sensation is detected by a pacinian corpuscle, which is innervated by a sensory neuron whose body is in the dorsal root ganglion. The axon of this neuron enters the cord via the dorsal horn and ascends the posterior column of the spinal cord all the way to the lower medulla, where it synapses in the gracile nucleus (because it was from the lower body). From there the second-order neuron sends its axons across the midline to ascend in the contralateral medial lemniscus all the way to the ventral posterolateral nucleus (VPL) of the thalamus. From the thalamus the information is relayed to the sensory cortex where it is processed. In addition to vibration, discriminative touch and joint proprioception follow the same pathway. The lateral lemniscus is a tract in the brainstem involved in the relay of auditory information.

18.2 **A.** The sensory organ responsible for registering light touch is the **Meissner corpuscle**. The pacinian corpuscle is responsible for registering vibration, the Golgi tendon organ measures muscle tension, and the muscle spindle registers muscle fiber length. All of these organs are part of the dorsal column–medial lemniscal system and are innervated by large-diameter myelinated nerve fibers.

18.3 **B.** In this scenario, the patient would lose **proprioception in the right leg**. The dorsal columns transmit the signals for discriminative touch, vibration, and proprioception. A lesion to one of these columns will cause a loss of all of these sensations below the level of the lesion on the ipsilateral side. The sensation loss will be ipsilateral because the fibers of the pathway do not decussate until after they have synapsed in the nuclei gracilis and cuneatus in the lower medulla.

NEUROSCIENCE PEARLS

▶ Friedreich ataxia is indicated by loss of position and vibration sense, kyphosis, heart failure (hypertrophic cardiomyopathy), and cerebellar ataxia.

▶ Friedreich ataxia is the most common autosomal recessive cerebellar ataxias.

▶ Proprioceptive information is carried from mechanoreceptors located in the muscles, joints, and skin.

▶ The medial lemniscal pathway is located in the dorsolateral region of the spinal cord and carries position, movement, touch, pressure, and vibratory information.

▶ Injuries to the lemniscal pathway can result in inability to recognize limb position, astereognosis, and a positive Romberg sign.

REFERENCES

Buck LB. The bodily senses. In: Kandel ER, Schwarz JH, Jessell TM, Siegelbaum SA, Hudspeth AJ, eds. *Principles of Neural Science*. 5th ed. New York, NY: McGraw-Hill; 2012.

Martin JH. The somatic sensory system. *Neuroanatomy: Text and Atlas*. 2nd ed. Stamford, CT: Appleton and Lange; 1996.

Ropper AH, Brown, RH. Degenerative diseases of the nervous system. In: *Adam and Victor's Principles of Neurology*. 8th ed. New York, NY: McGraw-Hill; 2005.

CASE 19

A 27-year-old woman was hit by a car while riding her bicycle to work. She was brought to the emergency room where she complained of severe neck pain. On physical examination she was noted to have spastic paralysis of the right upper and right lower extremities, loss of vibration and position sense on the right side, and loss of pain and temperature sensation on the left side. An MRI reveals an acutely herniated disk compressing the right side of the upper cervical spinal cord.

▶ What syndrome could explain all of these patient's symptoms?
▶ Where is the injury located?
▶ What is the explanation for the loss of pain and temperature on the opposite side of the injury?

ANSWERS TO CASE 19:

Spinothalamic Pathway

Summary: Twenty-seven-year-old woman involved in a bicycle accident. An MRI reveals an acutely herniated upper cervical disk creating compression of her right spinal cord. She has symptoms of ipsilateral spastic paralysis, ipsilateral loss of vibration and position sense, and contralateral loss of pain and temperature sensation.

- **Syndrome that explains this patient's symptoms:** Brown-Sequard syndrome is characterized by an ipsilateral paralysis and loss of proprioception and vibration and a contralateral loss of pain and temperature sensation.

- **Location of injury:** Brown-Sequard syndrome can be caused by hemicompression or hemisection of the spinal cord.

- **Cause of the contralateral loss of pain and temperature sensation:** Injury to the spinothalamic tract will result in contralateral pain and temperature loss from one or two levels below the injury, and ipsilateral pain and temperature loss at the level of the injury.

CLINICAL CORRELATION

Brown-Sequard syndrome is a disorder that results from a hemisection injury to the spinal cord. Three main neural systems are affected, producing the resulting symptoms: the disruption of the corticospinal tract carrying the upper motor neurons produces a spastic paralysis on the same side of the body; loss of one or both dorsal columns will result in an ipsilateral loss of vibration and proprioception, and loss of the spinothalamic tract leads to pain and temperature sensation being lost from the contralateral side of the body. The deficit begins one or two segments below the lesion, and all sensory modalities are lost on the ipsilateral side at the level of the lesion because of the fibers still being uncrossed. While this syndrome is usually the result of penetrating traumatic injury to the spinal cord, it can rarely be caused by a herniated cervical disc. In this case, surgical intervention to decompress the disc would be indicated.

APPROACH TO:

Spinothalamic Pathway

OBJECTIVES

1. Know the origins of the neurons that make up the anterolateral system.

2. Be able to describe the pathway of the spinothalamic and spinoreticular tracts.

3. Understand the central projection patterns of these pathways.

DEFINITIONS

ANTEROLATERAL SYSTEM: The ascending pathway that conveys pain and temperature from the periphery to the brain.

DERMATOME: An area of the skin which innervates afferent nerve fibers coming to a single dorsal spinal root.

NOCICEPTOR: A sensory receptor that sends signals that cause the perception of pain in response to potentially damaging stimuli.

RETICULAR FORMATION: The network of neurons in the brainstem involved in consciousness, regulation of breathing, and transmission of sensory stimuli to higher brain centers.

SPINORETICULAR TRACTS: The fibers of the anterolateral system that do not reach the thalamus directly, instead synapsing first in the reticular formation before ascending to the thalamic nuclei, hypothalamus, and limbic system.

DISCUSSION

The **anterolateral system** is comprised of the **spinothalamic tracts** and the **spinoreticular tracts,** which do not reach the thalamus and cannot therefore be termed spinothalamic (Figure 19-1). These pathways mediate the sensations of pain, itching, temperature, and simple touch.

Neurons of the anterolateral system originate in the contralateral **dorsal horn** of the spinal cord. Most of their axons cross the midline through the **ventral white commissure** before ascending in the spinal cord to give rise to a diffuse bundle of

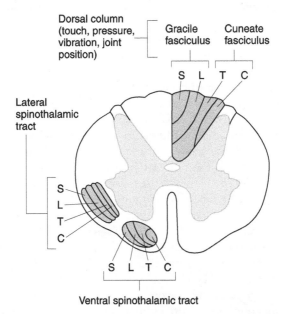

Figure 19-1. A cross-section of the spinal cord showing the spinothalamic and other ascending sensory pathways. (*With permission from* Morgan's Clinical Anesthesiology. *4th ed. New York: McGraw-Hill; 2006. Page 366, fig 18-4.*)

fibers projecting through the anterior and lateral funiculi. The originating neurons in the dorsal horn are activated by small-diameter, lightly myelinated, and unmyelinated dorsal root afferents, including Aδ(III) and C(IV) fibers, as well as larger myelinated cutaneous afferents. Aδ fibers have a faster conduction velocity and convey sharp pricking pain, and the C fibers convey dull aching pain.

A **dermatome** is the area of skin supplied by one dorsal root. The dorsal roots convey information from essentially all dermatomes below the face. The cell bodies of the dorsal root fibers are located in the **dorsal root ganglia.** Each ganglion cell possesses a single nerve process that divides into a central branch running to the spinal cord and a peripheral branch coming from a receptor organ or organs. The dorsal root fibers enter the spinal cord through the **dorsal root entry zone** in the region of the dorsolateral sulcus. The Aα and Aβ fibers are the most heavily myelinated and occupy the most medial position in this entry zone, while the small myelinated Aδ and unmyelinated C fibers occupy the most lateral position.

The peripheral pain receptors consist of the naked terminals of the small Aδ and C nerve fibers. Many of these are specialized chemoreceptors that are stimulated by tissue substances released in response to noxious and inflammatory stimuli, such as histamine, bradykinin, serotonin, acetylcholine, substance P, potassium, and cyclooxygenases. The concentration of hydrogen ions in these substances has been found to be critical in the activation of pain receptors. A stimulus that evokes pain is usually one that can cause damage or destruction to tissue.

As the axons of the Aδ and C fibers enter the dorsal root zone, they immediately divide into short ascending and descending branches that run longitudinally in the **posterolateral fasciculus,** or **Lissauer tract.** After traveling one or two segments, these fibers leave the tract to synapse with neurons in lamina IV in the dorsal horn. These neurons contain receptors for numerous neurotransmitters, such as excitatory amino acids and neuropeptides. The Aδ and C fibers release glutamate and substance P as their neurotransmitters. Interneurons in laminae II through IV project to the neurons in lamina V where they synapse with the cells of origin of the anterolateral system, including the spinothalamic tracts and the spinoreticular projections. Neurons in lamina I contribute fibers to the spinothalamic tracts.

The axons of the spinothalamic tract from laminae I and V of the dorsal horn decussate anterior to the central canal in the **ventral white commissure** and then ascend rostrally in the **anterolateral funiculus.** The spinothalamic and spinoreticular tracts ascend through the spinal cord and brain stem and supply information to other spinal cord segments, the **reticular formation,** the **superior colliculus,** and several **thalamic nuclei,** including the **intralaminar nuclei** and the **ventral posterolateral nucleus (VPL).** The VPL transmits information from the spinothalamic pathway to the lemniscal system through the **ventrobasal complex.** Projections of the anterolateral system to the VPL are carried somatotopically, and fibers carrying input from the upper body are located medially to those carrying input from the lower body. This somatotopic organization is maintained as fibers project from the VPL, conveying painful and thermal sensations to the **primary somatosensory cortex** of the **postcentral gyrus.** VPL axons end in the primary somatosensory cortex and provide information for accurate localization of sharp pricking pain. The fibers from the upper parts of the body project to cortical areas near the lateral fissure, while those from the lower parts terminate on the medial surface of the hemisphere in the **paracentral lobule.** The postcentral gyrus is interconnected with the posterior parietal lobe. These areas function together to localize pain stimuli and integrate the pain modality with other sensory stimuli.

The anterolateral system is a predominantly slow-conducting polysynaptic system. The fibers that do not reach the thalamus directly synapse in the reticular formation of the brain stem. These are the **spinoreticular tracts.** From the reticular formation, ascending fibers relay pain information to the **medial** and **intralaminar nuclei** of the thalamus as well as to the hypothalamus and limbic system. Additionally, the intralaminar nuclei project information to multiple cortical areas to ensure that pain sensation is not lost when there is damage to the primary somatosensory cortex. While most fibers of the anterolateral system cross the midline before ascending, many of the fibers of the spinoreticular tracts conveying visceral sensory information ascend ipsilaterally.

Painful sensations from the face, cornea, sinuses, and mucosal linings travel through fibers of the trigeminal nerve to its sensory **trigeminal ganglion.** After these fibers enter the brain stem in the pontine region, they form an ipsilateral descending tract, the **spinal tract of V,** which courses to the upper cervical segments of the spinal cord. The neurons synapse in the **spinal nucleus of V (spinal trigeminal nucleus),** and from there they decussate and form the **ventral trigeminal lemniscus,** which joins with the crossed main sensory trigeminal efferents and ascends to the **ventral posteromedial nucleus (VPM).** In this way, information from the spinal nucleus of V projects to the contralateral side and ascends to the **VPM** of the thalamus. The ascending trigeminal pathway also projects to the reticular formation and the medial and intralaminar thalamic nuclei, which also receive projections from the anterolateral system of the spinal cord. The VPM projects to the somatosensory cortex closest to the lateral fissure. The areas of the body most sensitive to somatosensory stimuli, the lips and fingers, have disproportionately large areas of neuronal representation in the somatosensory cortex. Thalamocortical fibers from neurons in the intralaminar thalamic nuclei and parts of the VPL and VPM nuclei relay pain information to the **secondary somatic sensory area** of the cerebral cortex.

This tract has also been known to respond to temperature. Naked nerve endings in the skin carry sensations of cold and warmth. The peripheral nerve fibers mediating these sensations consist of thinly myelinated Ad and some C fibers. Other types of C fibers mediate the painful components of the extremes of hot and cold sensations. The central nervous system pathway for thermal sensation follows the same course as the pain pathway. These two systems are so closely associated in the central nervous system that they cannot be distinguished anatomically, and injury to one usually affects the other.

The lateral spinothalamic tract can be purposefully transected in the spinal cord in an attempt to relieve intractable pain. This procedure, known as a **tractotomy,** is performed by making a cut in the anterior part of the lateral funiculus. There is usually some damage to the ventral spinocerebellar tract and possibly to some of the extrapyramidal motor fibers. However, no permanent symptoms are typically seen aside from a loss of pain sensitivity on the contralateral side, which begins one or two segments below the cut. The pain relief is sometimes only temporary, suggesting there are other routes in crossed and uncrossed tracts that mediate nociceptive sensations in the spinal cord.

CASE CORRELATES

- See Cases 18-23 (sensory systems), and Case 5 (spinal cord injury).

COMPREHENSION QUESTIONS

Refer to the following case scenario for questions 19.1-19.2:

A 12-year-old boy is brought into your office with a 5-cm laceration on his forehead that he got falling off his bicycle. Although he is trying to be brave, a few tears are running down his cheek, and when asked, he says it "really, really hurts."

19.1 What type of somatosensory receptor is involved in the initial detection of tissue injury as pain?

A. Merkel cell

B. Pacinian corpuscle

C. Unmyelinated free nerve ending

D. Heavily myelinated free nerve ending

19.2 Through what thalamic nucleus is the pain relayed to the primary sensory cortex?

A. VPL

B. VPM

C. Centromedian nucleus (CM)

D. Parafascicular nucleus (PF)

19.3 Where are the cell bodies of the primary sensory neurons involved in pain transmission located?

A. Dorsal horn of the spinal cord

B. Ventral horn of the spinal cord

C. Dorsal root ganglia

D. Near the site of sensation, in the subcutaneous tissue

19.4 A 72-year-old man comes into the emergency department complaining of vertigo, hoarseness, difficulty swallowing, and a droopy left eyelid. On examination you note a loss of painful sensation over the left half of his face and the entire right half of his body. Interestingly, however, he has intact discriminative touch in those areas. His history in remarkable for 60 pack-years of smoking and poorly controlled high blood pressure. What is the most likely location of the lesion in this man?

A. Lower cervical spinal cord

B. Medulla

C. Thalamus

D. Cerebral cortex

19.5 A 55-year-old man is noted to have severe pain from lung cancer metastases to the pelvic iliac crests. Long-acting morphine is injected into the epidural space at the L4-L5 interspace, inducing prolonged pain relief. What is the mechanism whereby the morphine acts on the postsynaptic area?

A. Activating the Na/H exchange mechanism

B. Increasing chloride influx into the cell

C. Increasing potassium efflux out of the cell

D. Blocking voltage-dependent potassium flow

E. Increasing calcium influx into the cell

ANSWERS

19.1 **C.** The "receptor cells" involved in the detection of pain are not discreet sensory organs, as they are for discriminative touch and proprioception, but rather the free nerve endings of the first-order neurons in the pain pathway. These neurons are unmyelinated, or very slightly myelinated, as opposed to the neurons innervating touch receptors, which are heavily myelinated. The free nerve endings of pain fibers respond to a variety of chemical stimuli, including potassium ions, serotonin, histamine, and bradykinin. There are also free nerve endings that respond to physical deformation (stretch-sensitive pain fibers).

19.2 **B.** This boy has a lesion on his face, so the pain sensation is traveling through the spinal trigeminal tract, and the thalamic relay for this tract (and all localized sensation for the face) is **VPM**, which projects to the face area of the primary sensory cortex. Had this been a lesion on the trunk or extremities, then the pathway would have been the spinothalamic pathway, which (like all localized sensation on the trunk and extremities) is relayed through VPL. It is true that CM and PF, which are also known as the intralaminar nuclei, are involved in the sensation of pain, but they relay more diffuse, poorly localized pain to diffuse areas of the cortex.

19.3   **C.** The cell bodies of pain-sensing fibers are located in the **dorsal root ganglia,** just like the cell bodies of touch and proprioception neurons. Unlike touch neurons, however, the axons of pain neurons do not ascend the spinal cord all the way to level of the brainstem. They synapse with second-order neurons of the pain pathway in the dorsal horn of the spinal cord at or one to two spinal levels above where the fibers enter the cord.

19.4   **B.** This man is experiencing what is known as Wallenberg syndrome, which is caused by a lesion in the **lateral medulla.** Often, this lesion results from occlusion of the posterior inferior cerebellar artery (PICA), which supplies this region. The sensory symptoms are explained by the fact that the spinal trigeminal tract descends from the sensory trigeminal ganglion to the spinal trigeminal nucleus in the ipsilateral lower medulla and upper cervical spine. The descending spinal trigeminal neurons (carrying pain from the ipsilateral face) are near the ascending spinothalamic tract (carrying pain from the contralateral body), and both of these tracts are supplied by the PICA, so both are damaged. The medial lemniscus is not in the same region, so is not damaged by an occlusion of PICA, accounting for the preservation of discriminative touch. A lesion in the lower cervical cord would not affect the face as this is below the level of the spinal trigeminal nucleus, and lesions of the thalamus and cortex would affect the body and face on the same side.

19.5   **C.** Morphine is an opioid agent which binds to the mu opioid receptor, which is a G-protein–linked receptor. G-protein–linked receptors in turn activate secondary messengers; one of the primary ones is potassium-channel activation, allowing for potassium efflux out of the cell. Hyperpolarization of the postsynaptic neuron leads to cessation of the pain transmission. Mu receptors are not thought to be related to sodium, chloride, or calcium conductance.

NEUROSCIENCE PEARLS

▶ The anterolateral area of the spinal cord (spinothalamic tracts and spino-reticular tracts) mediates pain, temperature, and simple touch.

▶ Most of the anterolateral system axons cross the midline through the ventral white commissure.

▶ Cell bodies of dorsal root fibers are located in dorsal root ganglia.

▶ Aδ fibers convey sharp pricking pain, while C fibers convey dull aching pain.

▶ Painful sensation from the face has a complicated neural pathway: (1) the trigeminal nerve to the (2) trigeminal ganglion, to the (3) spinal tract of V in the pons. The neural pathway then synapses in the (4) spinal nucleus of V, after which they decussate to form the ventral trigeminal lemnicus, and ascend to the (5) ventral posteriomedial nucleus (VPM) of the thalamus.

REFERENCES

Buck LB. Smell and taste: the chemical senses. In: Kandel ER, Schwarz JH, Jessell TM, Siegelbaum SA, Hudspeth AJ, eds. *Principles of Neural Science*. 5th ed. New York, NY: McGraw-Hill; 2012.

Martin JH. The somatic sensory system. *Neuroanatomy: Text and Atlas*. 2nd ed. Stamford, CT: Appleton & Lange; 1996.

Ropper AH, Brown RH. Other somatic sensation. *Adam's and Victor's Principles of Neurology*. 8th ed. New York, NY: McGraw-Hill; 2005.

A 75-year-old man came to his physician with complaints of a continuous severe, sharp, burning pain on the right hand side of his chest wall that extended around the right side onto his back. He stated that he initially felt numbness and tingling in this area before the pain began. The physician notes on examination that the area of skin corresponding to the pain is red, with several fluid-filled vesicles. The physician diagnoses the patient as having shingles (herpes zoster), a viral disease which causes inflammation.

▶ What receptors are involved in feeling pain?
▶ What factors activate these receptors?
▶ Are there any available treatment options?

ANSWERS TO CASE 20:

Nociception

Summary: A 75-year-old man has right chest wall, side, and back numbness and tingling that progresses to severe, sharp, burning pain. The corresponding area is red with fluid-filled vesicles. The affected area is sharply demarcated by dermatomal borders.

- **Receptor:** Nocireceptor.

- **Factors activating receptors:** Substance P, Histamine, and K^+ are released after tissue damage or injury and activate nocireceptors.

- **Available treatment options:** While there is no cure for herpes zoster, antiviral medications such as acyclovir can be effective in moderating the progress of symptoms and can be taken prophylactically by those at high risk of developing the disease. A vaccine is now available.

CLINICAL CORRELATION

Herpes zoster is a viral disease causing inflammation of the dorsal root ganglia along with severe pain and cutaneous eruption in the dermatomal distribution of the ganglion cells. The disease is caused by the varicella-zoster virus, which causes chickenpox during the primary infection. The virus becomes latent in the dorsal root ganglion cells and can reactivate later in life, frequently in people who are immunosuppressed. The severe pain results from the spread of the virus along the peripheral nerve fibers and the inflammation of neurons in the dorsal root ganglia that convey pain sensation.

APPROACH TO:

Nociception

OBJECTIVES

1. Know the different types of painful sensations.

2. Understand the importance of cerebral cortical processing of pain information in the perception of pain.

3. Be able to describe the endogenous analgesic system.

DEFINITIONS

OPIOIDS: A chemical substance that can bind to the opioid receptors of the central nervous system providing pain relief.

ANALGESIA: Blocking of the conscious perception of pain.

HYPERESTHESIA: Abnormal increase in sensitivity to stimulation.

HYPESTHESIA: Abnormal decrease in sensitivity to stimulation.

PARESTHESIA: Spontaneous sensation of prickling, tingling, or numbness.

RADICULAR PAIN: Pain distributed over an area that is consistent with the boundaries of a dermatome.

REFERRED PAIN: Pain that is perceived in a surface area of the body far removed from its actual source.

PHANTOM PAIN: Pain that is felt in a part of the body that either no longer exists because of amputation or is insensate caused by nerve severance.

DISCUSSION

Pain is an unpleasant sensation that occurs in response to either an externally perceived event or an internal cognitive event. The unpleasant sensory and emotional perception associated with actual or potential tissue damage functions to warn of avoidable injury. The sensations we describe as painful include pricking, burning, aching, stinging, and soreness. There are several aspects to pain: a distinctive sensation; the individual's reaction to this sensation; activity in both somatic and autonomic systems; and both reflex and volitional efforts of avoidance or escape.

Three types of pain sensation occur after an acute noxious event. **Fast pain** consists of a sharp, pricking sensation that can be localized accurately and results from activations of Ad myelinated fibers. **Slow pain** is a burning sensation that has a slower onset, greater persistence, and a less clear location, and results from activation of unmyelinated C fibers. Slow pain runs through archispinothalamic and paleospinothalamic tracts, while fast pain runs through the neospinothalamic tract. **Deep** or **visceral pain** can be described as aching, burning, or cramping. Visceral pain results from stimulation of visceral and deep somatic receptors as in joints or muscles. Visceral receptors are innervated by both unmyelinated C fibers and myelinated Ad fibers that pass through the sympathetic nerves. The cell bodies of both these fibers are located in the dorsal root ganglia.

The **spinothalamic pathway**, projecting to the VPL of the thalamus and from there to the primary somatosensory cortex, is essential for the spatial and temporal discrimination of painful stimuli. The **spinoreticular pathways**, which connect to the secondary somatosensory cortex, hypothalamus, and limbic system, mediate systemic autonomic responses to pain and probably the emotional and affective responses as well. The cingulate cortex is one of the limbic areas activated by pain.

The cerebral cortex is an important component to the processing of pain. Patients who have suffered complete destruction of the somatosensory areas of the cerebral cortex on one side of the brain may still detect painful stimuli on the contralateral side of the body as long as the thalamus and lower structures of the pain pathway remain intact. Intractable pain can be relieved by destruction of the posterior and intralaminar nuclei of the thalamus. A **prefrontal leukotomy** transects the fibers linking the dorsomedial and anterior nuclei of the thalamus with the frontal lobe and anterior cingulate cortex and can diminish the anguish of constant pain by changing the psychological response to painful stimuli. This lesion, however, is also accompanied by negative changes in personality and intellectual capacity. A **cingulotomy,** or bilateral section of the cingulum bundle, has been shown to relieve a patient's reaction to pain without causing the drastic personality changes that occur with prefrontal leukotomy.

There are no pain receptors in the parenchyma of the internal organs, including the brain. The pain receptors are found within the walls of arteries, the peritoneal surfaces, pleural membranes, and the dura mater covering the brain. All of these structures can respond to inflammation or mechanical deformation with severe pain. Any abnormal contraction or dilatation of a wall of a hollow viscus, such as a blood vessel, also causes pain. The pain fibers from the viscera project to the spinal cord as components of the sympathetic nerves, while fibers conveying nonpainful visceral sensations project to the central nervous system as components of the parasympathetic nerves.

The afferent pathway of visceral pain receptors follows the peripheral sympathetic nerves from the viscera to the sympathetic trunk. Here they pass to the thoracic and lumbar spinal nerves over the **white rami** and enter the spinal cord. Their cell bodies are located in the dorsal root ganglia of segments T1 to L2, and their axons terminate at synapses in the dorsal horn of the intermediate gray matter. These nuclei project axons bilaterally through the anterolateral system to the brain stem reticular formation, intralaminar thalamic nuclei, and hypothalamus. Some visceral pain is also mediated by neurons from the deep central spinal gray matter whose axons ascend in the dorsal midline with the dorsal columns. These fibers terminate on neurons in the nuclei of the dorsal columns, which then project to the ventral posterior thalamus.

Pain originating from the viscera tends to be vaguely localized. It may even be perceived in a surface area of the body far away from its actual source. This phenomenon is known as **referred pain.** A common example of referred pain occurs with pain of coronary heart disease, which may be felt in the chest wall, the left axilla, or down the inside of the left arm. Inflammation of the peritoneum covering the diaphragm may be felt over the shoulder. This occurs because the peripheral afferents that supply the skin area of the referred pain enter the same segment of the spinal cord as the visceral afferents conducting pain from the affected organ. Visceral sensory fibers discharge into the same pool of neurons in the spinal cord as fibers from the skin, and an abundance of impulses results in misinterpretation of the true origin of the pain. Noxious stimuli affecting deep somatic structures can also result in referred pain.

Radicular pain is pain distributed over an area that is consistent with the boundaries of a dermatome. Mechanical compression or local inflammation of dorsal nerve roots can irritate pain fibers and produce pain along the dermatomal distribution of the affected root. Sensory changes other than pain may be associated with dorsal root irritation. Localized areas of spontaneous prickling, tingling, or numbness are termed **paresthesias.** Zones of **hyperesthesia** may be present, in which tactile stimuli are grossly exaggerated. If the pathological process progresses to the point of destroying the nerve fibers, the dorsal roots will eventually lose their ability to conduct sensory impulses, resulting in **hypesthesia** or **anesthesia.** The majority of skin areas

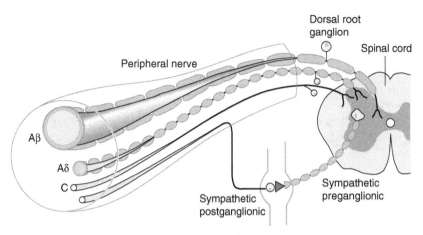

Figure 20-1. Components of a typical cutaneous nerve. (*With permission from* Harrison's Principles of Internal Medicine. *15th ed. New York: McGraw-Hill; 2001. Chap 11, fig 11-1.*).

receive innervation from fibers from more than one dorsal root; therefore, damage to a single dorsal root usually does not cause complete sensory loss (Figure 20-1).

Endogenous analgesia is an important component to understand in the treatment of pain. Stimulation of areas along the medial **periventricular** and **periaqueductal axis**, including the **midline raphe nuclei** of the brain stem, has been shown to produce **analgesia**. The raphe nuclei, found throughout the brain stem, are populated with neurons that produce **serotonin**. The axons of the **caudal raphe nuclei** descend to the spinal cord through the **dorsolateral fasciculus**. These axons terminate in the dorsal horn, where they attenuate the responses of spinothalamic cells to spinal nerve afferents mediating noxious stimuli from nociceptors. The release of opioid peptides is thought to primarily mediate endogenous analgesia. Other neurotransmitters such as serotonin, dopamine, and norepinephrine can also have analgesic effects. The three families of opioid receptors include enkephalin, β-endorphin, and dynorphin. Enkephalin interneurons are found in laminae I through III of the dorsal horn of the spinal cord. Inputs from descending serotonergic and noradrenergic fibers activate these interneurons and inhibit transmission of painful sensations at the first synaptic connection in the pain pathway.

CASE CORRELATES

- See Cases 18-23 (sensory systems).

COMPREHENSION QUESTIONS

20.1 A 62-year-old man with high blood pressure and a 50 pack-year history of smoking develops sudden onset of chest and left arm pain one hot afternoon while mowing the lawn. He is rushed to the emergency department by his wife, where he is diagnosed with an acute myocardial infarction. Therapy is begun immediately. What term is used to describe the pain this man felt in his arm during this event?

A. Central pain

B. Phantom pain

C. Radicular pain

D. Referred pain

20.2 The back pain experienced by a 50-year-old man with pneumonia can best be classified as follows:

A. Radicular pain

B. Phantom pain

C. Referred pain

D. Central pain

20.3 Following a thoracotomy and removal of a small lung nodule, one of your patients is complaining of severe pain at the incision site and underlying chest wall. The standard doses of pain medication do not seem to have the desired result, and you are afraid to give too much for fear of oversedating him. To more adequately control this patient's pain you decide to start him on a morphine PCA (patient-controlled analgesia) pump. This works very well, and by the following morning he has no further complaints of pain. Through what mechanism does morphine decrease pain?

A. Blocking sensory nerve action potentials

B. Modulating the brain and spinal cord's response to painful stimuli

C. Blocking pain transmission at the level of the thalamus

D. Preventing nerve endings from generating action potentials in response to normally painful stimuli

ANSWERS

20.1 **D.** This is an example of **referred pain**. The patient feels pain in his arm, but there is in fact no pathology affecting his arm. Instead, there is pathology affecting his heart, which is innervated by nerves from the same spinal levels as those innervating the arm. The afferent pain signal actually originating from the heart is erroneously interpreted as coming from the arm, resulting in referred pain. Central pain arises from a lesion in the thalamus or cortex that is interpreted as pain in the body part corresponding to the lesion. Phantom pain is "felt" in amputated limbs, and radicular pain is pain distributed over a spinal cord level, caused by compression of the corresponding nerve root.

20.2 **C.** **Referred pain** occurs when pain of visceral origin is localized to a part of the body far removed from the source. Inflammation of the pleura can frequently be referred to the thoracic dermatomes of the back.

20.3 **B.** Morphine and other opioid pain medications decrease pain by **modulating the response of the brain and spinal cord to painful stimuli.** These molecules bind to receptor sites for endogenous opioids which are located in diffuse areas of the CNS, including in the brain and spinal cord. In the spinal cord, binding to opioid receptors modulates the response of second- and third-order neurons in the pain pathway to painful stimuli coming in from the periphery. Opioid receptors in the brainstem activate descending tracts that further modulate pain transmission in the spinal cord. Through poorly understood mechanisms, opiates also alter the perception of pain; in other words, the sensation of pain is still there, but it is not so debilitating.

NEUROSCIENCE PEARLS

▶ Three types of pain sensations occur after an acute noxious event: (1) fast pain, (2) slow pain, and (3) visceral pain.

▶ Fast pain is an accurately localizable, immediate, sharp sensation resulting from activation of myelinated Ad fibers.

▶ Slow pain is a vaguely localizable, burning pain resulting from activation of unmyelinated C fibers.

▶ Visceral pain is an aching pain resulting from stimulation of deep somatic receptors in joints or muscles, which are innervated by both C and Ad fibers. Stimulation of areas along the medial periventricular and periaqueductal axes has been shown to produce analgesia.

REFERENCES

Basbaum AI. Smell and taste: the perception of pain. In: Kandel ER, Schwarz JH, Jessell TM, Siegelbaum SA, Hudspeth AJ, eds. *Principles of Neural Science*. 5th ed. New York, NY: McGraw-Hill; 2012.

Martin JH. The somatic sensory system. *Neuroanatomy: Text and Atlas*. 2nd ed. Stamford, CT: Appleton & Lange; 1996.

Ropper AH, Brown RH. Pain. *Adam's and Victor's Principles of Neurology*. 8th ed. New York, NY: McGraw-Hill; 2005.

A 22-year-old male is involved in a motorcycle accident and suffers a closed head injury with bilateral paramedian frontal lobe contusions. After a lengthy stay in the intensive care unit, he enters a rehabilitation facility and begins to make significant progress. He begins to notice that he is unable to taste any of his food or to smell his coffee. On examination he is noted to have complete absence of the ability to smell. An MRI is obtained 3 months after his injury, which demonstrates encephalomalacia in the paramedian frontal lobe regions corresponding to the area of his prior contusions. The patient is diagnosed with posttraumatic olfactory dysfunction.

▶ What is the mechanism of this patient's dysfunction?
▶ What bony anatomical structure is likely damaged?

ANSWERS TO CASE 21:

Olfaction

Summary: A 22-year-old man has a history of a closed head injury following a motor-cycle accident. After a period of neurological recovery, the patient notes an inability to smell. Neuroimaging reveals posttraumatic changes in the paramedian frontal lobes, which is in the region of the olfactory bulbs and tracts.

- **Mechanism of posttraumatic olfactory dysfunction:** Posttraumatic olfactory dysfunction may be caused by a shearing injury of the olfactory nerve filaments or by brain contusion and hemorrhage within the olfactory brain regions.

- **Bony anatomical site:** The cribriform plate of the naso-orbito-ethmoidal region is commonly affected in this case.

CLINICAL CORRELATION

Posttraumatic anosmia is a common finding in patients that have suffered head injuries. The axons of olfactory receptor cells are delicate and pass through small foramina of the cribriform plate at the base of the skull and synapse directly in the olfactory bulb. Any tearing or shearing of the axons during the trauma can result in olfactory dysfunction. It may occur with fractures of the naso-orbito-ethmoidal region, involving the cribriform plate, or with rapid translational shifts in the brain secondary to coup or contrecoup forces generated by blunt head trauma. Head trauma often results in traumatic brain injury in the form of cortical contusion or intraparenchymal hemorrhage. Contusion of the olfactory bulbs or cortical lesions at the olfactory brain regions (amygdala, temporal lobe region, frontal lobe region) can lead to posttraumatic anosmia. The treatment is expectant observation.

APPROACH TO:

Olfaction

OBJECTIVES

1. Know the anatomical structures involved in the olfactory system.

2. Be familiar with the cortical areas involved in olfactory associations.

3. Be able to name and describe some of the etiologies of olfactory dysfunction.

DEFINITIONS

MITRAL CELLS: Second-order neurons of the olfactory system. They receive information from olfactory receptor neurons through synapses in the olfactory bulb; their axons project as the olfactory tract.

ENTORHINAL CORTEX: An important memory center of the brain located in the medial temporal lobe. Mitral cells project to this region of the cortex and are thought to contribute to emotional associations with olfaction.

ANOSMIA: Absence of the sense of smell.

HYPEROSMIA: Abnormally acute smell function.

OLFACTORY AGNOSIA: Inability to recognize an odor sensation.

DYSOSMIA: Distorted smell perception.

DISCUSSION

Olfaction allows us to detect and discriminate between a great variety of odors within our environment. The olfactory system consists of several elements of the central nervous system: the **olfactory nerves, bulbs,** and **tracts,** the **olfactory tubercle,** the **primary olfactory cortex,** the **entorhinal cortex** of the parahippocampal gyrus, and the **amygdala.** The functions of this system include distinguishing isolated odors from other background odors, determining the concentration of the odor and being able to identify it at different concentrations, creating a representation of the odor, and pairing the odor with an associated memory. For many other species, olfaction is important for locating food and communication.

Olfactory stimuli are sensed by specialized peripheral olfactory receptors in a part of the nasal mucosa called the **olfactory epithelium.** This specialized tissue consists of pseudostratified columnar olfactory epithelium located on the superior concha, the roof of the nasal chamber, and the upper portion of the nasal septum. The receptor cells are **bipolar neurons** with **cilia** at their dendritic endings in the olfactory epithelium. Interestingly, olfactory neurons are the only special sensory receptors that are the nerve itself. The molecules that are smelled or tasted alter the membrane potential of the receptors through **ligand-receptor-mediated second messenger mechanisms** that result in **ion channel openings** in the ciliary membrane. There are more than **1000 different receptor proteins** in the olfactory cilia, with each neuron expressing a single receptor. Olfactory neurons are also unique in that they are short-lived neurons with an **average life span of 30-60 days (Figure 21-1).** Aside from perceiving and distinguishing odors, they contribute to the sensation of taste.

The ability to discriminate between different odorants is the result of differential expression of the receptor proteins in receptor cells across the surface of the mucosa, combined with selective convergence of axons from functionally related receptor cells to target cells in the olfactory bulb. These axons are grouped into fascicles before passing through the cribriform plate as the olfactory nerve. The olfactory axons terminate in the olfactory bulb in what are called the glomeruli. The olfactory bulbs, which rest on the cribriform plates, contain several types of neurons, including inhibitory interneurons and mitral cells. The mitral cells receive direct synaptic input from olfactory nerve fibers and project their axons in the lateral olfactory tract. Each mitral cell receives input from olfactory nerves expressing the same odorant receptor. Inhibitory interneurons also form synapses with mitral cell dendrites and inhibit mitral cells in surrounding glomeruli, producing lateral inhibition.

The **anterior olfactory nucleus** is located in the **olfactory stalk** and contains groups of neurons. The olfactory portion of the anterior commissure originates in the olfactory stalk as one of these groups. Information from the ipsilateral olfactory bulb is received and transmitted to the contralateral olfactory bulb by way of the anterior commissure through this group of neurons. The olfactory stalk is positioned

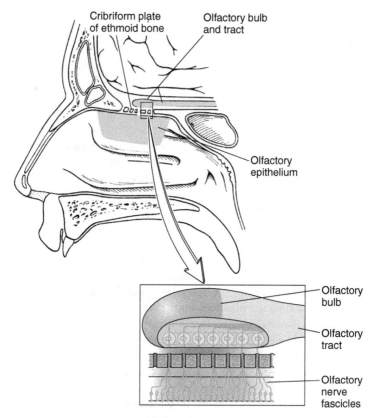

Figure 21-1. Basic neural circuits in the olfactory bulb. (*With permission from Martin JH.* Neuroanatomy: Text and Atlas. *3rd ed. New York, NY: McGraw-Hill; 2003: 217.*)

in the **olfactory sulcus** of the frontal lobe, directly lateral to the gyrus rectus of the **orbitofrontal cortex.** The olfactory tract bifurcates here into **medial** and **lateral striae.** A portion of the fibers in the medial olfactory stria contain the axons of the anterior olfactory nucleus neurons which cross the anterior commissure to the contralateral olfactory bulb. The remaining fibers consist of the mitral cell axons and terminate in the ipsilateral olfactory tubercle within the anterior perforated substance. Mitral cell axons only project ipsilaterally.

The **lateral stria**, or lateral olfactory tract, consists primarily of mitral cell axons and projects to the lateral margin of the **anterior perforated substance**, the **piriform cortex,** a small rostral portion of the **entorhinal cortex**, and the **corticomedial amygdala**. This pattern distinguishes the olfactory system as being the only sensory system in which second-order neurons (the mitral cells) project directly to the cerebral cortex and not to the thalamus.

The primary olfactory cortex, or **piriform cortex**, is the key region involved in the **conscious perception of smell.** Lesions in this region have been shown to result in failure to discriminate between various odorants. The piriform cortex is also unique in that it consists of a **three-layered cortex**, making it phylogenetically older than the six-layered cortex of the visual, auditory, and somatosensory systems.

The olfactory cortex has **reciprocal thalamocortical connections** used for discriminative functions. Projections travel directly from the primary olfactory cortex to the **lateral orbitofrontal cortex** and indirectly through the magnocellular portion of the **dorsomedial nucleus** of the **thalamus**. Other communications important in olfactory discrimination include corticocortical connections between the temporal lobe and the orbitofrontal cortex.

Olfactory impulses that reach the corticomedial **amygdala** are important in many species for the control of social behaviors. In humans, however, the behavioral significance of olfactory projections to the amygdalae has not been clarified. The function of the olfactory input to the hippocampus through the **entorhinal cortex** is also unclear. These pathways, however, are thought to integrate olfactory information with visual, auditory, and somatosensory impulses arriving from other association cortices. Through connections with the amygdala, the hippocampus is thought to participate in the integration of multisensory inputs into appropriate **emotional and physiological responses** to external stimuli. This may contribute to the development of emotional responses to specific odors.

Anosmia, the loss of sense of smell, can result from several etiologies. **Traumatic injuries** that injure the frontal lobes can affect the primary olfactory cortex. Fractures of the cribriform plate can result in injury to the olfactory bulbs or tracts. **Infections** such as the common cold, viral hepatitis, syphilis, bacterial meningitis, cerebral abscesses, or osteomyelitis of the frontal or ethmoid regions can also injure the olfactory system. The most frequent cause of smell loss in adults is a severe upper respiratory tract infection (URTI). The smell loss is most commonly partial. Direct insult to the olfactory neuroepithelium is the primary cause of the problem. Viruses can cause edema and hyperemia of nasal membranes, necrosis of cilia, and cellular damage. They can also produce varying degrees of destruction. Biopsies of olfactory epithelia from patients with post-URTI anosmia showed greatly reduced numbers of olfactory receptors. If the infection resolves, the surviving receptors can regenerate. Anosmia results when there is a lack of regeneration following severe destruction of the epithelium. If there is patchy destruction, patients may have **hyposmia**. If the regeneration of receptor neurons and their central attachments are "misguided" to reach abnormal locations in the brain, patients may experience **dysosmia**, or distorted smell. In addition to direct insults, many viruses can invade the CNS via the olfactory neuroepithelium to cause subsequent dysfunction. No effective treatment has been found for post-URTI hyposmia. Even though spontaneous recovery in some patients is theoretically possible, meaningful recovery is rare when marked loss has been present for some time. Less common causes of anosmia include olfactory groove meningiomas, frontal lobe gliomas, metabolic diseases, amphetamine or cocaine use, and Parkinson or Alzheimer disease. A complete anosmia will result in loss of the ability to recognize flavors as the olfactory and taste systems function together in the perception of flavors. **Hyperosmia**, an increase in olfactory sensitivity, frequently occurs in early pregnancy and can also occur in conversion disorders and some psychoses.

CASE CORRELATES

- See Cases 18-23 (sensory systems).

COMPREHENSION QUESTIONS

Refer to the following case scenario for questions 21.1-21.2:

A 37-year-old man comes into your office with the complaint that he cannot smell. He states that several months ago he was involved in a bar fight in which his nose and several other facial bones were fractured. After being discharged from hospital following this incident, he has not been able to smell.

21.1 Olfactory neurons are vulnerable to damage as they pass through which opening in the skull?

A. Inferior orbital fissure

B. Foramen ovale

C. Cribriform plate

D. Internal nasal valve

21.2 Which of the following statements best describes this man's olfactory neurons prior to his injury?

A. The maximal number of olfactory neurons is present at birth, and the number slowly decreases throughout life as neurons are damaged.

B. Each olfactory neuron expresses receptors for only one odorant.

C. Olfactory neurons project directly to the olfactory cortex.

D. Olfactory neurons sensing different but complimentary odorants synapse with the same mitral cell.

21.3 A 33-year-old woman comes into your office with the complaint that she smells things that others do not smell. This occurs at unpredictable times, typically lasts for several minutes, and then goes away. The smell is typically something burning. You suspect that this woman is having simple partial seizures and order an EEG. An abnormal discharge in which brain region would confirm your findings?

A. Occipital lobe

B. Superior parietal lobe

C. Medial temporal lobe

D. Anterolateral frontal lobe

ANSWERS

21.1 **C.** As the olfactory neurons pass through the tiny holes in the **cribriform plate,** they are susceptible to injury, as occurred in this case. Olfactory neurons are small, bipolar neurons that run a very short distance from the olfactory epithelium of the superior nasal cavity through the cribriform plate to the olfactory bulb in the olfactory groove of the frontal lobe. Depending on the degree of damage to the olfactory epithelium and the cribriform plate, the neurons may be able to regenerate, resulting in a return of the sense of smell.

21.2 **B.** Although there are thousands of odorant receptors expressed in humans, **each olfactory neuron expresses only one odorant receptor**. The neurons expressing the same receptor project through the cribriform plate and synapse with mitral cells in the olfactory bulb. Each mitral cell receives input from neurons expressing the same odorant. Mitral cells then project their axons down the olfactory tract directly to the olfactory cortex. The neurons in the olfactory epithelium have a lifespan of just a few months, after which time they die and are replaced by new olfactory neurons derived form the basal cell layer of the olfactory epithelium.

21.3 **C.** The olfactory cortex, made up of the piriform cortex and the periamygdaloid cortex, is located on the **medial aspect of the temporal lobe**. Abnormal EEG readings from this area during the experience of an abnormal smell would confirm that seizure discharge is responsible for this woman's symptoms. The occipital lobe houses the primary visual cortex; discharge here would result in abnormal vision. The superior parietal lobe contains the primary sensory cortex, and a discharge here would result in abnormal sensation. The anterolateral frontal lobe is not a sensory area, but is involved in executive function.

NEUROSCIENCE PEARLS

▶ Olfactory receptor neurons detect odorants in the olfactory epithelium in the nose and transmit information through the cribriform plate to mitral cells in the olfactory bulbs.

▶ The olfactory system is the only sensory system in which second-order neurons (the mitral cells) project directly to the cerebral cortex.

▶ All sensory pathways except the olfactory pathway have a relay nuclei through the thalamus.

▶ The projection of olfactory information to the limbic area explains why certain smells can be evocative of memory and emotion.

▶ Anosmia can result from traumatic, infectious, and neoplastic etiologies.

REFERENCES

Buck LB. Smell and taste: the chemical senses. In: Kandel ER, Schwarz JH, Jessell TM, Siegelbaum SA, Hudspeth AJ, eds. *Principles of Neural Science*. 5th ed. New York, NY: McGraw-Hill; 2012.

Martin JH. The olfactory system. *Neuroanatomy: Text and Atlas*. 2nd ed. Stamford, CT: Appleton & Lange; 1996.

Ropper AH, Brown RH. Disorders of smell and taste. *Adam's and Victor's Principles of Neurology*. 8th ed. New York, NY: McGraw-Hill; 2005.

A 25-year-old woman has been having difficulty for several months understanding what people are saying over the phone when she holds the receiver to her left ear. She has no problem with her right ear and no other symptoms. On examination, when a vibrating tuning fork is placed in the center of her forehead, she hears the sound louder in her right ear than in her left ear. The base of a vibrating tuning fork is then placed on her left mastoid process. When she is no longer able to hear the sound through the bone, the vibrating tuning fork is held close to her left ear. She is still able to hear the vibrating tuning fork in the air after she can no longer hear it on bone. Based on the examination findings and further workup, the patient is diagnosed with an acoustic neuroma.

▶ Is the hearing loss more likely to be conductive or sensorineural in nature?
▶ What are other possible diagnoses?

ANSWERS TO CASE 22:

Audition

Summary: A 25-year-old woman has a several month history of left-sided hearing loss.

- **Interpretation of examination findings:** The Weber test indicates a sensorineural etiology for the patient's hearing loss. The Rinne test is not suggestive of a conductive hearing loss. Both of these findings are consistent with the diagnosis of an acoustic neuroma, which affects the eighth cranial nerve.

- **Other possible diagnoses:** Sensorineural hearing loss can be caused by a variety of etiologies in addition to a tumor involving the eighth cranial nerve or the cerebellopontine angle. Drugs such as aminoglycosides and salicylates can have toxic side effects to the hair cells of the cochlea. Infectious, immune-mediated, or traumatic nerve injury can also lead to hearing deficits similar to those that the patient exhibits.

CLINICAL CORRELATION

The Weber test is performed by holding the tuning fork to the center of her forehead. Normally, the sound is heard in the midline. In air conduction hearing loss, the sound is lateralized toward the abnormal side. In sensorineural hearing loss, the sound is lateralized away from the abnormal side. This patient's examination lateralized away from her abnormal side. The Rinne test was performed by comparing bone with air conduction on the affected side. Normally, hearing by air conduction will continue after hearing by bone conduction ceases. A person with conductive hearing loss will hear by bone conduction better than by air conduction. This patient was able to hear by air conduction after hearing by bone conduction had ceased. An MRI was obtained which revealed an acoustic neuroma of the left eighth cranial nerve localized at the cerebellopontine angle.

The capture and interpretation of sound is a key element used in communication and interpretation of our surroundings. Sound waves travel through the external auditory canal, where they are translated into mechanical energy in the middle ear. Auditory information is then transmitted to the fluid-filled inner ear where it is detected and organized tonotopically by the hair cells of the cochlea. This information is then transmitted through the cochlear nerve to the brain stem and auditory cortex, where the information is processed and interpreted. An acoustic neuroma impairs the ability of the cochlear nerve to function properly, resulting in a sensorineural hearing loss. Other types of hearing loss can be congenital or acquired and can be because of conductive or sensorineural etiologies. The ability to communicate and interact socially can be significantly impaired with diminished or absent hearing. Treatment options for acoustic neuromas include surgical resection and stereotactic radiosurgery.

> # APPROACH TO:
> ## Audition

OBJECTIVES

1. Understand the basic anatomical structures involved in audition.

2. Review the central auditory pathways.

3. Be familiar with the diagnostic tests used to differentiate between conductive and sensorineural hearing loss.

DEFINITIONS

HERTZ: A unit of frequency equal to one cycle of a wave per second. The frequency of a sound wave determines the pitch of the sound.

DECIBEL: A unit of measurement of the amplitude of a sound wave. The amplitude determines the loudness of a sound.

OVAL WINDOW: Oval opening in the cochlea which separates the air-filled middle ear from the fluid-filled inner ear cavity. Vibrations of the stapes bone against the oval window transmit sound to the inner ear.

ROUND WINDOW: Round opening in the cochlea connecting the inner ear to the middle ear. Movement of the membrane covering the round window allows for sound waves to be dissipated into the air-filled middle ear.

SENSORINEURAL HEARING LOSS: Partial or complete deafness that occurs as a result of injury to the cochlear nerve or the sensory structures of the inner ear.

CONDUCTIVE HEARING LOSS: Partial or complete deafness that occurs as a result of disruption of auditory conduction through the bones of the middle ear or mechanical obstruction of the external ear.

DISCUSSION

The structures of the auditory system can be divided into three components: the external, middle, and inner ear. The external ear consists of the auricle and external auditory canal, and is separated from the middle ear by the **tympanic membrane**. The middle ear is filled with air and contains three bony ossicles: the **malleus, incus,** and **stapes**. The malleus attaches to the tympanic membrane, and the stapes attaches to the **oval window** by a ligamentous membrane. The oval window separates the middle ear cavity from the inner ear space, which is filled with fluid. Sound waves traveling through air strike the tympanic membrane and create motion, which is conveyed through the ossicles to the oval window. This chain of ossicles functions not only as an amplifier but also as an impedance matcher. The impedance of the tympanic membrane is matched to the higher impedance of the oval window, thus miminizing the energy loss as the sound waves travel from air to fluid. Sound consists of sinusoidal waves of air molecules. The frequency of a wave is measured in **hertz**

(Hz) and determines the pitch of a sound. The amplitude of a wave is measured in **decibels (dB)** and determines the loudness of a sound. The human ear can detect sound frequencies between 20 and 20,000 Hz and loudness between 1 and 120 dB.

The oval window opens up into the vestibule of the inner ear, which is filled with **perilymph**. The **cochlea** and the semicircular canals lie on either side of the vestibule. The cochlea is a perilymph-filled tube, which wraps around itself approximately 2.5 times. The semicircular canals are also filled with perilymph and are important for providing information about the position and movement of the head. These three chambers of the inner ear lie within the temporal bone and form the **bony labyrinth**. The **membranous labyrinth**, suspended within the bony labyrinth, is filled with **endolymph**, and contains the sensory organs.

The bony cochlea turns around a central bony **modiolus**, which forms the axis of the cochlear turns. The **spiral lamina** is a ridge of bone that divides the cochlear cavity into two chambers: the **scala vestibuli** and the **scala tympani**. The **scala media**, or membranous cochlear duct, lies between the scala vestibule and scala tympani. The **organ of Corti** is suspended in the endolymph within the scala media. It rests on the **basilar membrane** as it spirals within the cochlear turns.

As the footplate of the stapes moves against the oval window, a pressure wave is produced in the perilymph of the scala vestibuli. This pressure wave travels within microseconds to the apical connection between the scala vestibuli and scala tympani, called the **helicotrema**. The vibrations from this pressure wave are transmitted through the fluid to the basilar membrane of the organ of Corti. The cytoarchitecture of the organ of Corti is such that specific portions resonate harmonically with each audible frequency, allowing for **tonotopical organization of sound.** The width of the basilar membrane is greater and more flexible toward the apex than at the base, where it is stiffer. This allows lower frequencies to resonate near the apex and helicotrema and higher frequencies near the base and oval window.

The organ of Corti contains two types of receptors: **inner and outer hair cells**. The base of each inner hair cell is indirectly attached to the basilar membrane and functions as an auditory receptor cell. **Stereocilia** extend above the apical surface of the cell and lie just below the tectorial membrane, which is attached separately to the wall of the scala media. As sound waves travel within the cochlea, the basilar and tectorial membranes move independently of each other. The stereocilia touch the tentorial membrane and bend, opening ionic channels that create changes in the potential of the hair cell membrane. Specifically, the stereocilia bend when they come into contact with the tectorial membrane and produce a change in the K^+ conductance inward. The inward K+ current depolarizes the cell and activates Ca^{2+} conductance that further contributes to the depolarization. Transmitter is released at the base of the hair cell and binds to **spiral ganglion cells** to trigger an action potential. The base of each hair cell synapses with dendritic processes from up to ten spiral ganglion cells. These ganglion cells are bipolar cells which synapse with only one inner hair cell each. They form the **spiral ganglion**, and their axons form the cochlear division of the **eighth cranial nerve**. The outer hair cells contain stereocilia embedded in the tectorial membrane. They have contractile properties which allow them to control the apposition of the tectorial membrane to the inner hair cells, and therefore the sensory response properties of the organ of Corti.

The pressure waves that initially travel in the scala vestibuli traverse the scala media, vibrate the basilar membrane, and induce pressure waves in the scala tympani. These then induce movements of the elastic diaphragm covering the round window within the scala tympani. The round window opens up into the air-filled middle ear where the pressure waves are finally dissipated.

The cochlear nerve travels to the brain stem, where it enters at the medullary pontine junction. Each entering nerve fiber bifurcates and synapses with neurons in both the **dorsal and ventral cochlear nuclei**. These nuclei contain tonotopically organized neurons. The dorsal, intermediate, and ventral acoustic striae project from the cochlear nuclei and relay information to central and rostral structures. The **dorsal acoustic stria** projects from the dorsal cochlear nucleus to the contralateral lateral lemniscus. The **intermediate acoustic stria** projects from the ventral cochlear nucleus, taking a similar course as the dorsal stria. The **ventral acoustic stria** projects from the ventral cochlear nucleus as well and travels to the ipsilateral and contralateral nuclei of the trapezoid body and superior olivary nuclei. These nuclei then project to the ipsilateral and contralateral lateral lemnisci. The superior olivary nucleus is the first point where information from both ears converges. It is sensitive to small changes in time of arrival and intensity of stimulus, which helps to localize the sound.

The **monaural central auditory pathway** carries information about the frequency of sound. It consists of neuronal fibers projecting from the dorsal and ventral cochlear nuclei through the dorsal and intermediate striae to the contralateral inferior colliculus. The **binaural pathway** ascends bilaterally with the ventral acoustic stria and carries information about the location of origin and direction of auditory stimuli. The pathway includes synapses in the trapezoid body, superior olivary complex, and lateral lemniscus nuclei before ending in the inferior colliculus. Axons then travel to the **medial geniculate nucleus** via the brachium of **the inferior colliculus**. The medial geniculate nuclei are the final sensory relay stations of the hearing pathway. The auditory radiation is formed by the efferent projections from the medial geniculate nucleus to the transverse temporal gyri and the adjacent planum temporale in the temporal lobe. The **primary and secondary auditory** cortices are located within these areas of the temporal lobe.

Sound is perceived when auditory impulses arrive at the primary auditory cortex. The processing and interpretation of sound, however, involves a combination of structures involved in audition. Information about the location of a sound is initially processed in the superior olive and inferior colliculus; however, precisely locating a sound requires processing up to the auditory association areas in the superior temporal gyrus and posterior parietal cortex. The interpretation of combinations of different frequencies in sequence begins in the cochlear nuclei and continues to the inferior colliculus and medial geniculate nucleus and primary auditory cortex. Tonotopic organization is continuous throughout these structures.

Representation of sound is achieved bilaterally in each temporal lobe. The ascending pathways from the brain stem to the auditory cortex consist of both crossed and uncrossed fibers (Figure 22-1). Each lateral lemniscus conducts stimuli from both ears, and an ipsilateral lesion above the level of the cochlear nuclei does not usually interfere with impulses from both ears. Because of this, deafness in one ear usually implies a lesion below the brain stem at the level of the cochlear nerve, cochlea, or middle ear.

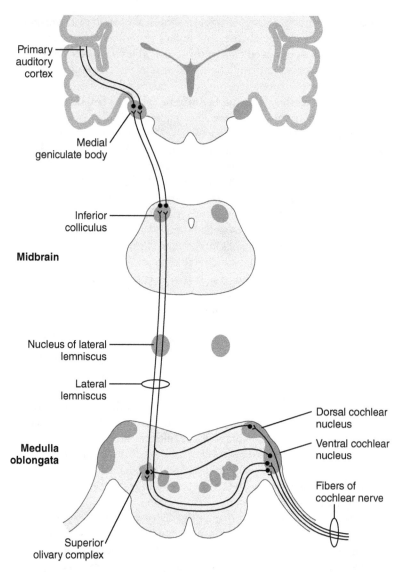

Figure 22-1. The ascending auditory pathways.

CASE CORRELATES

* See Cases 18-23 (sensory systems).

COMPREHENSION QUESTIONS

22.1 A 75-year-old man comes into your office complaining of increasing difficulty hearing, particularly understanding conversation in noisy places. On audiometric testing you note a loss of high-frequency hearing from the upper end of the normal spectrum down into the speech range. The hearing loss is bilaterally symmetric. Further workup for this patient is negative and you diagnose him with presbyacusis, an age-related degeneration of cochlear hair cells. Based on his pattern of hearing loss, in which area of the cochlear duct would you expect the hair cells to be most degenerated?

A. At the base of the cochlear duct

B. At the apex of the cochlear duct

C. In the middle of the cochlear duct

D. Evenly throughout the cochlear duct

22.2 A 27-year-old man presents to your office following an industrial accident in which a large explosion occurred on his right side. He complains of pain and loss of hearing in his right ear. On examination you note a completely destroyed tympanic membrane on that side, and audiometric testing reveals near complete hearing loss in all frequencies on that side. Which of the following best describes the thalamic response to sound in this patient?

A. Stimulation of the right lateral geniculate body only

B. Stimulation of the right medial geniculate body only

C. Stimulation of bilateral medial geniculate bodies

D. Stimulation of bilateral lateral geniculate bodies

22.3 A 68-year-old man with a prior stroke 3 years ago is brought into your office by family members who state that he had a sudden, complete hearing loss that occurred yesterday evening. Prior to the event he had been well, and did not have any difficulties hearing. On examination he has a complete failure to respond to any auditory stimuli. Otological examination is normal, as are otoacoustic emissions and auditory brainstem responses. All of this information indicates that the peripheral and central auditory pathways are intact up to the level of the thalamus. Which location is the most likely site of this patient's lesion?

A. Bilateral parietal lobes

B. Right superior temporal lobe

C. Left superior temporal lobe

D. Bilateral superior temporal lobes

ANSWERS

22.1   **A.** Loss of high-frequency hearing results from damage to the hair cells located **at the base of the cochlear duct**. Presbycusis is an age-related degeneration of hair cells in the organ of Corti, beginning at the base of the cochlear duct and slowly progressing toward the apex. The tonotopic organization of the cochlear duct predicts the frequencies of hearing affected. The cochlear duct is stiffer at the base, causing it to resonate at a high frequency, and more lax at the apex where it responds to lower frequency vibrations. Loss of hair cells at the base of the cochlear duct therefore cause a loss of high-frequency hearing, and as the degeneration progresses up the duct, the hearing loss descends the normal hearing spectrum, until it reaches the speech range (roughly 200-6000 Hz).

22.2   **C.** Because of the decussation in the ascending auditory pathway, sound will continue to **stimulate bilateral medial geniculate bodies** in this patient. The thalamic nucleus involved in the ascending auditory pathway is the medial geniculate body, which receives input via the brachium of the inferior colliculus. Additionally, sound from one ear is represented bilaterally in the CNS, so even though this man has no response from his right ear, sound waves affecting his left ear will cause stimulation of bilateral structures above the cochlear nuclei. The point of initial decussation in the ascending auditory pathway is the trapezoid body. The lateral geniculate body is the thalamic relay for vision.

22.3   **D.** The primary auditory cortex is located on the **superior aspect of the temporal lobe bilaterally**. This area is also known as the transverse temporal gyrus of Heschl. Each lobe receives input from both ears, although the contralateral ear is more highly represented. Since the auditory cortex receives bilateral input, a lesion must affect both temporal lobes in order to cause cortical deafness.

NEUROSCIENCE PEARLS

▶ Sound is transmitted through the external, middle, and inner ear via air, bony ossicles, and fluid, respectively.

▶ The organ of Corti within the cochlea is organized tonotopically with higher pitches at the base and lower pitches at the apex, enabling us to differentiate sound tones.

▶ The stereocilia touch the tentorial membrane and bend, opening ionic channels that create changes in the potential of the hair cell membrane.

▶ The upper auditory pathways contain both crossed and uncrossed fibers creating bilateral representation of information from each ear.

REFERENCES

Hudspeth AJ. Hearing. In: Kandel ER, Schwarz JH, Jessell TM, Siegelbaum SA, Hudspeth AJ, eds. *Principles of Neural Science*. 5th ed. New York, NY: McGraw-Hill; 2012.

Martin JH. The auditory and vestibular systems. *Neuroanatomy: Text and Atlas*. 2nd ed. Stamford, CT: Appleton & Lange; 1996.

Ropper AH, Brown RH. Deafness, dizziness, and disorders of equilibrium. *Adam's and Victor's Principles of Neurology*. 8th ed. New York, NY: McGraw-Hill; 2005.

A 50-year-old man has a traffic accident after driving through an intersection without noticing an oncoming car from his right. He has also noticed a tendency to bump into walls when rounding corners. His physician performs visual field testing and finds bitemporal visual field deficits indicating bitemporal hemianopia. An endocrinology workup is negative for any abnormalities. An MRI of the head is obtained which reveals a mass in the region of the sella turcica. A diagnosis of pituitary adenoma is made.

- ▶ Compression of what structure would produce this patient's visual symptoms?
- ▶ What neural bundles follow this structure in the visual pathway?
- ▶ How would compression of these structures affect the patient's visual symptoms?
- ▶ What treatment options are available?

ANSWERS TO CASE 23:

Vision

Summary: A 50-year-old man has visual deficits resulting in an impaired ability to drive and a tendency to bump into things.

- **Compressed structure in the visual pathway:** Bitemporal hemianopia results from a lesion affecting the optic chiasm. A mass in the sella turcica can compress the overlying optic chiasm, resulting in deficits of both temporal visual fields.

- **Structures associated with the optic chiasm in the visual pathway:** The optic chiasm splits to form the left and right optic tracts. Information from the left visual fields of both eyes travels in the right optic tract, while information from the right visual fields of both eyes travels in the left optic tract. Thus, a lesion of either optic tract will lead to deficits in the contralateral visual field.

- **Mass effect:** Damage to the brain due to pressure of a tumor, which often causes blockage or excess accumulation of fluid within the skull.

- **Treatment options:** A number of surgical, radiotherapeutic, and pharmacological treatment options are available for pituitary adenomas. Endocrine-deficient patients will require appropriate replacement therapy. Surgical removal is most often performed by transsphenoidal craniotomy.

CLINICAL CORRELATION

Visual perception begins with rays of light reflected from an object reaching the eye. The light is refracted as it passes through the cornea and lens. The optical properties of the lens cause the image to be inverted and reversed as it is projected onto the retina. Light from the left half of the visual field will therefore project onto the right half of the retina, and light from the superior half of the visual field will project onto the inferior half of the retina. Information from the right half of the retina is carried along the visual pathway to the right cerebral hemisphere, and information from the left half of the retina is carried to the left cerebral hemisphere. Lesions along the visual pathway from the eye to the visual cortex create deficits that correspond to the associated visual field. A pituitary mass can compress the optic chiasm, impeding nerve fibers carrying information from the nasal half of each retina, as they cross at the chiasm. Surgical or medical decompression of the mass, if done early enough, can lead to improvement of symptoms.

APPROACH TO:
Vision

OBJECTIVES

1. Know the anatomic structures involved with visual perception.

2. Be able to describe the visual pathway.

3. Understand the effects of lesions interrupting the visual pathway.

DEFINITIONS

SCOTOMA: An isolated area within the visual field in which vision is absent or diminished (a blind spot).

HOMONYMOUS HEMIANOPIA: A loss of vision in the same visual field of both eyes.

BITEMPORAL HEMIANOPIA: A loss of vision in the outer half of the visual field of each eye.

RECEPTIVE FIELD: A specific area of the retina that maximally stimulates or inhibits firing of its corresponding ganglion cell when stimulated by light.

RETINOTOPIC PATTERN: The topographic representation of the retina created by ganglion cells from adjacent areas of the retina projecting onto adjacent neurons in the lateral geniculate nuclei (LGN) and from there to adjacent neurons in the visual cortex.

VISUAL ACUITY: The size of the smallest dark object that can be correctly identified in a light background.

DISCUSSION

The retina is formed as an extension of the central nervous system. It contains two types of photoreceptors: **rods** and **cones**. Rods mediate light perception and are important for nocturnal vision but provide low visual acuity. Cones mediate color vision and provide high visual acuity. The overall ratio of rods to cones in the retina is 20:1. The periphery of the retina primarily contains rods, while the **fovea centralis** within the **macula** only contains cones and functions as a specialized region of the retina adapted for high visual acuity.

Photoreceptors in the retina contain visual pigments that can trap photons of light. Rods contain the pigment **rhodopsin**, whereas cones contain three forms of the pigment **iodopsin**. Each form of iodopsin absorbs light maximally in different parts of the visible light spectrum: red, green, and blue. The absorption of light by the visual pigments initiates a chemical reaction that results in hyperpolarization of the cell membrane. This results in a graded potential which can then be used by retinal cells to transmit information.

The circuitry of the retina is formed from six basic cell types: rods, cones, horizontal cells, bipolar cells, ganglion cells, and amacrine cells. As many as 1500 rods will converge onto a single bipolar cell, which can then affect a ganglion cell through the amacrine cell interneurons. In contrast, cone cells synapse directly with ganglion cells.

The receptor cells and bipolar cells transmit excitatory signals, whereas the intervening interneurons, the horizontal cells, and the amacrine cells transmit inhibitory signals. This retinal circuitry processes information about the color and contrast of the images projected onto the retina. The retinal ganglion cells provide information important for detecting the shape and movement of objects. There are two types: P-cells, which are color-sensitive detectors, and M-cells, which are color-insensitive motion detectors. The axons of the ganglion cells converge to form the **optic nerve**. Optic nerve fibers from both eyes combine to form the optic chiasm, which lies superior to the sella turcica and directly above the pituitary gland. A partial crossing of fibers (decussation) occurs at the optic chiasm. After the decussations in the optic chiasm, each optic tract represents the contralateral visual fields. The fibers from the nasal half of each retina cross to the contralateral side. Fibers from the temporal half of each retina approach the chiasm but do not decussate. At the level of the optic chiasm, some ganglion cell axons terminate in the suprachiasmatic nucleus of the hypothalamus, where information is provided to regulate circadian rhythms. The remainder of the axons continue past the optic chiasm as the **optic tracts**. The optic tracts terminate primarily in the **lateral geniculate nuclei** (LGN) of the thalamus, the **superior colliculus**, and the **pretectal area**.

Input from the optic tract to each LGN is received in a retinotopic pattern representing the contralateral visual half-field. This means that ganglion cells in adjacent areas of the retina will project onto adjacent neurons in the LGN.

The superior colliculus receives retinotopically organized input directly from the ipsilateral optic tract. Neurons in the superior colliculus carry visual input to the pons by way of the **tectopontine tract** and to the spinal cord by way of the **tectospinal tract**. The tectopontine tract also relays visual information to the cerebellum and aids in the control of eye movements through the paramedian pontine reticular formation. The tectospinal tract mediates reflex control of head and neck movements in response to visual input. The superior colliculus also has reciprocal connections with neurons in the visual cortex.

The neurons leaving the lateral geniculate nucleus form the **optic radiations** and project to the **primary visual cortex** of the occipital lobes. The inferior radiations carry information regarding superior visual fields, while the superior radiations carry information regarding inferior visual fields. The radiating fibers from the lateral aspect of the LGN course downward and forward before bending back in a sharp loop through the temporal lobe. They then course in the lateral wall of the inferior horn of the lateral ventricle and then to the occipital lobe. The radiating fibers from the medial aspect of the LGN travel adjacent to the lateral fibers, but take a more direct course over the top of the inferior ventricular horn and then to the occipital lobe.

The optic radiations travel to the cortex surrounding the **calcarine fissure** on the **medial occipital lobe**. The gyrus above the calcarine fissure is called the **cuneus** and receives visual impulses from the ipsilateral upper quadrant of both retinas, which corresponds to the lower quadrant of the contralateral visual field. The gyrus below the calcarine fissure is called the **lingual gyrus** and receives impulses from the lower quadrant of both retinas. The primary visual cortex is called the **striate cortex** because of a stripe of myelinated fibers named the **line of Gennari** running horizontally through the cortex. The visual cortex is divided into the dorsal stream and ventral stream. The dorsal (or "where") stream is involved in spatial awareness and recognizing where objects are in space. The ventral (or "what") stream is involved in object recognition and form representation.

The optic fibers maintain their topographic arrangement from the LGN to the cortex, maintaining the retinal map (Figure 23-1). Thus, information from the upper quadrant of the left visual field projects to the lower right quadrant of the retina, then travels to the lateral portion of the right LGN and then to the right visual cortex below the calcarine sulcus. Information from the fovea in the center of the retina projects to the occipital pole. The visual association areas have connections with the frontal lobes and brainstem and influence visually guided saccades, ocular pursuit movements, accommodation, and convergence.

There are a number of differences in the visual fields. A monocular visual field occurs when only one eye is used, as opposed to a binocular visual field, which involves both eyes. In monocular vision the field of view is increased, while depth perception is limited. The central visual field operates best under high illumination and has the greatest visual acuity and color sensitivity. The peripheral visual field on the other hand, is more sensitive to dim light, operates under low illumination, and has little color sensitivity.

Lesions in different parts of the visual pathway can produce distinctive visual field defects. An injury to the optic nerve fibers carrying information from the **macula** results in loss of vision at the center of the visual field, creating a **central scotoma**, as well as loss of **visual acuity** and color vision. Complete injury to an optic nerve results in complete blindness in the ipsilateral eye. The optic chiasm can be affected by a pituitary tumor compressing the inferior aspect of the chiasm, or from a craniopharyngioma compressing the superior aspect. The fibers decussating at the chiasm carry information from the nasal retina and temporal visual fields and a lesion at this location results in **bitemporal hemianopia.** A lesion of the optic tract will affect fibers from the ipsilateral nasal retina and the contralateral temporal retina, resulting in a **homonymous hemianopia**. A lesion of the optic radiations will also result in a homonymous hemianopia. If the optic radiations are only affected in the anterior temporal lobe, a predominantly superior visual field deficit will result.

A lesion of the occipital lobe that affects the entire primary visual cortex can also create a homonymous hemianopia. **Macular sparing,** preservation of the central 5-10 degrees of vision in an otherwise blind visual field, is often present because of the extensive macular representation in the occipital cortex.

Visual input from the upper halves of the retina carries information from the lower visual field to the cuneus, whereas input from the lower halves of the retina carrying information from the upper visual field travels to the lingual gyrus. A lesion of the lingual gyrus will create a deficit of the contralateral upper field called a **superior quadrantanopia.**

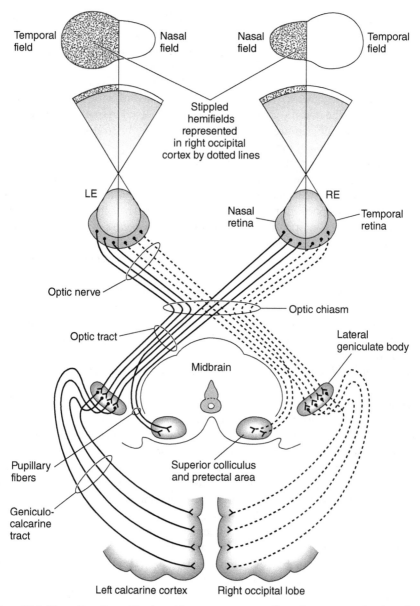

Figure 23-1. The optic pathway. The dotted lines represent nerve fibers that carry visual and pupillary afferent impulses from the left half of the visual field. (*With permission from* Vaughan & Asbury's General Opthalmology. *New York, NY: McGraw-Hill; 2000. 16th ed. Chapter 14, figure 14-2.*)

> CASE CORRELATES
>
> - See Case 18 (proprioception), Case 19 (spinothalamic pathway), Case 20 (nociception), Case 21 (olfaction), and Case 22 (audition).

COMPREHENSION QUESTIONS

23.1 A 7-year-old boy is referred to you by the school nurse because of concerns about his vision. He apparently has been having difficulty in certain classes and has been noted to have erroneously colored some assignments. As a quick screening test, you show the child some Ishihara color plates, and he is unable to see the numbers on some of them, indicating that he has dyschromatopsia specifically discriminating between red and green. What cell type is most likely abnormal in this child's retina?

A. Rod cells

B. Cone cells

C. Bipolar cells

D. Retinal ganglion cells

23.2 A 43-year-old woman presents to your clinic complaining of vision loss. The visit is prompted by an auto accident where she pulled out in front of a car approaching from the left that she claims she never saw. On complete visual field testing you find that this woman has a left homonymous hemianopia. You perform an MRI and find a tumor impinging on part of her visual pathway. Based on her visual field defect, which of the following is the most likely location of her tumor?

A. Left optic nerve

B. Optic chiasm

C. Right optic tract

D. Right inferior optic radiations (Meyer loop)

23.3 A 64-year-old man with hypertension and an extensive history of smoking presents to the clinic with complaints of loss of peripheral vision. He notes that this vision problem occurred abruptly several days ago, and he was hoping it would go away, but it has not. On visual field testing, he has a right homonymous hemianopia with macular sparing. Where is the most likely location of this man's lesion?

A. Left occipital lobe

B. Left lingual gyrus

C. Left cuneus

D. Left superior optic radiations

ANSWERS

23.1 **B.** Color blindness most often occurs because of a defect in one of the three types of **cone cells** present in the human eye. Each type of cone detects a different color—red, blue, or green—and defect in any one of these types results in difficulty with color discrimination. Rod cells respond to all colors of light (monochrome vision) and are effective in low-light situations. Bipolar cells function to consolidate information from many rod cells and transmit it to retinal ganglion cells, which project their axons down the optic nerves and tracts to the thalamus.

23.2 **C.** This patient is most likely to have a defect of the **right optic tract**. Vision is represented contralaterally in the CNS, but it is not divided by eye, rather by visual field. In other words, objects to the right of midline are represented on the left side of the brain and vice versa. This woman has a complete loss of vision to the left of midline, indicating a right-sided lesion behind the optic chiasm, where the pathways become fully segregated by side. The only lesion of the ones listed that can cause this is a lesion of the right optic tract. A lesion of the right inferior optic radiations would only cause a defect in half of the left visual field, that coming from the inferior half of the retina, which would cause a left superior quadrantopia. Lesions of the optic chiasm typically cause bitemporal hemianopia caused by damage to the crossing nasal fibers from both eyes. A lesion of the left optic nerve would cause complete blindness in the left eye.

23.3 **A.** Because this patient has a right homonymous hemianopia with macular sparing, the most likely location for the lesion is the **left occipital lobe**. The right visual field projects to the left cortex, with the inferior half represented superior to the calcarine sulcus in the cuneus, and the superior half represented inferior to the calcarine sulcus in the lingual gyrus. A lesion in either of these gyri results in a corresponding quadrantopia. Extensive damage to the occipital pole, including to both of these gyri, however, can cause vision loss in the entire visual field. An interesting phenomenon in large occipital lesions, however, is that of macular sparing. It is believed that because the macula is represented so heavily in the visual cortex, even large lesions leave some areas of cortex representing the macula intact.

NEUROSCIENCE PEARLS

▶ The optic fibers carry information topographically to the visual cortex, maintaining the retinal map.

▶ Lesions along the visual pathway will correspond to distinctive visual field deficits.

▶ The macular region of the retina is adapted for high visual acuity, contains only cones, and has an extensive representation in the visual cortex.

▶ A lesion of the macula leads to a central scotoma.

REFERENCES

Martin JH. The visual system. *Neuroanatomy: Text and Atlas*. 2nd ed. Stamford, CT: Appleton & Lange; 1996.

Ropper AH, Brown RH. Disturbances of vision. *Adams and Victor's Principles of Neurology*. 8th ed. New York, NY: McGraw-Hill; 2005.

Wurtz RH, Kandel ER. Central visual pathways. In: Kandel ER, Schwarz JH, Jessell TM, Siegelbaum SA, Hudspeth AJ, eds. *Principles of Neural Science*. 5th ed. New York, NY: McGraw-Hill; 2012.

A 65-year-old man develops sudden weakness in his right arm and leg. His wife notes that the lower right half of his face appears to droop as well. He is taken to the emergency room, where he is noted to have a medical history significant for diabetes and poorly controlled hypertension. On physical examination he is noted to have a flaccid paralysis of the right side and paralysis of the right lower side of his face. An MRI is immediately obtained, which reveals an acute, left-sided ischemic infarct.

► Which infarcted area of the brain is responsible for the patient's symptoms?
► What treatment is available?
► High levels of what lipid greatly increases the risk of stroke?

ANSWERS TO CASE 24:

Movement Control

Summary: A 65-year-old man with diabetes and poorly controlled hypertension develops acute-onset right-sided hemiparesis and a partial right facial droop. The MRI is consistent with a left-sided ischemic stroke.

- **Affected brain area responsible for patient's symptoms:** The left internal capsule contains the corticobulbar and corticospinal fibers innervating the patient's contralateral face, arm, and leg as they travel to the brainstem.

- **Treatment options:** If an ischemic stroke is diagnosed within a 3-hour window, tissue plasminogen activator can be used to attempt to dissolve the clot in the blood vessels perfusing the internal capsule.

- **Stroke risk factors:** High levels of cholesterol increase risk for ischemic stroke. Other risk factors include advanced age, hypertension, smoking, heart disease, and diabetes.

CLINICAL CORRELATION

Cerebral stroke is characterized by the sudden loss of blood supply to an area of the brain, resulting in a loss of corresponding neurological function. The majority of strokes are ischemic, caused by a thrombosis or embolism to a cerebral blood vessel. They can present as hemorrhagic in quality as well, and an ischemic stroke can convert to a hemorrhagic strokes. The symptoms of a stroke will depend on the area of the brain that is affected by the lost blood supply. The most common location for a stroke is the posterior limb of the internal capsule, which carries the descending corticospinal and corticobulbar fibers and results in purely motor symptoms.

APPROACH TO:

Control of Movement

OBJECTIVES

1. Know the structures of the central nervous system involved in motor control.

2. Describe the descending motor pathways and their functions.

3. Be aware of the interactions and influences of the sensory systems on motor function.

DEFINITIONS

HOMUNCULUS: A figure of the human body superimposed on the cortical surface of the brain to represent the motor or sensory regions of the body represented there.

UPPER MOTONEURONS: Upper-level neurons from the cerebral cortex, cerebellum, and basal ganglia that control descending motor pathways either directly or indirectly.

LOWER MOTONEURONS: Motor neurons originating from nuclei of the spinal cord and brainstem that innervate skeletal muscle and provide the final direct link from the nervous system through the neuromuscular junctions.

PREMOTOR CORTEX: An area of motor cortex, located between the primary motor cortex and the prefrontal cortex, responsible for sensory guidance of movement and control of proximal and trunk musculature of the body.

BETZ CELLS: Large pyramidal cells located within the primary motor cortex, which give rise to the portion of the corticospinal tract which synapses directly with lower motoneurons in the anterior horn cells of the spinal cord.

SOMATOTOPIC: The maintenance of spatial organization within the central nervous system. For instance, fibers innervating the foot will travel next to fibers innervating the lower leg.

DISCUSSION

The motor system is organized in a functional hierarchy, with each level responsible for a specific task. Movement must be planned, have a purpose, respond to sensory input, and function with coordination using spatiotemporal details of muscle positions. There are several anatomical pathways that project to the spinal cord from higher motor centers. Most of these are organized **somatotopically**, with movements of adjacent body parts being controlled by contiguous areas of the brain at each level within the motor hierarchy.

The **primary motor cortex** lies in the precentral gyrus and paracentral lobule of the frontal lobe and is responsible for controlling simple movement. It extends from the lateral fissure upward to the dorsal border of the hemisphere and beyond to the paracentral lobule. The left motor strip controls the right side of the body, and the right strip controls the left side. Neurons in the lowest lateral part of the motor strip influence the larynx and tongue, followed in upward sequence by neurons affecting the face, thumb, hand, arm, thorax, abdomen, thigh, leg, foot, and perineal muscles. The areas for the hand, tongue, and larynx are disproportionately large, given the elaborate motor control needed for these muscle groups. This functional representation is called a **homunculus**.

The **premotor cortex** lies immediately rostral to the primary motor area on the lateral surface of the hemispheres. The premotor area also contains a homunculus representation. The medial aspect contains the **supplementary motor area.** The **postcentral gyrus** and the **secondary motor cortex** located where the pre- and postcentral gyri are continuous at the base of the central sulcus are also cortical regions that influence movement. The **frontal eye fields**, located in the middle frontal gyrus, initiate voluntary saccades and contain neurons that influence eye movement.

Movements result from the actions of neuronal networks at many different levels of the nervous system. The brainstem and spinal cord contain pattern generator for rhythmic activities and complex movements such as locomotion. The descending pathways interact with and control the lower level neuronal patterns of discharge in a **hierarchical manner (Figure 24-1).** The cerebral cortex can control contractions of individual muscles and can determine the force of these contractions. Populations of motor cortical neurons, however, must act together to specify the **direction** and **force** of movements. The premotor and supplementary motor areas are important in planning movements. The supplementary motor area also functions to integrate movements performed simultaneously on both sides of the body.

The **lower motoneurons** innervate skeletal muscle through **neuromuscular junctions.** Their cell bodies reside in the spinal cord and cranial nerve nuclei of

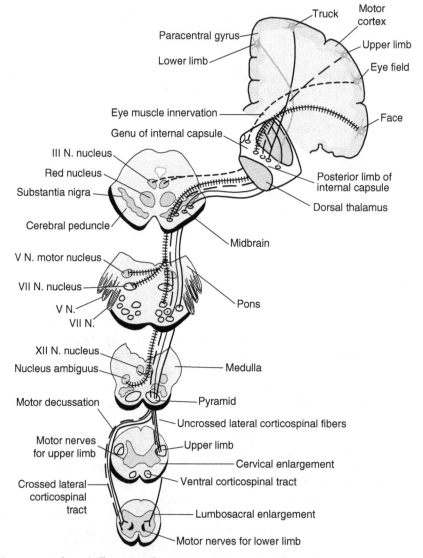

Figure 24-1. Schematic illustration of upper motor neuron pathways. (*With permission from* Adam and Victor's Principles of Neurology. *New York: McGraw-Hill; 2000. 7th ed. Page 51, figure 3-2.*)

the central nervous system, where they are influenced by the higher nervous system structures to stimulate or inhibit voluntary movement. The lower levels of the motor system coordinate simple reflexes and control the amount of force and velocity generated by a single muscle. They coordinate movements and changes in posture. The **upper motoneurons** technically include the cerebral cortex, cerebellum, and basal ganglia, which all regulate lower motoneuron activity either directly or indirectly through interneurons. The higher levels of the motor systems are involved in more global tasks and coordinate and calculate the activity of many limbs or muscle groups, and evaluate the appropriateness of a particular action.

The **corticospinal tract**, or **pyramidal tract**, **controls skilled movements** of the distal limbs and influences the **distal flexor muscles**. One-third of the axons in the corticospinal tract originate in the primary motor cortex, one-tenth of these cells originating from **Betz cells**, which are large pyramidal cells located in the fifth cortical layer. One-third of the corticospinal tract axons arise from the premotor and supplementary motor regions and the remainder of the fibers arise from the parietal lobe, primarily the postcentral gyrus. The areas of the cortex that contribute to the corticospinal tract are collectively termed the **sensorimotor cortex**. After passing through the **posterior limb of the internal capsule** and the middle of the cerebral pedicle, or **crus cerebri**, the corticospinal tract splits into bundles in the pons prior to reforming as a discrete bundle to form the **medullary pyramid**. Approximately 90% of the fibers cross to the other side at the **level of the pyramidal tract decussation** in the lower medulla and continue descending as the **lateral corticospinal tract.** This tract travels to all levels of the spinal cord, synapsing in the lateral aspects of laminae IV through VIII. Most of the synapses in these layers are with interneurons which then synapse directly with motoneurons in lamina IX. Some fibers in the lateral corticospinal tract, however, synapse directly with the motoneurons in lamina IX. The remaining 10% of corticospinal fibers that do not decussate in the medulla descend in the anterior funiculus of the cervical and upper thoracic cord as the **ventral corticospinal tract**. Most of these fibers decussate through the **ventral white commissure** at their level of termination prior to synapsing with interneurons and motoneurons of the contralateral side. The number of fibers in both lateral and ventral corticospinal tracts successively decreases in lower spinal cord segments as more and more fibers reach their terminations.

The corticospinal tract fibers that synapse with the interneurons of the dorsal horn influence both local reflex arcs and originating cells of ascending sensory pathways. This system allows the cerebral cortex to **control reflex motor output and to modify the sensory input that reaches the brain.** Cortical **excitatory signals usually result from monosynaptic connections with motoneurons** and are facilitated by the neurotransmitter **glutamate. Inhibitory signals occur through synaptic connections in inhibitory interneurons** and are mediated by **glycine.** The activation of the corticospinal tract generally results in **excitatory input to motoneurons of flexor muscles** and **inhibitory input to those of extensor muscles.**

The **corticobulbar tract** arises primarily from the ventral portion of the sensorimotor cortex on the lateral surface of the hemisphere and from the frontal eye fields. The axons diverge from the corticospinal tract at the level of the mid-brain and terminate in the brainstem nuclei of cranial nerves III, IV, V, VI, VII, IX, X, XI, and XII.

Fibers projecting from the frontal eye fields indirectly influence eye movements by synapsing with cells in the pontine reticular formation that then project to the nuclei of cranial nerves III, IV, and VI. The cranial nerve motor nuclei receive innervation from both cerebral hemispheres, creating symmetric movements on both sides of the face. The lower facial nucleus and hypoglossal nucleus receive far heavier innervation from the ipsilateral cortex, allowing muscles controlled by these groups (lower face and tongue) to be controlled independently on the two sides. Similar to the corticospinal tract, the corticobulbar tract contains fibers that terminate on ascending sensory neurons, allowing for mediation of sensory information from the nucleus gracilis, nucleus cuneatus, sensory trigeminal nuclei, and nucleus of the solitary tract.

The **corticotectal tract** contains fibers that project from cortical areas of the occipital and inferior parietal lobes to the superior colliculus, the interstitial nucleus of Cajal, and the nucleus of Darkschewitsch. Axons project from here to the **pontine reticular formation** and from there to the **medial longitudinal fasciculus** (MLF) to synapse in the oculomotor, trochlear, and abducens nuclei. This input allows for cortical influence on extraocular muscle activity. The corticotectal tract also connects with neurons in the superior colliculus that give rise to the **tectospinal tract**. This tract crosses at the **dorsal tegmental decussation** and descends to the cervical spinal cord, where the fibers become incorporated with the MLF. Input from the tectospinal tract influences neurons innervating the muscles of the neck and is concerned with reflexive turning movements of the head and eyes.

The cortical areas that give rise to the corticospinal tract also form the **corticorubral tract**. These neurons project to the **ipsilateral red nucleus** in the tegmentum of the midbrain. Neurons of the red nucleus then give rise to **the rubrospinal tract**, which crosses at the **ventral tegmental decussation** and descends through the lateral tegmentum of the pons, midbrain, and medulla. Descending just anterior to the lateral corticospinal tract, the fibers synapse in laminae V, VI, and VII of the spinal gray matter. The rubrospinal tract facilitates flexors and inhibits extensor motoneurons.

The **corticoreticular fibers** travel with the corticospinal and corticobulbar tracts to the reticular formation. The reticular formation of the brainstem receives sensory information from numerous systems and communicates heavily with the cerebellum and limbic systems. Two corticoreticular fibers originate from both the **sensorimotor cortex**, and other regions of the brain such as the **medial prefrontal cortex**, the **limbic lobe**, and the **amygdala**. These cortical areas integrate somatic and visceral components of complex reflex systems such as micturition and genital function, as well as controlling the complex emotional and social behaviors associated with them.

The pontine reticulospinal tract is important for the control of both posture and locomotion. It originates from the pontine reticular formation and travels with the MLF through the medulla and cervical spinal cord before terminating in the thoracic spinal cord. Its fibers synapse in laminae VII and VIII with excitatory signals for extensor motoneurons innervating the midline musculature of the body and the proximal extremities.

Several **raphe nuclei** within the reticular formation have an important function in the modulation of the responsiveness of the motor system to reflex or corticospinal inputs. The **caudal raphe nuclei** in the reticular formation give rise to fibers that project to the spinal cord, where they influence incoming sensory signals as well as motor responsiveness. Fibers from the **nucleus raphe magnus** exert important influences on the transmission of pain stimuli from peripheral nerves. The **nucleus locus ceruleus** and nucleus subceruleus give rise to noradrenergic projections to the spinal cord through the ventrolateral funiculus. While these raphe-spinal connections do not evoke movement, they are important in producing excitatory or general inhibitory effects that influence the overall motoneuron responsiveness and modulate the motor system in different phases of sleep-wake cycles and with changing emotional states.

The **vestibulospinal tracts** are important pathways for the control of postural tone and postural adjustments of the body that accompany head movements. They arise from neurons in the vestibular nuclei of the medulla and descend as **the lateral vestibulospinal tract** over the entire length of the cord, and the **medial vestibulospinal tract** through the upper thoracic levels. They descend in the anterior funiculus and synapse in laminae VII and VIII of the spinal gray matter to evoke excitatory postsynaptic potentials in extensor motoneurons innervating the neck, back, forelimb, and hindlimb muscles.

The integration of sensory input with movement allows us to continually interact with our environment through varied and purposeful motor behaviors. Motor systems are continuously refined by repetition and learning because of constant influences from the complex cortical association areas. Lesions along the motor hierarchy can lead to both negative (paralysis) and positive (spasticity) sequelae caused by the combination of excitatory and inhibitory input to lower motor systems.

CASE CORRELATES

- See Cases 24-27 (movement systems).

COMPREHENSION QUESTIONS

24.1 A 68-year-old woman presents to your clinic with the complaint of gradually worsening weakness in her right leg and a recent onset of weakness of her left leg. She states that the weakness in her right leg is now so profound that she is nearly incapable of moving it. On examination she has normal muscle bulk in both her lower extremities, hyperreflexia of her right patellar and Achilles reflexes, and 20% strength in the right lower extremity. You suspect a brain tumor could be causing her symptoms and send her for an MRI, which most likely shows a tumor in which of the following locations?

   A. On the convexity of the left hemisphere over the fronto-parietal region

   B. On the convexity of the right hemisphere over the fronto-parietal region

   C. In the midsagittal plane near the paracentral lobule

   D. On the convexity of the left hemisphere over the parieto-temporal region

Refer to the following case scenario to answer questions 24.2–24.3:

A 25-year-old man is hospitalized following a motorcycle accident in which he was not wearing a helmet. He remains unconscious in critical condition in the ICU for several weeks but eventually regains consciousness. When he does so, it is noted that while he is able to move all of his muscle groups with full strength while testing them individually and can voluntarily make simple movements without problem, he has great difficulty performing complex movements.

24.2 Which of following areas in the motor system is most likely damaged in this patient?

A. Primary motor cortex

B. Premotor and supplemental motor areas

C. Internal capsule

D. Anterior horn cells

24.3 At what level in the nervous system do the majority of the descending fibers of the corticospinal tract decussate to become contralateral?

A. Internal capsule

B. Medullary pyramids

C. Ventral white commissure of the spinal cord

D. Corticospinal neurons do not decussate

ANSWERS

24.1 **C.** The lesion described is a **parasagittal meningioma,** a generally benign tumor of the meninges. In this case, it is growing in the midsagittal plane, and causing compression of the left paracentral lobule and also the right paracentral lobule to a lesser extent. Recall that the somatotopic organization of the primary motor cortex places control of the contralateral leg and foot in the paracentral lobule, and control of more superior body parts in progressively more inferior aspects of the precentral gyrus as you progress toward the Sylvian fissure. A lesion over the lateral convexity of the fronto-parietal region would compress one of the primary motor strips, located in the precentral gyrus, resulting in weakness of the contralateral hand or arm or face, depending on the exact location of the tumor.

24.2 **B.** The deficit in executing complex movements can be attributed to a dysfunction in his **supplemental and premotor areas.** The motor system is arranged in a hierarchical fashion, with each higher level adding complexity to possible movements. At the top of this hierarchy are the supplemental motor area and premotor cortex, which are involved in planning and execution of complex motor behaviors. Neurons in the primary motor cortex are involved in simple movements and can determine the speed and strength with which muscle groups contract. Descending axons from both these areas travel through the internal capsule and the corticospinal tracts to synapse with anterior horn cells, which control the actual contraction of individual muscle groups. Since this patient has full voluntary movement and strength in all of his muscle groups, his primary motor cortex seems to be intact.

24.3 **B.** Approximately 90% of the descending axons of the corticospinal tract decussate **at the level of the medullary pyramids** in the pyramidal decussation. These fibers then travel in the contralateral lateral corticospinal tract in the lateral column of the spinal cord. They end on alpha motor neurons and interneurons at the appropriate spinal level, resulting in control of movement by the contralateral motor cortex. The remaining 10% of the corticospinal neurons do not cross in the medullary pyramids, but travel down the ipsilateral cord in the anterior corticospinal tract, and eventually cross in the ventral white commissure of the spinal cord at their target spinal level.

NEUROSCIENCE PEARLS

▶ The motor system is somatotopically organized in a functional, hierarchical pattern.

▶ Cortical motor areas are important for planning movements and integrating motor output with sensory input.

▶ Descending motor pathways carry excitatory and inhibitory input to the spinal cord, allowing for purposeful, controlled movements.

REFERENCES

Ghez C, Krakauer J. The organization of movement. In: Kandel ER, Schwartz JH, Jessell TM, eds. *Principles of Neural Science.* 4th ed. New York, NY: McGraw-Hill; 2000 .

Martin JH. Descending projection systems and the motor function of the spinal cord. In: *Neuroanatomy: Text and Atlas.* 2nd ed. Stamford, CT: Appleton & Lange; 1996.

Ropper AH, Brown RH. Disorders of motility. In: *Adams and Victor's Principles of Neurology.* 8th ed. New York, NY: McGraw-Hill; 2005.

A 62-year-old man presents to his general practitioner complaining of a steady tremor in his hands, which has slowly progressed for the past 6 weeks. The patient also states that walking has become increasingly difficult although he attributes this to old age. Upon physical examination, the patient has increased muscle tone with a notably hunched posture and a resting tremor. When asked to make purposeful movements, the patient is slow to initiate the movement; however, the tremor is alleviated while moving. Examinations of cranial nerve function and deep tendon reflexes are normal. The patient shows no symptoms of dementia, Alzheimer, or any other cognitive disorders. An MRI is ordered which shows mild brain atrophy appropriate for his age, but otherwise unremarkable. You make the diagnosis of Parkinson disease.

▶ What microscopic structures are found in the neurons of patients with this disorder?
▶ What are the treatment options available?

ANSWERS TO CASE 25:

Basal Ganglia

Summary: A 62-year-old man with bradykinesia and a steady resting tremor in both hands. His tremor is temporarily mitigated while making purposeful movements. The patient has increased muscle tone and a hunched posture. When asked to walk, the patient's gait is disturbed. Examinations of cranial nerve function and deep tendon reflexes are normal. MRI reveals mild global brain atrophy appropriate for the patient's age but is otherwise normal.

- **Microscopic pathology:** Lewy bodies are eosinophilic cytoplasmic inclusions with a halo of radial fibrils and are composed mainly of the protein alpha-synuclein. These pathological structures are thought to accumulate over time and disrupt normal intracellular functions of nerve cells.

- **Treatment options:** While no current treatment options are curative, administration of **levodopa** (a precursor of dopamine) and other parkinsonism medications help alleviate the symptoms of Parkinson disease. Some patients may also be candidates for neurosurgical interventions such as **thalamotomy, subthalamotomy,** or **pallidotomy** procedures to alleviate movement symptoms once medical treatments have become ineffective. Also, current research shows that embryonic stem cell transplantation into the striatum may one day become an effective treatment option for patients with Parkinson disease.

CLINICAL CORRELATION

Parkinson disease (PD) is a common neurological disorder of the basal ganglia arising from neural cell degeneration within the pars compacta region of the substantia nigra. The loss of these pigmented dopaminergic neurons reduces the amount of dopamine synthesized by the substantia nigra. Without the appropriate amount of dopamine in the striatum, there is an antagonistic effect on the direct pathway and agonistic effect on the indirect pathway of the nigrostriatal projection. This often leads to hypokinesia. The direct pathway has the net effect of exciting thalamic neurons, which in turn make excitatory connections with cortical neurons. The indirect pathway has the net effect of inhibiting thalamic neurons, thereby inhibiting cortical neurons. The pathological cause of cell death in PD is still unclear, and the majority of cases are idiopathic. Other causes of parkinsonism are related to drug-induced, toxic, genetic, and traumatic etiologies. Parkinsonism symptoms are also observed in individuals after long-term use of antipsychotic medications such as haloperidol or after ingestion of the neurotoxin MPTP.

APPROACH TO:

Basal Ganglia

OBJECTIVES

1. Know the anatomical structures and cellular projections of the basal ganglia.

2. Be able to describe the different neurological functions of the basal ganglia.

3. Know common disorders associated with the basal ganglia.

DEFINITIONS

STRIATUM: The caudate nucleus and the putamen. Together, the two nuclei act as the "receiving portion" of the basal ganglia.

PALLIDUM: Comprised of the internal (GP_i) and external (GP_e) portion of the globus pallidus, the pallidum plays a modulatory role between the striatum and the thalamus. The efferent connections from the substantia nigra and the pallidum act as a mediator between the basal ganglia and the rest of the nervous system.

MEDIUM SPINY NEURONS: The most common cell type found in the striatum. Contain the inhibitory neurotransmitter gamma-aminobutyric acid (GABA).

HEMIBALLISMUS: Involuntary, violent movements of the contralateral side of the body resulting from lesions involving the subthalamic nucleus.

CHOREA: A movement disorder of the basal ganglia characterized by a rapid, irregular flow of motions as well as grimacing movements of the face.

ATHETOSIS: A movement disorder of the basal ganglia characterized by slow, writhing movements of a wormlike character involving the extremities, trunk, and neck.

DISINHIBITION: Inhibition of an inhibiting projection pathway.

DISCUSSION

The basal ganglia are composed of the **caudate nucleus, putamen,** and **globus pallidus.** Because of the high degree of cellular connections with these three structures, the **substantia nigra** and **subthalamic nucleus** are also considered components of the basal ganglia. The basal ganglia have significant connections with both the thalamus and the cortex. It functions primarily in the modification and elaboration of movements initiated by the primary motor cortex. A lesion in one of the nuclei within the basal ganglia will therefore produce a disruption in movement and muscle tone but no pareses.

Other significant roles of the basal ganglia include cognitive functions such as verbal working memory, motor planning, repetitive movement learning, and the association of motivation and emotions to the execution of movements.

The caudate nucleus and the putamen together are termed the **striatum.** Together, these two nuclei act as the "receiving portion" of the basal ganglia. The main afferent connections travel from the cerebral cortex, the intralaminar thalamic nuclei, and the dopamine-containing cell groups in the mesencephalon. The putamen largely receives information from the primary motor and somatosensory cortex, while the caudate nucleus receives the majority of its inputs from the association areas of the cortex. The **intralaminar thalamic nuclei** project cells to the striatum in large part from the centromedian nucleus. The dopaminergic afferent connections of the striatum arise from the **substantia nigra** and the **ventral tegmental area.**

The primary **efferent pathways** from the basal ganglia begin in the globus pallidus and substantia nigra and project to the **thalamus, mesencephalic tegmentum, and superior colliculus.** The efferent connections from the substantia nigra and the pallidum act as a mediator between the basal ganglia and the rest of the nervous system.

The globus pallidus is composed of internal and external segments, each receiving afferents input from the striatum and subthalamic nucleus. The actions of the striatal fibers on the internal segment of the globus pallidus (GP_i) are inhibitory while the effects from the subthalamic nucleus are excitatory. The activity of the GP_i is determined from the summation of inhibitory and excitatory input. The efferent output from the GP_i acts on the substantia nigra and the thalamus. While the external portion of the globus pallidus (GP_e) projects to the subthalamic nucleus, the substantia nigra has efferent connections with the thalamus and superior colliculus. The thalamus functions as a relay center for information projecting in both directions between the basal ganglia and the cerebral cortex. The basal ganglia additionally regulate muscle movements and tone through its efferent input to the reticulospinal tract and reticular formation. Six different cell types have been identified in the striatum, the most common of which are **medium spiny neurons.** These contain the inhibitory neurotransmitter **GABA.**

The substantia nigra has two components: the pars compacta and the pars reticula. The **pars compacta** is densely packed with neuromelanin-containing neurons (giving the nucleus its dark color). The afferent connections of the substantia nigra come from many different cell groups, the majority being from the striatum. The inhibitory afferents of the substantia nigra contain the neurotransmitter GABA. Excitatory afferents originate from the **pedunculopontine nucleus** and **subthalamic nucleus** (transmitting glutamate), the **raphe nucleus** (transmitting serotonin), the **locus coeruleus** (transmitting norepinephrine), and the **basal forebrain.**

The efferent connections of the substantia nigra primarily project to the striatum and thalamus, with some association to the superior colliculus and reticular formation (Figure 25-1). The pars compacta contains the dopaminergic nigrostriatal neurons, making it a key location in PD. The nigrostriatal pathway has the dual effect of exciting the direct pathway and inhibiting the indirect pathway—when this pathway is destroyed in PD, the indirect pathway cannot be inhibited. The pars reticula contains inhibitory GABAergic nigrothalamic neurons.

Dopamine is one of the key modulatory neurotransmitters in the striatum. Dopamine receptors can be divided into D_1 and D_2 types, each of which possess many of their own subtypes. The receptors influence many types of ion channels

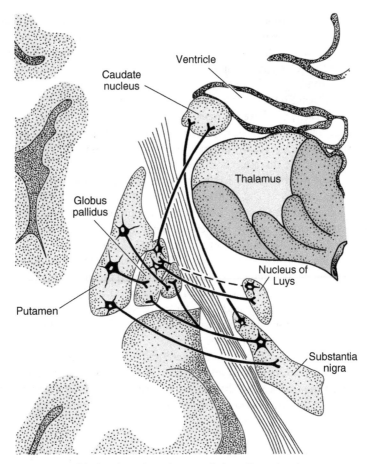

Figure 25-1. Diagram of the basal ganglia in the coronal plane, illustrating the main interconnections. (*With permission from* Adam and Victor's Principles of Neurology. *7th ed. New York: McGraw-Hill; 2000. Page 303, figure 15-2.*)

on both presynaptic and postsynaptic terminals. This allows the neurotransmitter to affect the stabilization of the membrane potential and contribute to keeping that potential in a range where the cell is apt to fire in bursts, that is, in a state suited for efficient signal transmission.

Excitatory afferent connections are received from the motor cortex by the subthalamic nucleus. The subthalamic nucleus sends the majority of its efferent connections to the pars reticula and the globus pallidus. The effects of the subthalamo-pallidal efferents are excitatory. The excitatory influences of the striatum determine the activity of the GABA-containing neurons within the GP_i. The subthalamic nucleus is believed to control or stop ongoing movements. Because of the inhibitory effects that the GP_i has on the thalamocortical neurons, increased activity in the subthalamic nucleus would result in an inhibition of voluntary movement as a result of the inhibition of the motor cortex. The clinical presentation of **hemiballismus,** involuntary, violent movements of the contralateral side of the body resulting from lesions involving the subthalamic nucleus, may be because of increased activity of the thalamocortical neurons. Several diseases have been linked to the basal ganglia. **Parkinson disease** is a neurodegenerative disorder of the dopaminergic neurons of the substantia nigra. The disease is characterized by **tremor, rigidity, akinesia,** and **postural instability.** Current therapies for PD are limited to the treatment of the movement symptoms.

Huntington disease is an inherited disease that results from progressive neurodegeneration of primarily GABAergic neurons within the striatum. It is characterized in its early stages by **forgetfulness, depression, sudden falls, irritability,** and **choreiform movements,** which gradually progress until the patient is no longer ambulatory. In the late stages of the disease, continued cell deterioration eventually results in **dementia.** A locus for the disease has been identified on the short arm of chromosome 4 allowing carriers of the disease to be identified before symptoms present. The disease is caused by a selective loss of striatal neurons in the indirect pathway. Without the inhibition from the indirect pathway, the thalamic neurons can fire spontaneously, causing the motor cortex to execute motor programs with no control by the person. **Tourette syndrome** is a disorder involving a functional disturbance between the basal ganglia and the frontal cortex. The syndrome presents with multiple **tics,** often associated with **uncontrollable vocal outbursts.** Patients with Tourette syndrome are often treated with dopamine antagonists to relive the motor symptoms. **Tardive dyskinesia** is an iatrogenic disorder that results from the long-term administration of neuroleptic medications such as haloperidol, chlorpromazine, and thioridazine. The disorder is characterized by **rigidity, and uncontrollable movements of the head, face, lips, and tongue.** The symptoms are caused by sensitization of dopamine receptors in the mesolimbic pathway. This sensitization results in an imbalance in communication between the nigrostriatal pathway and the basal nuclear motor loop.

CASE CORRELATES

- See Cases 24-27 (movement systems), and cases 7 and 15 (Parkinson disease).

COMPREHENSION QUESTIONS

Refer to the following case scenario to answer questions 25.1-25.3:

A 38-year-old man is brought into your office by his family because he has begun to behave somewhat strangely recently. He has been increasingly irritable and has withdrawn from activities he normally enjoys. Additionally, he has been having strange fidgeting movements of his hands and fingers nearly all the time that he did not have in the past. Based on this history and your physical examination findings, you perform an MRI, which is suggestive of Huntington disease, a disorder that affects the basal ganglia.

25.1 With which of the following processes are the basal ganglia most associated?

A. Directly controlling input to the alpha motor neurons

B. Planning and execution of complex motor activities

C. Modification of movements initiated by the motor cortex

D. Integration and smoothing of multiple movements

25.2 Which of the following structures serves as the primary input to the basal ganglia?

A. Striatum

B. Globus pallidus

C. Subthalamic nucleus

D. Substantia nigra

25.3 Which of the following structures serves as the primary output from the basal ganglia?

A. Striatum

B. Internal segment of the globus pallidus

C. External segment of the globus pallidus

D. Subthalamic nucleus

ANSWERS

25.1 **C.** The primary function of the basal ganglia is thought to be modulation of cortical output, including **modification of movements initiated by the motor cortex.** The major outputs of the basal ganglia are inhibitory to the thalamus, and the variety of basal ganglia circuits alters these outputs to further inhibit or disinhibit the thalamus. Since the thalamic outputs to the cortex are for the most part excitatory, the basal ganglia are capable of inhibiting or disinhibiting the cortex. It is thought that these actions serve to somehow select the appropriate motor programs at the right time, while suppressing other programs that are not needed at the time. The primary motor cortex directly controls alpha motor neurons, the premotor and supplemental motor areas are involved in the planning of complex movements, and the cerebellum serves to integrate and smooth out multiple movements.

25.2 **A.** The **striatum**, composed of the caudate and putamen, serves as the primary input structure of the basal ganglia. Motor and somatosensory information primarily is received by the putamen, while input from the association cortex is primarily received by the caudate. The striatum then projects to both segments of the globus pallidus and the substantia nigra, primarily with inhibitory action. Stimulation of striatal neurons leads to the activation of complex basal ganglial circuitry that ultimately modulates thalamic outflow to the cortex.

25.3   **B. The internal segment of the globus pallidus** projects inhibitory neurons from the basal ganglia to the thalamus. The thalamic nuclei involved are the ventral anterior and ventrolateral nuclei for motor control and the dorsomedial nucleus for output to the association cortex. The thalamic output to the cortex is excitatory, so the various basal ganglial circuits inhibit or disinhibit thalamic outflow, thereby affecting motor control. The pars reticularis of the substantia nigra also has inhibitory outflow to the thalamus that behaves very similarly to the internal segment of the globus pallidus.

NEUROSCIENCE PEARLS

► The primary function of the basal ganglia is the modification and elaboration of motor movements initiated by the primary motor cortex.

► Dopamine has an important function as a modulatory neurotransmitter in the basal ganglia.

► The striatum acts as the "receiving portion" of the basal ganglia.

► The primary efferent connections of the basal ganglia begin in globus pallidus and substantia nigra.

REFERENCES

Brodal P. The basal ganglia. *The Central Nervous System: Structure and Function.* 3rd ed. New York, NY: Oxford University Press; 2004.

DeLong M. The basal ganglia. In: Kandel ER, Schwarz JH, Jessell TM, Siegelbaum SA, Hudspeth AJ, eds. *Principles of Neural Science.* 5th ed. New York, NY: McGraw-Hill; 2012.

Melrose RJ, Poulin RM, Stern CE. An fMRI investigation of the role of the basal ganglia in reasoning. *Brain Res.* April 2007;1142:146–158.

A 56-year-old man presents to the emergency room with complaints of difficulty walking and frequent falls over the past 48 hours. He has a 30-year history of chronic alcoholism. His nutritional status is poor and he admits to spending all of his money on alcohol instead of buying food to eat. Prior to his presentation, he describes an alcoholic binge and several days of anorexia. On physical examination he is noted to walk with a wide-based, irregular gait. He has poor coordination on tests requiring rapid leg movements.

Based on the patient's history of severe alcoholism and malnutrition, alcoholic cerebellar degeneration was thought to be the most likely diagnosis. After treatment with vitamin B$_1$ (thiamine), aggressive hydration with intravenous fluids, and good nutrition, the patient's symptoms gradually improved and he was discharged to alcohol treatment program.

► Which part of the cerebellum is most likely affected?
► If left untreated, what disease is he at risk for?

man with 2 days of difficulty walking, frequent falls, and
gait with poor coordination of his lower extremities. He has
ic alcoholism and poor nutrition. He was treated with vitamin B_1
ggressive hydration with intravenous fluids, and good nutrition. His
gradually improved and he was discharged to alcohol treatment program
normal gait.

- **Area of the cerebellum affected:** Degenerative changes appear in the anterior and superior parts of the cerebellar vermis and are associated with ataxia of gait but preservation of speech and coordination of upper extremities.

- **Disease at risk for if symptoms untreated:** Wernicke-Korsakoff syndrome is a far more serious disease that also results from malnutrition and frequently chronic alcoholism. It results from severe thiamine deficiency, which leads to a degeneration of the eye movement nuclei and the cerebellar vermis and cortex, particularly the Purkinje cell layer. The symptoms include ataxia, ophthalmoplegia, and confusion. If untreated, the symptoms can progress to coma and death.

CLINICAL CORRELATION

Alcoholic cerebellar degeneration is caused by toxic degeneration of Purkinje cells and is clinically characterized by impaired gait, tremor, and truncal ataxia. The midline cerebellar structures are predominantly involved with resulting effect of the lower extremities. The disease is thought to be caused by a combination of alcohol neurotoxicity and nutritional deficiency. The treatment of alcoholic cerebellar degeneration is alcohol abstinence, adequate calorie intake, and thiamine supplementation. While the effects on the cerebellum are frequently permanent, early treatment can lead to an improvement in symptoms and prevent the development of more serious pathology, such as Wernicke-Korsakoff syndrome.

The neurological structures most affected by thiamine deficiency (giving rise to Wernicke encephalopathy or Korsakoff psychosis) are the mammillary bodies. Administration of dextrose without thiamine can precipitate Wernicke's encephalopathy since thiamine is a coenzyme for pyruvate dehydrogenase, important in glucose metabolism; sudden glucose administration exacerabates thiamine deficiency, leading to mammillary body necrosis. The mammillary bodies are part of the Papez circuit and are involved in emotion and memory.

Another cause of cerebellar ataxia is ataxia-telengiectasia syndrome, an autosomal recessive disorder. Children with this disorder have cerebellar atrophy in the first year of life, which leads to ataxia and oculocutanous telangiectasia (abnormally dilated blood vessels). These individuals have severe immune deficiency and frequent infections, especially of the respiratory tract. They also have an increased risk of cancer due to inefficient repair mechanisms for DNA damage caused by UV light (DNA hypersensitivity to ionizing radiation).

APPROACH TO:
The Cerebellum

OBJECTIVES

1. Know the key components of cerebellar anatomy.

2. Understand the functions of the cerebellum.

3. Know the afferent and efferent connections of the cerebellum.

DEFINITIONS

ATAXIA: Problems with movement resulting from the combined effects of dysmetria and decomposition of movement.

DYSDIADOCHOKINESIA: Reduced ability to perform rapidly alternating movements.

DYSMETRIA: Disturbance of the trajectory or placement of a body part during active movements.

MOSSY FIBERS: Originating neurons from spinal and brainstem nuclei (except the inferior olive) that form synapses with granule cells in the cerebellar cortex and have excitatory input to many Purkinje cells.

NYSTAGMUS: Rapid and involuntary oscillatory movement of the eye.

SPINOCEREBELLUM: Comprises the vermis and the intermediate zones of the cerebellar cortex, the fastigial and interpose nuclei. Receives major input from the spinocerebellar tract. Major output to rubrospinal, vestibulospinal, and reticulospinal tracts.

VESTIBULOCEREBELLUM: Flocculonodular lobe and its connection with the lateral vestibular nuclei. Involved in vestibular reflexes and maintenance of posture.

DISCUSSION

The cerebellum participates in the execution of a wide variety of movements. It maintains the fine control and coordination of both simple and complex movements. It is essential for coordinating posture and balance in walking and running, executing sequential movements, producing rapidly alternating repetitive movements and smooth-pursuit movements, and for controlling the trajectory, velocity, and acceleration of movements.

The cerebellar cortex is comprised of gray matter arranged in slim, folded layers called **folia.** Three layers of cells constitute the cortex: the **molecular layer,** the **Purkinje cell layer,** and the **granular layer.** The Purkinje cells mediate all of the outgoing signaling of the cerebellum. Below the cortical layers, the white matter of the cerebellum houses the afferent and efferent fibers. Within the white matter are the cerebellar nuclei. These nuclei play an important mediating role in the cerebellar efferent connections.

The cerebellum includes a midline structure called the vermis and two large lateral structures known as the **cerebellar hemispheres.** Together, the vermis and hemispheres can be transversely divided into the **flocculonodular lobe,** the **anterior lobe,** and the **posterior lobe.** The flocculonodular lobe is the phylogenetically oldest portion of the cerebellum. It is composed of two parts connected by a thin stalk: the **flocculus** and the **nodulus** (itself a portion of the vermis in the midline). The remaining portion of the vermis and the cerebellar hemispheres are termed the **corpus cerebelli.** The medial-most portions of the cerebellar hemispheres are defined as the **intermediate zones.** The primary fissure divides the remaining functional areas of the cerebellum into the posterior and anterior lobes.

Information pertaining to the movement and relative position of the head is relayed primarily to the vestibular nuclei. Connections also occur, however, between the flocculonodular lobe and posterior portion of the vermis, allowing for cerebellar mediation of eye movements and movements related to balance. The cerebellar cortex receives constant information from the skin, joints, and muscles of the limbs and trunk through the **dorsal, ventral,** and **rostral spinocerebellar** tracts and through the **cuneocerebellar tract.** All of this information is integrated with input from the auditory, vestibular, and visual sensory systems in order to determine the progress of motor movements within the cerebellum. The spinocerebellar tracts are **somatotopically organized** and provide the physiological basis for the cerebellar somatotopic organization.

The pontine nuclei provide the cerebellum with its greatest number of afferent connections via the **pontocerebellar tract.** These nuclei function as mediators of information between the cerebellum and cerebrum. The cerebrum sends information to the pontine nuclei via the **corticopontine tract.** The majority of the fibers of the corticopontine tract begin in the motor cortex and somatosensory cortex, with significant connections also coming from the premotor area, supplemental motor area, posterior parietal cortex, and prefrontal and visual cortices. These connections allow the cerebellum to evaluate and coordinate motor movements. In addition, the pontine nuclei receive cortical projections from the cingulate gyrus and hypothalamus, providing the physiological basis for the influence of emotion on motor movements. The pathways involving the cerebellar cortex, pontine nuclei, and cerebellum are referred to as the **corticopontocerebellar pathway.**

The medial portion of the cerebellum, the **vermis,** is characterized by afferents from the spinal cord. The lateral portions of the cerebellar lobes receive their afferents mainly from the cerebrum. The area between the vermis and the lateral portions of the cerebellar lobes, termed **the intermediate zone,** receives afferents from both the spinal cord and the cerebral cortex.

The cerebellar cortex is made up of three layers: the molecular layer, the Purkinje cell layer, and the granular layer. The molecular layer possesses few cell bodies and is primarily composed of axons and dendrites whose cell bodies lie in deeper layers. The Purkinje cell layer, as its name suggests, consists of mostly Purkinje cells, which are arranged in a single layer. The dendrites of the Purkinje cells are very dense and receive numerous synaptic connections with the parallel fibers (around 200,000 per cell). The Purkinje cells are GABAergic and thus inhibitory. The Purkinje cells are also the only cells to project out of the cerebellar cortex. The granular layer is the deepest layer of the cerebellar cortex. The granule cells project their axons to the molecular layer, where they split in two directions and run parallel with the cortical surface. These cells are termed parallel fibers and synapse with many Purkinje cells. The granule cells release glutamate and create an excitatory effect on the Purkinje cells.

The cerebellar cortex receives two forms of afferent connections: **mossy fibers** and **climbing fibers.** Because each mossy fiber forms synapses with many granule cells which then synapse with many Purkinje cells, the mossy fibers have the ability to excite many Purkinje cells. The effect on a Purkinje cell from a mossy fiber is, however, relatively weak. Mossy fibers originate from almost all nuclei except the inferior olive and generally fire at a high frequency of about 50–100 synapses per second. The fibers from the inferior olive consist of climbing fibers, which project to the molecular layer. As the climbing fibers travel, each one forms many synapses with the dendrites of the Purkinje cells it "climbs" alongside. Whereas each Purkinje cell receives synapses from only one climbing fiber, one climbing fiber may synapse with many Purkinje cells, allowing a single action potential from a climbing fiber to create an excitatory effect in numerous Purkinje cells. The frequency of climbing fiber firing is very low, normally firing less than once per second. These different forms of afferent connections allow the cerebellar cortex to receive very precise signals. The mossy fibers mainly convey information of the force, velocity, direction, and individual muscles involved in motor movements. Climbing fibers provide "error signals" regarding motor movement and may also be involved in certain aspects of motor learning.

The majority of cerebellar efferents arise from the cerebellar nuclei. Four different bodies constitute the cerebellar nuclei: **the fastigial nucleus, dentate nucleus, globose nucleus**, and **emboliform nucleus.** They are located within the deep white matter in each hemisphere. The cerebellar hemispheres exert their effect on the ipsilateral portion of the body. A cerebellar lesion will therefore clinically manifest symptoms on the same side of the body as the lesion. The cerebellar hemispheres send their cortical efferents mainly to the motor cortex, with some minor connections also reaching the supplemental premotor area and lateral premotor area. These minor connections are the basis of the cerebellar effect on cognition. The efferents from the dentate nucleus cross the midline after exiting the cerebellar peduncles ending, primarily, in the **ventrolateral nucleus of the thalamus.** The inferior and middle cerebellar peduncles convey mainly inputs to the cerebellum, and the superior cerebellar peduncle conveys the outputs of the cerebellum. The efferent output from the intermediate zone projects to the interposed nuclei, where it influences motor neurons through either the **rubrospinal** or **pyramidal tracts.** The **fastigial nucleus** sends fibers to the **reticular formation** and **vestibular nuclei,** allowing cerebellar control of motor neurons via **reticulospinal** and **vestibulospinal tracts.** These tracts facilitate cerebellar influence on posture and autonomic movements.

Damage to different areas of the cerebellum (Figure 26-1) creates characteristic symptoms. A lesion of the flocculonodular lobe or midline will usually result in problems with stance and gait, titubation, head posture, and ocular-motor disorders resulting in nystagmus. Anterior lobe lesions typically result in **gait ataxia.** Disease infiltrating the neocerebellum often presents with ataxia involving voluntary movements. Lesions of the lateral hemispheres can also result in **asynergia, gait ataxia,** hypotonia, dysarthria and **speech ataxia, dysmetria, dysdiadochokinesia, asynergia**

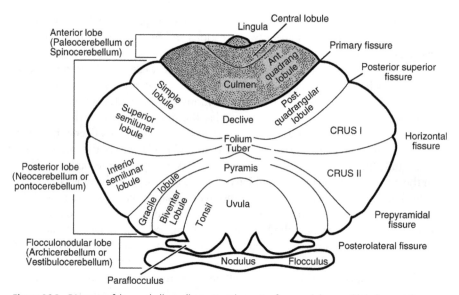

Figure 26-1. Diagram of the cerebellum, illustrating the major fissures, lobes, and lobules and the major phylogenetic divisions. (*With permission from* Adam and Victor's Principles of Neurology. *7th ed. New York: McGraw-Hill; 2000. Page 87, figure 5-1.*)

or the decomposition of complex movements, and **intentional tremors** as well as **static tremors.**

CASE CORRELATES

- See Cases 24-27 (movement systems), and Case 18 (ataxia).

COMPREHENSION QUESTIONS

Refer to the following case scenario to answer questions 26.1-26.2:

A 4-year-old girl is brought into your clinic by her parents because of headaches, vomiting, and lethargy for 3 weeks. She is also having difficulty using silverware with her right hand. She has no problem performing finger-to-nose pointing with her left side, but consistently past points on the right and her movements on that side are very coarse. On funduscopic examination you note bilateral papilledema and order an MRI. It shows a tumor in the posterior fossa.

26.1 Based on her symptoms, which is the most likely location of this tumor?

    A. Left cerebellar hemisphere

    B. Right cerebellar hemisphere

    C. Cerebellar midline

    D. Within the fourth ventricle

26.2 Through which cerebellar nucleus is the outflow from the lateral cerebellar hemispheres mediated?

A. Dentate

B. Globose

C. Fastigal

D. Emboliform

26.3 A 35-year-old concert pianist is involved in an auto accident in which he sustains a head injury, resulting in damage to the left side of his cerebellum among other injuries. Following a lengthy hospital stay and recovery period, he discovers that he is no longer able to play the piano with the same ease and grace that he did prior to the accident. Even after extensive practice and retraining, he is still unable to play at the same level as before. Which of the following inputs to the cerebellum is thought to play a key role in motor learning?

A. Excitatory input from mossy fibers

B. Inhibitory input from mossy fibers

C. Excitatory input from climbing fibers

D. Inhibitory input from climbing fibers

26.4 A 25-year-old woman with a 12-month history of headaches and a 3-month history of "clumsiness walking" is being evaluated. An MRI scan shows that the cerebellar tonsils are in a low position and impinging into the vertebral canal. The patient asks how this condition developed. You explain it is most likely due to:

A. Congenital abnormality

B. Autoimmune disease

C. Genetic disorder

D. Traumatic injury

E. Brain tumor

F. Vascular disorder

ANSWERS

26.1 **B.** This patient has **right-sided cerebellar** symptoms: difficulty with fine movements and past pointing on finger-to-nose testing. Since the cerebellar hemispheres affect movement in the limbs, and the cerebellum acts ipsilaterally, the defect must be in the right cerebellar hemisphere. A left-side lesion would cause left-sided symptoms, and a lesion in the midline would cause truncal and gait ataxia. The rest of this girl's symptoms are explained by mass effect from the tumor causing increased intracranial pressure (ICP), which causes headaches, vomiting, lethargy, and papilledema.

26.2 **A.** The primary outflow from the cerebellum is through the deep nuclei (dentate, emboliform, globose, and fastigial). The lateral cerebellar hemispheres project to the **dentate nucleus**; the intermediate zone of the cerebellar cortex projects to the globose and emboliform nuclei, which are also known as the interposed nuclei; and the cerebellar vermis projects to the fastigial nucleus. From the dentate nucleus, cerebellar outflow projects through the superior cerebellar peduncle to the red nucleus and the ventral anterior and ventrolateral nuclei of the thalamus.

26.3 **C.** It is thought that the firing of **excitatory climbing fibers** represents some sort of error signal to the cerebellum, and is integral in the process of motor learning. The two types of input to the cerebellum are mossy fibers and climbing fibers, both of which are excitatory. Climbing fibers originate in the inferior olivary nucleus, enter the cerebellum via the inferior cerebellar peduncle, and synapse directly with Purkinje cells with a rather strong excitatory effect. Mossy fibers affect Purkinje cells less directly and, therefore have a relatively weak effect. Mossy fibers synapse with granule cells, which send out parallel fibers that interact with Purkinje cells.

26.4 **A.** This cerebellar description is likely the Chiari malformation, which is a **congenital underdevelopment** of the posterior fossa. The underdeveloped fossa then causes the cerebellum to herniated through the foramen magnum. Chiari type I can be asymptomatic in childhood and manifests as headaches and ataxia in the adult. Chiari type II is more severe and symptomatic at birth and is associated with other CNS anomalies such as hydrocephalus and lumbosacral meningomyelocele.

NEUROSCIENCE PEARLS

▶ Alcoholic cerebellar degeneration is caused by toxic degeneration of Purkinje cells and is clinically characterized by impaired gait, tremor, and truncal ataxia.

▶ In a patient with chronic thiamine deficiency due to alcoholism, thiamine should be given BEFORE glucose to avoid precipitation of Wernicke encephalopathy.

▶ A child with ataxia-telangeictasia has ataxia, prominent vasculature on the skin (telangiectasia), pulmonary infections, and increased risk of cancer.

▶ The symptoms of cerebellar lesions manifest on the same side of the body as the lesion.

▶ The Purkinje cells mediate all of the efferent projections from the cerebellum.

▶ Afferent input to the cerebellum is provided by the mossy fibers from spinal cord and brainstem nuclei and from climbing fibers from the inferior olive.

REFERENCES

Brodal P. The cerebellum. *The Central Nervous System: Structure and Function.* 3rd ed. New York, NY: Oxford University Press; 2004.

Ghez C, Thach WT. The cerebellum. In: Kandel ER, Schwarz JH, Jessell TM, Siegelbaum SA, Hudspeth AJ, eds. *Principles of Neural Science.* 5th ed. New York, NY: McGraw-Hill; 2012.

Ropper AH, Brown RH. Incoordination and other disorders of cerebellar function. *Adams and Victor's Principles of Neurology.* 8th ed. New York, NY: McGraw-Hill; 2005.

A 6-month-old female infant is brought to the pediatrician by her mother. The mother states she has noticed the child moving her eyes back and forth for no apparent reason over the past several weeks and as a result is concerned that her daughter may have vision problems. The patient's chart shows that she had an uncomplicated pregnancy and delivery and has been meeting her developmental milestones. On examination the pediatrician notes involuntary, rhythmic, horizontal eye movements. The remainder of the examination is normal. Based on the patient's history, the physician decides that the patient's disorder is most likely congenital.

► What are the eye movements called?
► Are there any treatment options available?

ANSWERS TO CASE 27:

Eye Movements

Summary: A 6-month-old female infant presents with several weeks of abnormal eye movements. Physical examination demonstrates horizontal nystagmus and is otherwise normal.

- **Eye movements: Nystagmus,** which are involuntary, rhythmic, horizontal eye movements.

- **Treatment options:** In most cases there is no treatment for congenital nystagmus. Some patients may, however, qualify for surgical intervention intended to mitigate the effect the nystagmus has on visual acuity.

CLINICAL CORRELATION

Congenital nystagmus is the most common form of nystagmus. Typically, the patient's visual acuity is only mildly affected, and the patient has little awareness of the movements. The etiology of congenital nystagmus is unknown, although it may be caused by a disruption in the nuclei which influence eye movements. Nystagmus may also be acquired, presenting after neurological trauma, ischemic episodes, or cerebrovascular accidents.

APPROACH TO:

Eye Movements

OBJECTIVES

1. Know the muscles that move the eye and the types of movements they control.

2. Describe the various types of eye movements.

3. Know the centers for eye movement control found in the brainstem, cerebellum, and cortex.

DEFINITIONS

CONVERGENCE: The medial rectus muscles contract to move both eyes toward the midline in order to keep the image in each eye focused.

SACCADES: Rapid conjugate eye movements that move the visual axis from one point of fixation to another.

SMOOTH-PURSUIT MOVEMENTS: Slow conjugate eye movements that track a moving object and keep it on the fovea.

OPTOKINETIC REFLEX MOVEMENTS: Stabilizing movements which occur after a retinal slip, when the entire visual field moves with respect to the head.

VESTIBULAR-OCULAR REFLEX: Eye movements that occur in response to vestibular signals from movement of the head that move the eyes in an equal and opposite direction in order to keep a stationary image.

VERGENCE: Converging or diverging the eyes to keep both foveae aligned to a target that moves closer or farther away.

DISCUSSION

Several different types of movements are used to provide the delicate control of the eyes necessary to keep the image from a viewed object placed on the macula. The macula lies near the center of the retina and has the fovea at its center. This is the area of the retina with the highest visual acuity.

The extraocular muscles facilitate three basic movements: horizontal eye movements, vertical eye movements, and rotary eye movements. Horizontal eye movements are principally controlled by the **medial rectus** and **lateral rectus** muscles. Vertical eye movements are mainly mediated by the **inferior and superior rectus** muscles. Rotary eye movements are primarily controlled by the **inferior and superior oblique** muscles. These muscles are innervated by cranial nerves leaving the brainstem. The **abducens nerve** or sixth cranial nerve innervates the lateral rectus muscle for abduction of the eye. The **trochlear nerve** or fourth cranial nerve innervates the superior oblique muscle, pulling the eye down and out. The remainder of the eye muscles are innervated by the **oculomotor nerve,** or third cranial nerve.

All normal eye movements are synchronized, allowing both eyes to focus on the same image. To achieve this level of control, a delicate balance of excitatory and inhibitory stimuli must be sent to the synergistic and antagonistic muscle groups. The varying degrees of tension between the extraocular muscles permits smooth, coordinated movement. Damage to the extraocular muscles or the neurological mechanisms which control their movement can lead to a deficit between the coordination of the eyes, resulting in **diplopia,** or double vision.

There are five types of eye movements that interact in order to place a viewed object on the fovea for best visual resolution and to keep it there as either the object or the observer moves. These five eye movements may be either voluntarily or reflexively controlled and include the following: saccade movements, smooth-pursuit movements, vergence movements, vestibuloocular movements, and optokinetic movements.

Saccades are rapid, conjugate movements that place the desired image on the fovea. These movements are rapid and accurate to allow for object fixation. Saccades allow the entire visual field to be assessed, compensating for the decrease in visual acuity that occurs as an image moves away from the fovea. There are several types of saccades. **Volitional saccades** direct the gaze to either a remembered location or a location where a target will likely appear. A **reflexive saccade** can be stimulated by a nonvisual stimulus such as a sound. The **frontal eye fields** in the precentral sulcus initiate volitional and reflexive saccades. The **parietal eye fields** in the posterior parietal cortex mediate visually guided saccades of both volitional and reflexive types. There are, however, interconnections between the frontal and parietal eye fields, with each influencing the other. Input from the cerebral cortex travels to the contralateral **paramedian pontine reticular formation (PPRF).** These inputs come directly from the frontal eye fields, and come by way of the **superior colliculus** from the parietal eye fields.

During **horizontal saccades,** the PPRF sends signals to the abducens nucleus of cranial nerve VI to activate both abducens neurons and interneurons. These signals then travel to the lateral rectus muscle and through the interneurons to the **medial longitudinal fasciculus** in order to signal the contralateral **oculomotor nucleus** of cranial nerve III for coordinated movement of the medial rectus muscle. The **nucleus prepositus hypoglossi** projects information to the abducens nucleus about the current position of the head and eyes in order to hold the eyes on target until the end of the saccade.

Vertical saccades also use the PPRF pathway; however, impulses from the PPRF relay through eye movement centers in the midbrain before reaching the motor nuclei. The tegmentum of the rostral midbrain contains the **rostral interstitial nucleus** of the **medial longitudinal fasciculus** and the **interstitial nucleus of Cajal.** These nuclei regulate vertical and torsional saccades through connections to the oculomotor and trochlear motor nuclei.

Large, acute lesions of the cerebral hemispheres can affect saccade movements in the direction contralateral to the side of the lesion. The eyes will often deviate toward the side of the lesion. These deficits are usually temporary because of the ability of other pathways to compensate. Lesions of the pons will result in similar deficits, but affect eye movements toward the ipsilateral side of the lesion. A cerebellar lesion can cause small dyssymmetric saccades and difficulty in maintaining off-center gaze.

Smooth-pursuit movements allow an object to be tracked as it moves slowly through a visual field. These movements keep the desired image on the retina. The movements themselves are involuntarily; however, they do require the observer's attention to be focused on the object. If pursuit movements fail, saccades are needed to catch up with the target. Visual inputs to the temporo-occipital junction carry information about the speed and direction of movement of a target and initiate pursuit movements. Information is projected via corticopontine fibers to the ipsilateral **dorsolateral pontine nuclei** and from there to the cerebellar vermis, flocculus, **vestibular nuclei,** and **nucleus prepositus hypoglossi.** From here the contralateral abducens nucleus and the medial rectus subnucleus of the oculomotor nerve are activated via the medial longitudinal fasciculus.

Vestibulo-ocular movements are reflexive movements of the eye in the opposite direction of a head movement, preventing an object from moving away from the fovea as the head moves. These eye movements occur with the same velocity as the head, allowing the image to remain stable on the retina. These movements can occur along any visual axis. The pathway for these movements begins with signals from the **semicircular canals** that travel to the **vestibular nuclei.** The fibers of the vestibular nerve enter the medulla oblongata and pass between the inferior peduncle and the spinal tract of the trigeminal. They then divide into ascending and descending fibers. The latter end by arborizing around the cells of the medial nucleus. The ascending fibers either end in the same manner or in the lateral nucleus.

Some of the axons of the cells of the lateral nucleus, and possibly also of the medial nucleus, are continued upward through the inferior peduncle to the roof nuclei of the opposite side of the cerebellum, to which also other fibers of the vestibular root are prolonged without interruption in the nuclei of the medulla oblongata. A second set of fibers from the medial and lateral nuclei end partly in the tegmentum, while the remainder ascends in the medial longitudinal fasciculus to arborize around the cells of the nuclei of the oculomotor nerve.

Vergence movements allow an image to remain in focus when the object's depth is changing relative to the observer. They occur together with pupillary contraction and accommodation and change the visual axis of the eyes relative to each other through either convergence or divergence. The activation of the vergence system results from a blurred image or from an image falling on non-corresponding retinal areas. Signals from the posterior temporal and prefrontal cortex project to vergence cells in the brainstem and from there to cranial nerves III and VI.

Optokinetic movements occur in response to retinal slip during prolonged head movement at a constant velocity. Initially, the vestibulo-ocular system will produce compensatory eye movements in response to head acceleration, but these movements will fade as the motion of the endolymph in the semicircular canals reaches equilibrium and the vestibular input ceases. The optokinetic movements will then begin to compensate for retinal slip. The **indirect optokinetic pathway** consists of signals between the temporo-occipital cortex and the **accessory optic system nuclei.** The **direct optokinetic pathway** consists of input from retinal ganglion cell axons to the **nucleus of the optic tract** and **nuclei of the accessory optic system,** which then project to the cerebellum and vestibular nuclei.

CASE CORRELATES
- See Case 24 (movement control), Case 25 (basal ganglia), and Case 26 (cerebellum).

COMPREHENSION QUESTIONS

27.1 A 29-year-old woman comes into the emergency department after falling off a ladder. She did not lose consciousness at the time, and does not recall hitting her head as she fell. Nonetheless, you do a screening neurological examination to make sure she does not have any gross deficits. As part of this examination, you ask her to follow your finger with her eyes as you draw an invisible H in front of her. In addition to testing cranial nerve III, IV, and VI, and the integrity of the extraocular muscles, what type of eye movement is this testing?

A. Vergence

B. Visual saccades

C. Vestibulo-ocular movements

D. Smooth pursuit

27.2 In addition to having your patient follow as you trace an invisible H in the air, you ask her to keep looking at your fingers as you move them closer to her face in the midline. Her eyes appropriately cross as your finger approaches her nose. What is the stimulus for this type of eye movement?

A. Activation of the frontal eye fields

B. Blurred images perceived by the visual cortex

C. Angular motion detected in the semicircular canals

D. Movement detected by accessory optic nuclei

27.3 A 43-year-old woman comes into your office with the complaint of inability to look to the left. This problem has been going on for several months, with increasing frequency. She is now totally incapable of looking at objects on her left side without turning her head. On examination, she is unable to voluntarily direct either of her eyes toward the left side. If she follows a finger with her eyes, however, beginning in the right visual field, she can track the object as it crosses midline all the way to the lateral extent of normal ocular movement. She finds this quite distressing. Which of the following structures is most likely damaged resulting in this woman's symptoms?

A. Right frontal eye field

B. Left frontal eye field

C. Right PPRF

D. Left rostral interstitial nucleus of the MLF

ANSWERS

27.1 **D.** Tracking a moving object with the eyes with a stationary head is known as **smooth pursuit.** This seemingly simple action is in fact quite complicated, requiring the appropriate interaction of numerous levels of the brain and spinal cord. Smooth pursuit movements are initiated in the visual association areas of the parieto-occipital cortex, which transmit information about the speed and direction of object motion to dorsolateral pontine nuclei. These in turn project to the cerebellum (particularly the flocculus) and the vestibular nuclei, which project to the abducens and oculomotor nuclei for final control of the extraocular muscles.

27.2 **B.** The type of eye movement described is called vergence, and it is initiated by **blurred images perceived by the visual cortex** or by images falling on noncorresponding parts of the retinas. The exact pathway responsible for vergence has not been elucidated, but it is thought to involve the visual association cortex sending projections to vergence nuclei in the midbrain which then project to the oculomotor nuclei to control the final eye movements. Frontal eye fields initiate voluntary saccades, angular motion is the stimulus for vestibulo-ocular reflexes, and the accessory optic nuclei control optokinetic movements.

27.3 **A.** This woman has a defect in voluntary saccadic movements toward left, which are initiated by the **right frontal eye field.** That she can track an object past midline indicates that the nuclei responsible for ocular movement and the extraocular muscles are all intact and that the defect is at a higher level. The frontal eye field projects to the contralateral PPRF, directly and via the parietal eye field and the superior colliculus. The PPRF then sends signal to the ipsilateral abducens nucleus, which control the lateral rectus and sends impulses to the contralateral oculomotor nucleus to control the medial rectus. This causes both eyes to deviate away from the frontal eye field initiating the movement. The right PPRF is involved in voluntary saccades toward the right side, and the rostral interstitial nucleus of the MLF is involved in vertical saccades.

NEUROSCIENCE PEARLS

▶ Six muscles control the movement of the eye within the orbit: the inferior rectus, superior rectus, medial rectus, lateral rectus, inferior oblique, and superior oblique.

▶ The muscle groups are innervated by the third, fourth, and sixth cranial nerves.

▶ Damage to the extraocular muscles, or to the neurological mechanisms which control their movement, can lead to a deficit between the coordination of the eyes, resulting in diplopia, or double vision.

▶ The five types of voluntary and reflexive eye movements are saccades, smooth-pursuit movements, optokinetic reflexive movements, vergence movements, and vestibulo-ocular movements.

REFERENCES

Brodal P. Eye movements. *The Central Nervous System: Structure and Function.* 3rd ed. New York, NY: Oxford University Press; 2004.

Haines DE. Visual motor systems. *Fundamental Neuroscience.* 2nd ed. Philadelphia, PA: Churchill Livingstone; 2002.

Martin JH. The olfactory system. *Neuroanatomy: Text and Atlas.* 2nd ed. Stamford, CT: Appleton & Lange; 1996.

Ropper AH, Brown RH. Disorders of ocular movement and pupillary function. *Adams and Victor's Principles of Neurology.* 9th ed . New York, NY: McGraw-Hill; 2009.

A 25-year-old African American woman presented to the neurosurgery clinic with an 8-month history of amenorrhea and bitemporal hemianopia. Following laboratory and imaging studies, she was diagnosed with a prolactinoma. After failing bromocriptine therapy for her disease, she elected to undergo transsphenoidal surgical resection of her tumor. Her operation went well, and she was doing fine postoperatively until she developed insatiable thirst and began voiding large amounts of urine. Based on her history and symptoms, the diagnosis of neurogenic diabetes insipidus (DI) is made.

▶ What are the laboratory findings associated with neurogenic DI?
▶ What are the other causes of neurogenic DI?

ANSWERS TO CASE 28:

Regulatory Functions of the Hypothalamus

Summary: A 25-year-old woman with a pituitary prolactinoma has polydipsia and polyuria following successful transsphenoidal resection of her tumor.

- **Laboratory abnormalities:** Urine specific gravity less than 1.005 and urinary output greater than 250 mL/h.

- **Other causes of neurogenic DI:** Head trauma, meningitis, encephalitis, autoimmune disease, familial, idiopathic, neoplastic, and following intracranial procedures.

CLINICAL CORRELATION

The hypothalamus is responsible for various aspects of homeostasis, one of which is water balance. Vasopressin, or antidiuretic hormone (ADH), is produced in the paraventricular and supraoptic nuclei of the hypothalamus. The neuron cell bodies in these nuclei have axons that descend into the median eminence, pituitary stalk, and then into the posterior lobe of the pituitary gland where ADH is stored. Plasma osmolality changes stimulate vasopressin release directly from the supraoptic and paraventricular nuclei and indirectly from osmoreceptors in other hypothalamic nuclei. Volume-sensitive pathways can also regulate vasopressin release. Baroreceptors and mechanoreceptors in the aortic arch, carotid sinus, and right atrium can signal for vasopressin release in times of volume depletion.

Vasopressin stimulates aquaporin channels to open in the luminal membranes of the cortical and medullary collecting tubules of the kidney to promote water reabsorption. Treatment of DI consists of administering synthetic vasopressin (eg, DDAVP) to compensate for low levels of vasopressin.

APPROACH TO:

Regulatory Functions of the Hypothalamus

OBJECTIVES

1. Know the names and principal functions of the substances formed by the hypothalamus.

2. Know the principal functions of each hypothalamic nucleus.

3. Differentiate between DI, Syndrome of Inappropriate secretion of ADH (SIADH), and cerebral salt wasting.

DEFINITIONS

THYROTROPIN-RELEASING HORMONE (TRH): Stimulates release of thyroid-stimulating hormone (TSH) from the anterior pituitary.

GROWTH HORMONE–RELEASING HORMONE (GHRH): Stimulates growth hormone release from the anterior pituitary.

CORTICOTROPIN-RELEASING HORMONE (CRH): Stimulates adrenocorticotropic hormone release from the anterior pituitary.

GONADOTROPIN-RELEASING HORMONE (GnRH): Stimulates luteinizing and follicle-stimulating hormone release from the anterior pituitary.

PROLACTIN-RELEASING HORMONE: Stimulates prolactin release from the anterior pituitary.

OXYTOCIN: Produced by the paraventricular and supraoptic nuclei and migrates into the posterior pituitary.

DISCUSSION

The hypothalamus is a bilateral structure that resides on the sides and floor of the third ventricle. It has three general functions: **autonomic** regulation, **endocrine** regulation, and **circadian** regulation. The hypothalamus is comprised of three different zones. The periventricular zone, which borders the third ventricle, has neuroendocrine functions. The medial zone, which is bound laterally by the fornix, functions in autonomic and neuroendocrine control of the enteric system. The lateral zone, which is bound medially by the fornix and laterally by the internal capsule, functions in autonomic and neuroendocrine control of the cardiovascular system. The circuitry of the hypothalamus is very complex, and many of its neuronal connections are bidirectional. It sends and receives information via the blood stream—through the pituitary by way of the hypophyseal-portal system and through circumventricular organs, sites at which the blood-brain barrier is highly permeable and allows passage of chemical stimuli into the brain. The hypothalamus is the key brain site for the integration of multiple biological systems to maintain homeostasis.

These functions are executed by the various nuclei of the hypothalamus (Figure 28-1).

The **paraventricular** and **supraoptic nuclei** have already been discussed. The **anterior nucleus** is responsible for regulating the dissipation of heat from the body and stimulates the parasympathetic nervous system. A lesion will cause hyperthermia. The **posterior nucleus** regulates heat conservation and stimulates the sympathetic nervous system. A lesion will cause hypothermia. A lesion of the **dorsomedial nucleus** will result in "savage behavior." The **ventromedial (VM) nucleus** regulates satiety. A lesion in the VM nuclei will result in obesity and savage behavior. The **preoptic nucleus** is responsible for regulating gonadotropin release. The **suprachiasmatic nucleus** controls circadian rhythms. A lesion will result in disruption of sleep-wake cycles. The **lateral nucleus** regulates the desire to eat. A lesion will result in starvation. The **mamillary body** receives input from the hippocampal formation by way of fibers in the fornix. Hemorrhagic lesions are found in the mamillary nuclei

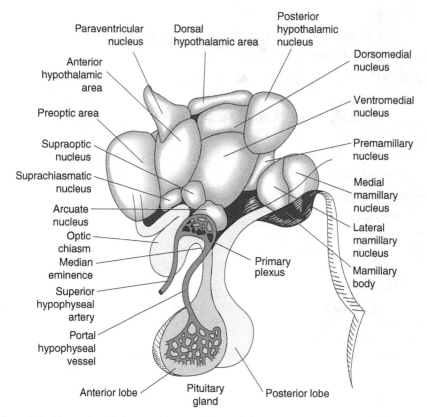

Figure 28-1. Human hypothalamus, with a superimposed diagrammatic representation of the portal hypophyseal vessels. (*With permission from* Ganong's Review of Medical Physiology. *22nd ed. New York: McGraw-Hill; 2005. Page 223, figure 14-2.*)

in Wernicke encephalopathy. The **arcuate nucleus** has axons projected back to the **hypophyseal portal system**, where they release dopamine to inhibit prolactin release from the anterior pituitary gland.

Many authors divide hypothalamic dysfunction into **global** and **partial**. In global dysfunction, many or all of the hypothalamic functions are deranged. Common causes are usually systemic in nature. Sarcoidosis, metastatic cancer, idiopathic inflammatory diseases, and germ cell cancers all may cause global hypothalamic dysfunction. Partial hypothalamic dysfunction results when only one of the hormones is affected, as in the case of DI.

Syndrome of inappropriate secretion of antidiuretic hormone (SIADH) occurs when there is an unnecessary release of ADH. Common triggers include neurosurgical procedures, head trauma, neoplasms, CNS infections, and autoimmune disorders like Guillain-Barré syndrome. Often the initial finding is dilutional hyponatremia without edema in the setting of low plasma osmolality and high urine specific gravity (>300 mOsm/L). Treatment is simple: free water restriction or, in the case of seizures secondary to hyponatremia, 3% sodium chloride with sodium correction no greater than 12 mEq/L in the first 24 hours and no greater than 20 mEq/L in 48 hours. This avoids the debilitating sequela of central pontine demyelinolysis. Another common cause of hyponatremia in neurosurgical patients is **cerebral salt wasting**. In SIADH, in which the mechanism is excessive water that dilutes the sodium in the serum; in contrast, in cerebral wasting, there is a volume contraction because of salt loss in the urine. Thus, patients with cerebral salt wasting will have a suboptimal fluid balance and high levels of urine sodium, which is the opposite of SIADH patients. Thus, it is crucial to assess a patient's volume status and urine sodium before instituting treatment. For example, fluid restriction in a patient with presumed SIADH could have detrimental effects on the renal, cardiovascular, and neurological systems of a patient who is actually volume-depleted in the setting of cerebral salt wasting.

Diabetes insipidus is a disease characterized by excretion of large amounts of highly diluted urine, which cannot be controlled by reducing fluid intake. It denotes inability of the kidney to concentrate urine. **Diabetes insipidus** is caused by a deficiency of vasopressin (also known as ADH), or by an insensitivity of the kidneys to that hormone. The regulation of urine production occurs in the hypothalamus, which produces vasopressin in the supraoptic and paraventricular nuclei. After synthesis, the hormone is transported in neurosecretory granules down the axon of the hypothalamic neuron to the posterior lobe of the pituitary gland, where it is stored for later release. In addition, the hypothalamus regulates the sensation of thirst in the ventromedial nucleus by sensing increases in plasma osmolality and relaying this information to the cortex. The main effector organ for fluid homeostasis is the kidney. Antidiuretic hormone acts by increasing water permeability in the collecting ducts. Specifically, it acts on proteins called aquaporins which open to allow water into the collecting duct cells. This increase in permeability allows for reabsorption of water into the bloodstream, thus concentrating the urine.

CASE CORRELATES

- See Cases 28-36 (regulatory systems).

COMPREHENSION QUESTIONS

28.1 A 32-year-old man presents to your office with a complaint of sleep disturbance. He works on a rotating shift at a chemical plant and changes from days to evenings to nights every week. He states that his problem is that he often cannot fall asleep when he needs to, and is frequently very tired at work. You diagnose him with circadian rhythm sleep disorder and discuss treatment options with him. Which hypothalamic nucleus is normally responsible for controlling the circadian rhythm?

   A. Supraoptic nucleus

   B. Posterior nucleus

   C. Suprachiasmatic nucleus

   D. Lateral nucleus

28.2 A morbidly obese patient comes into your office for an annual checkup. As a responsible physician, you inform this patient of the risks associated with excess body fat and discuss some methods for losing weight. The patient states that she has tried "more diets than I can count" and that none of them worked. She states that there must be some underlying problem with her body that prevents her from losing weight. If she had a hypothalamic lesion responsible for her excess weight, where would it be?

A. Paraventricular nucleus

B. Lateral nucleus

C. Dorsomedial nucleus

D. Ventromedial nucleus

28.3 A 60-year-old woman comes into your office complaining of fatigue, weight gain, and feeling cold all the time. On further questioning she complains of constipation and states that she hasn't been eating as much as usual, despite the gain in weight. You check her thyroid status and find low thyroxine and low TSH, indicating a defect either in the pituitary or in the hypothalamus. If the problem is in the hypothalamus, which nucleus would it be affecting?

A. Arcuate nucleus

B. Paraventricular nucleus

C. Anterior nucleus

D. Mamillary bodies

ANSWERS

28.1 **C.** The **suprachiasmatic nucleus**, located just superior to the optic chiasm, is responsible for controlling circadian rhythm. The free running time of the cycle generated by the suprachiasmatic nucleus is about 25 hours, but it receives input directly from the retina and therefore alters its generated frequency to match the actual length of the day.

28.2 **D.** The **ventromedial nucleus** is considered the "satiety center" of the hypothalamus, and at least in laboratory animals, ablating this nucleus bilaterally results in animals that eat uncontrollably and become morbidly obese. Because the hypothalamus is so small in humans, lesions of a single nucleus, let alone bilateral lesions of a nucleus, are extremely rare. The lateral nucleus is considered the "feeding center" of the hypothalamus, and bilateral ablation of these nuclei in laboratory animals results in animals that have no interest in eating and will actually starve to death with food accessible.

28.3 **A.** The **arcuate nucleus** of the hypothalamus is responsible for secreting factors that stimulate or inhibit release of pituitary hormones. Among these
factors is TRH (thyrotropin-releasing hormone), which travels through the
hypophyseal portal system to the anterior pituitary where it stimulates thyrotropes to release TSH (thyroid-stimulating hormone). Low TSH in the face of
low levels of thyroid hormone either indicates a defect in the thyrotropes of
the pituitary, in the arcuate nucleus of the hypothalamus, or in the connection
between the two. The paraventricular nucleus sends axons down the pituitary
stalk to the posterior lobe of the pituitary where it releases vasopressin and
oxytocin directly into the bloodstream.

NEUROSCIENCE PEARLS

▶ The hypothalamus is crucial to autonomic, endocrine, and circadian
regulation.

▶ A lesion of the VM nuclei results in obesity and aggressive, savage behavior.

▶ Vasopressin is produced in the paraventricular and supraoptic nuclei of
the hypothalamus.

▶ Baroreceptors and mechanoreceptors in the aortic arch, carotid sinus, and
right atrium signal for vasopressin release in times of volume depletion.

▶ Diabetes insipidus is a disease characterized by excretion of large
amounts of highly diluted urine, which cannot be controlled by reducing
fluid intake.

▶ Common causes of SIADH include neurosurgical procedures, head
trauma, neoplasms, CNS infections, and autoimmune disorders like
Guillain-Barré syndrome.

REFERENCES

Bear MF, Connors BW, Paradiso MA, eds. *Neuroscience: Exploring the Brain.* 3rd ed. Baltimore, MD: Lippincott Williams & Wilkins; 2006.

Purves D, Augustine GJ, Fitzpatrick D, et al, eds. *Neuroscience.* 3rd ed. Sunderland, MA: Sinauer Associates Inc; 2004.

Squire LR, Berg D, Bloom FE, du Lac S, eds. *Fundamental Neuroscience.* 4th ed. San Diego, CA: Academic Press; 2012.

A 51-year-old previously healthy man visits his primary care physician with the complaint of changes in the appearance of his face. Specifically, he is unhappy with recent increases in the size and shape of both his nose and ears. Later, in his presentation he states that he recently had to have his wedding ring removed because it was compressing his finger to the point of causing a sensory disturbance. Notably, he states that he has not regained full sensation of several of the fingers in his right hand following removal of the ring. On physical examination, the physician finds a man who appears his age with a distinct facial morphology including a large, bulbous nose, enlarged auricles, and a protruding jaw. His hands are disproportionately large with sausage-shaped fingers. On neurological examination, the only abnormal finding is decreased vision in both temporal fields. Based on these findings, the patient is diagnosed with acromegaly secondary to a growth hormone–releasing neoplasm.

► What is the most likely location of this neoplasm?
► How would the physician confirm his diagnosis?

ANSWERS TO CASE 29:

The Neuroendocrine Axis

Summary: A 51-year-old man is noted to have physical findings of growth hormone (GH) excess, known as acromegaly.

- **Location of neoplasm:** Growth hormone–releasing neoplasms most frequently occur in the anterior pituitary gland. Here, somatotropes are activated by GH-releasing hormone (GHRH) secreted from the hypothalamus. The presence of bitemporal homonymous hemianopsia also confirms that the lesion is in the pituitary gland.

- **Confirmation of diagnosis:** There are a series of diagnostic tests that should be performed on this patient. Insulin-like growth factor 1 (IGF-1) levels seem to be the most sensitive and useful laboratory test for the diagnosis of this condition. An MRI of the head should also be performed to evaluate the extent of tumor growth and compression of structures surrounding the sella turcica. Because assessing random levels of GH does not provide accurate information due to the inconsistent secretion of GH, levels should be measured after administration of a glucose load (normally, GH is suppressed by glucose).

CLINICAL CORRELATION

This patient exhibits the typical features of **acromegaly:** a large bulbous nose, enlarged auricles, a protruding jaw, and sausage-shaped fingers. Because this patient is an adult with fused epiphyseal plates, he does not exhibit features of gigantism that children with GH-releasing neoplasms exhibit. Most commonly, acromegaly is caused by a GH-producing tumor derived from somatotrophic cells in the anterior pituitary gland. The excess GH stimulates the liver to secrete even more IGF-1, the primary mediator of the growth-promoting effects of GH. Research shows that many of these pituitary tumors contain a mutation involving the alpha subunit of a stimulatory guanosine triphosphate–binding protein, which leads to a persistent elevation of cyclic adenosine monophosphate (cAMP) in the somatotrophs, resulting in excessive GH secretion. Diagnosis of this condition is critical as the disease can lead to early death if unchecked. Diagnosis involves measuring IGF-1 levels, an MRI of the brain focusing on the sella turcica, and measuring GH levels after administration of glucose.

APPROACH TO:

The Neuroendocrine Axis

OBJECTIVES

1. Be able to recognize the symptoms which cause acromegaly.
2. Be able to decide on the mode of treatment for the disease.

DEFINITIONS

THYROID-STIMULATING HORMONE (TSH): Released from the anterior pituitary gland and stimulates the thyroid to produce thyroid hormone.

ADRENOCORTICOTROPIC HORMONE (ACTH): Released from the anterior pituitary and stimulates the adrenal gland to produce glucocorticoids.

GH: Released from the anterior pituitary and stimulates the peripheral tissues to release IGF-1.

LUTEINIZING HORMONE (LH) AND FOLLICLE-STIMULATING HORMONE (FSH): Released from the anterior pituitary gland and stimulates estrogen/progesterone production in females and testosterone production in males. Both hormones have roles in gametogenesis.

PROLACTIN-INHIBITING FACTOR (PIF): Most likely dopamine; secreted from the hypothalamus and results in inhibition of prolactin release from the anterior pituitary.

PROLACTIN (PRL): Released from the anterior pituitary gland and stimulates lactation at the breast.

VASOPRESSIN (VP): Also known as ADH. This hormone is created in the hypothalamus and released from the posterior pituitary to cause water reabsorption in the collecting ducts.

DISCUSSION

The hypothalamic nuclei responsible for production and secretion of the releasing hormones include the **periventricular, preoptic, arcuate,** and **paraventricular nuclei.** The axons of neurons that produce the releasing hormones targeted for the anterior pituitary terminate at the base of the hypothalamus in a region of tissue called the **median eminence.** Here, a microvascular network of capillaries and veins formed by penetrating branches of the **superior hypophyseal artery** forms a portal system for hormone delivery to the anterior pituitary. The portal system carries the releasing hormones from the **tuberoinfundibular** system and bathes them around the cells of the anterior pituitary. The cell bodies that produce **VP** and **oxytocin** project their axons through the median eminence and into the posterior pituitary, where they terminate around capillaries formed by the inferior hypophyseal artery. The anterior and posterior lobes of the pituitary subsequently secrete stimulating hormones that act on various end organs.

Interestingly, the secretion of both releasing and stimulating hormones occurs in a pulsatile fashion. Disruption of this rhythmicity can lead to corruption of the regulatory functions of the neuroendocrine axis. Although there are multiple levels of internal regulation within the neuroendocrine axis, the principal factor governing hormone secretion is negative feedback inhibition by end-organ hormones on the hypothalamic-releasing factors.

TRH and TSH

Neurons in the **paraventricular nucleus** of the hypothalamus are responsible for production of TRH. TRH stimulates TSH secretion, which results in thyroid hormone production within the thyroid gland itself. Thyroid hormone affects protein synthesis and metabolic activity in all organ systems. Thyroid hormone is essential for fetal and neonatal development. Hypothyroidism can cause devastating central nervous system developmental abnormalities during the first 3 months of fetal development. Thyroid hormone circulates the body and will negatively inhibit both TRH and TSH release by acting on both the hypothalamus and thyroid gland.

CRH and ACTH

This subsection of the neuroendocrine axis is intimately involved in the stress response as well as establishing basal levels of cortisol. CRH is synthesized in the paraventricular nucleus and descends into the anterior pituitary through the portal system, where it stimulates ACTH release. ACTH will enter the systemic circulation and bind to cells in the cortical layer of the adrenal gland. This binding will activate glucocorticoid production. Although this process can occur in response to stresses such as sepsis, there is a rhythmic release of these hormones resulting in peak cortisol levels just before waking and lowest levels before midnight. Similarly to thyroid hormone, cortisol circulates the body and negatively inhibits both the hypothalamus and pituitary gland. Interestingly, acute stress phases result in improved memory and learning while chronically elevated cortisol levels result in poorer hippocampal function.

GHRH and GH

The hypothalamus produces GHRH and **somatostatin**, which stimulate and inhibit GH release, respectively. The mode of action and regulation of GH differ from TSH and ACTH in several ways. Firstly, GH lacks a specific organ target. It acts on a variety of tissues, stimulating IGF-1 production, and both of these hormones stimulate anabolic drive. Secondly, GH negatively inhibits its own secretion at the hypothalamus only, rather than at both the hypothalamus and the anterior pituitary. However, IGF-1 does negatively inhibit both the hypothalamus and pituitary.

GnRH and FSH/LH

Gonadotropin-releasing hormone (GnRH) is produced diffusely throughout the hypothalamus and travels to the anterior pituitary along the hypophyseal portal system. In males, LH stimulates Leydig cells to produce testosterone. Testosterone and FSH act in concert to stimulate spermatogenesis. Testosterone also results in the development of secondary sexual characteristics in males. In females, FSH stimulates ovarian follicle formation. Both FSH and LH control the timing of ovulation as well as estrogen and progesterone production. Analogous to males, these sex steroids are responsible for the secondary sexual characteristics in females.

PIF and PRL

Prolactin (PRL) is different from the other hormones secreted by the anterior pituitary in that it does not have a secreting factor. Instead, it is tonically inhibited by PIF, which is most likely dopamine. Suckling causes decreased levels of PIF, allowing PRL to be secreted and thus stimulating lactation.

Oxytocin

Oxytocin is synthesized in the **paraventricular** and **supraoptic nuclei** of the hypothalamus and released directly into the systemic circulation. Its major functions are stimulation of uterine contractions and assisting the flow of milk during lactation.

Vasopressin

Vasopressin is another hormone produced in the hypothalamus and directly released into the circulation at the posterior pituitary. It has multiple effects, including vasoconstriction and renal regulation of water homeostasis.

Treatment for growth hormone–releasing tumors is usually multimodal, utilizing a combination of medical and surgical therapies. First-line therapy often is surgical resection of the tumor by a neurosurgeon. Medical therapies, including use of a dopamine agonist, somatostatin, and GH receptor antagonists, are indicated.

CASE CORRELATES

* See Cases 28-36 (regulatory systems).

COMPREHENSION QUESTIONS

29.1 A 33-year-old man with ulcerative colitis who is taking chronic steroid therapy for control of his symptoms comes into your office for a routine checkup. You do some blood work, as usual, and find that his steroid level is therapeutic and his cortisol and ACTH levels are undetectable. The suppression of ACTH by exogenously administered steroids is an example of what basic neuroendocrine principle?

   A. Positive feedback
   B. Negative feedback
   C. Stimulated release
   D. Pulsatile release

29.2 A 27-year-old woman comes into your office with the complaint that she has not had a period for 3 months. She also has noticed a whitish discharge from both of her nipples. The first test you perform is a pregnancy test, which is negative. You follow that with a prolactin level, which is very elevated, at 150 ng/mL. You are concerned that this woman has a prolactinoma, a functional tumor of PRL-producing cells of the anterior pituitary. By what mechanism is the secretion of PRL normally controlled?

   A. Release stimulated by dopamine
   B. Release inhibited by dopamine
   C. Release stimulated by GnRH
   D. Release stimulated by CRH

29.3 A 32-year-old woman is being evaluated for infertility. Her husband's sperm count and function was normal, and she has no apparent abnormalities with her ovaries or reproductive tract. On testing, however, it is found that she has very low levels of FSH and LH. Consideration is being given to administration of GnRH to induce FSH/LH secretion from the pituitary. Which of the following is critical to remember in the administration of GnRH for this purpose?

A. It must be administered at the same time each day.

B. It must be administered at an increased dose each day.

C. It must be administered in a pulsatile fashion.

D. It must be administered intracerebrally.

ANSWERS

29.1 **B.** This is an example of **negative feedback**. In the normally functioning neuroendocrine system, the hypothalamus releases CRH, which causes the pituitary to release ACTH, which causes the adrenal glands to secrete cortisol. When the level of cortisol in the blood becomes high enough, it exerts an inhibitory effect on both the hypothalamus and pituitary through a process known as negative feedback. When a patient is taking exogenous steroids, they are not measurable as cortisol but they act in the same way on the hypothalamus and pituitary gland, suppressing CRH and ACTH secretion. Most of the hypothalamus-pituitary-endocrine organ systems work in a similar manner.

29.2 **B.** PRL is tonically **inhibited from being released by dopamine**. For this reason, when patients are placed on antidopaminergic drugs (eg, some antipsychotics) they can develop gynecomastia and nipple discharge. One of the interesting things about PRL is that its regulation is different than that of the other anterior pituitary hormones. The rest are under stimulatory control, released from the adenohypophysis when the releasing factor is present in the hypophyseal portal system.

29.3 **C.** All of the hypothalamic-releasing hormones with the exception of TRH are released in a **pulsatile fashion**. If they are not pulsatile, they have decreased efficacy. Constant, high levels of GnRH actually serve to inhibit pituitary release of FSH and LH, producing the opposite effect that we are looking for in this patient. The normal pulse frequency of GnRH is approximately once every hour, so in order to induce ovulation in this patient, the medication must be administered that often via IV. This is a difficult regimen to adhere to, and for this reason women are often treated with FSH supplements rather than GnRH supplements; it is easier and more convenient and achieves similar results.

NEUROSCIENCE PEARLS

▶ The periventricular, preoptic, arcuate, and paraventricular nuclei of the hypothalamus are responsible for secretion of hormones.

▶ Axons carrying antidiuretic hormone (ADH) and oxytocin travel through the median eminence into the posterior pituitary, and terminate around capillaries from the inferior hypophyseal artery.

▶ The hypothalamus produces somatostatin, which inhibits GH release from the anterior pituitary.

REFERENCES

Bear MF, Connors B, Paradiso M, eds. *Neuroscience: Exploring the Brain.* 3rd ed. Baltimore, MD: Lippincott Williams & Wilkins; 2006.

Kandel ER, Schwarz JH, Jessell TM, Siegelbaum SA, Hudspeth AJ, eds. *Principles of Neural Science.* 5th ed. New York, NY: McGraw-Hill; 2012.

Squire LR, Berg D, Bloom FE, du Lac S, eds. *Fundamental Neuroscience.* 4th ed. San Diego, CA: Academic Press; 2012.

CASE 30

A confused 55-year-old female marathon runner presents to the emergency room complaining of a headache and muscle cramps. Her skin has turned red and is quite wet. After being asked to follow the nurse, the patient stands up, vomits, and subsequently faints. A nearby doctor immediately checks her pulse and temperature to find both elevated, with temperature approaching 39°C. The patient is immediately diagnosed with hyperthermia, and treatment is started.

- What is the normal range for human body temperature?
- What region of the brain regulates body temperature?
- What are the treatment options available to her?

ANSWERS TO CASE 30:

Thermoregulation

Summary: A 55-year-old female marathon runner is complaining of headache and cramps, and experiencing confusion, nausea, elevated pulse, and red skin.

- **Normal body temperature:** 36°C-37°C.

- **Brain region regulating temperature:** Body temperature regulation is controlled primarily by the hypothalamus.

- **Treatment options:** Rehydration with water; rest in cool conditions.

CLINICAL CORRELATION

Hyperthermia is an acute condition that occurs when the body creates more heat than it can dissipate, causing abnormally high body temperatures. Temperatures above 40°C are life threatening, and brain death begins at 41°C. Common signs include confusion, headache, muscle cramps, and nausea. Severe confusion may lead patients to become hostile and act as if intoxicated. High body temperatures will cause excessive perspiration until dehydration occurs. Acute dehydration causes blood pressure to drop significantly, possibly leading to dizziness or even fainting, especially when standing suddenly. To counter the drop in blood pressure, increased heart rate (tachycardia) and respiration rate (tachypnea) occur to increase the oxygen supply to the body. Cutaneous blood vessels dilate to increase heat dissipation, resulting in a red skin color. As heat stroke progresses, blood vessels constrict to help increase blood pressure, then causing a paler, bluish skin color. Eventually, body organs begin to fail until unconsciousness and coma result. Demographically, studies have shown that women and the elderly are at greater risk for heat stroke. Proper treatment of heat stroke includes immediate hospitalization. The patient's body temperature must be lowered quickly. Methods include moving patient to cool areas with fans or air conditioning, removing patient's clothes, and submerging the patient in cold water bath. Additionally, rehydration is critical and is achieved by drinking large amounts of water and isotonic beverages (eg, Gatorade).

APPROACH TO:

Thermoregulation

OBJECTIVES

1. Know the ways and means to prevent hypothermia.

2. Know the temperature-increasing and temperature-decreasing mechanisms.

DEFINITIONS

HOMEOTHERM: An organism, such as a mammal or bird, having a body temperature that is constant and largely independent of the temperature of its surroundings; an endotherm; "warm-blooded."

POIKILOTHERM: An organism, such as a fish or reptile, having a body temperature that varies with the temperature of its surroundings; an ectotherm; "cold-blooded."

PREOPTIC AREA: A region of the brain that is situated immediately below the anterior commissure, above the optic chiasma, on the anterior side of the hypothalamus that regulates certain autonomic activities often with other portions of the hypothalamus.

THERMORECEPTOR: A sensory receptor that responds to heat and cold.

PILOERECTION: Involuntary erection or bristling of hairs owing to a sympathetic reflex usually triggered by cold, shock, or fright, or caused by a sympathomimetic agent.

THYROXINE (T4): An iodine-containing hormone ($C_{15}H_{11}I_4NO_4$), produced by the thyroid gland, that increases the rate of cell metabolism and regulates growth.

DISCUSSION

Thermoregulation is the ability of an organism to maintain its body temperature within certain boundaries, even when the environmental temperature is different. Like all mammals, humans are **homeothermic**, or "warm-blooded." Human core temperature is about 37°C. Many nonmammals, such as reptiles and fish, are **poikilotherms**, or "cold-blooded" organisms.

Thermoregulation is controlled primarily by nervous feedback mechanisms operating through the **hypothalamus**. The thermostatic center is found in the anterior portion of the hypothalamus in the **preoptic area**. This area has a large number of heat-sensitive neurons (warm receptors) and about a third as many cold-sensitive neurons (cold receptors).

There are two general types of **thermoreceptors:** warm receptors and cold receptors. The axons of warm receptors are unmyelinated, slow-conducting C fibers, while the axons of cold receptors are lightly myelinated, faster-conducting Ad fibers. The receptive fields of thermoreceptors are small spots with diameters of approximately 1 mm in glabrous skin and 3-5 mm in hairy skin. About three to four spots are innervated by a single axon. At normal body temperature, both warm and cold spots discharge, but as temperatures increase, cold spots reduce their firing frequency while warm spots increase their firing. Activation of warm-sensitive receptors (at temperatures above 37°C) results in activation of neurons in the paraventricular nucleus and lateral hypothalamus to increase parasympathetic outflow and increase heat dissipation. Warm-sensitive receptors have inhibitory connections to cold-sensitive neurons. Increased discharge in cold-sensitive neurons results in activation of neurons in the **paraventricular nucleus** and the posterior hypothalamus to increase sympathetic outflow in order to generate and conserve heat. Cold-sensitive neurons do not have intrinsic temperature receptors but instead increase their discharge by the decrease in the discharge rate of warm-sensitive neurons.

Unlike the preoptic area, the skin has far more cold receptors than warm receptors. This signifies that peripheral temperature detection is mainly concerned with distinguishing cool and cold rather than detecting warm temperatures, probably in order to prevent **hypothermia**.

The temperature sensory signals from the peripheral thermoreceptors in the skin and mucous membranes and those from the internal central thermoreceptors in the hypothalamic preoptic area are transmitted bilaterally to a specific area in the posterior hypothalamus at the level of the mamillary bodies. Here, the signals are combined to control the heat-conserving and heat-producing mechanisms of the body.

The hypothalamus works together with higher cortical centers to keep core body temperature constant. Responses range from involuntary (mediated by the autonomic nervous system and neurohormones) to semivoluntary and voluntary behavioral responses.

Responses to Cold: Temperature-Increasing Mechanisms

Cold environments cool blood flowing into the skin and stimulate the skin's cold receptors. When blood temperature drops below normal, the neurons stimulate regions in the caudal hypothalamus responsible for mechanisms of heat conservation, initiating a variety of responses to promote heat gain and inhibit heat loss. Involuntary responses, which involve activation of the sympathetic nervous system, include the following:

1. **Vasoconstriction.** Release of norepinephrine from the sympathetic fibers constricts cutaneous blood vessels, thereby reducing blood flow and heat loss to the cold air.

2. **Piloerection.** Contraction of the arrectores pilorum (stimulated by α_1 adrenoreceptors) causes piloerection trapping warm air close to the skin.

3. **Increased heat production.** Sympathetic excitation and shivering allow for increased thermogenesis. Cellular metabolism can be increased via sympathetic stimulation, or by the actions of circulating epinephrine and norepinephrine in the blood. This is called **chemical thermogenesis**. Epinephrine and norepinephrine have the ability to uncouple oxidative phosphorylation, allowing excess foodstuffs to be oxidized, releasing heat energy. The amount of brown fat, a type of fat that contains large numbers of special mitochondria where uncoupled oxidation occurs, is proportional to the amount of chemical thermogenesis in an organism. Infants, who have small amounts of brown fat, are able to use this process to double heat production. However, adults, who have almost no brown fat, are only able to increase heat production by 10%-15% via this process.

Another form of chemical thermogenesis involves increased cellular metabolism caused by thyroxine. Cooling the hypothalamic preoptic area causes an increase in production and secretion of thyrotropin-releasing hormone (TRH) by the hypothalamus. TRH travels through the portal veins to the anterior pituitary, where it stimulates secretion of thyroid-stimulating hormone (TSH). TSH then stimulates the thyroid to release **thyroxine (T4)**, which increases the body's **metabolic rate**.

Shivering is controlled by the primary motor center for shivering, found in the **dorsomedial posterior hypothalamus**. This center is usually excited by cold signals from the skin and spinal cord and inhibited by warm signals from the hypothalamic preoptic area. When body temperatures fall too low, the motor center will transmit signals through bilateral tracts down the brain stem and the lateral columns of the spinal cord, and eventually to the anterior motor neurons, stimulating an increase in the tone of skeletal muscles around the body. Shivering begins when muscle tone reaches a critical level and can increase heat production up to four or five times that of normal.

More voluntary responses include increased physical activity, such as pacing and hand rubbing, and behavioral changes, such as wearing extra clothing and huddling in groups. These voluntary behaviors are activated mainly by the cerebral cortex and limbic system.

Responses to Heat: Temperature-Decreasing Mechanisms

Body temperature rises when the body is exposed to excess heat. The skin's warm receptors and higher blood temperature signal the hypothalamus to inhibit adrenergic activity of the sympathetic nervous system, leading to metabolic rate reduction and cutaneous vasodilation. Responses to heat include the following:

1. **Vasodilation.** Inhibition of the sympathetic centers in the posterior hypothalamus causes cutaneous vasodilation, allowing for increased blood flow and greater heat loss via the skin.

2. **Sweating.** In particularly warm conditions, the hypothalamus signals the cholinergic sympathetic fibers to release **acetylcholine** in order to stimulate the muscarinic cholinergic receptors on the **eccrine sweat glands** to induce sweat. Eccrine sweat glands are found all around the body but are most prevalent on the palms, soles, and forehead. Heat is then lost via the evaporation of sweat from the skin at a rate of about 0.58 kcal/mL.

3. **Decrease in heat production.** All mechanisms involved in heat production and conservation are severely inhibited.

Voluntary behavioral responses to heat, such as resting, wearing less clothing, fanning, and drinking cold fluids, help with heat loss as well.

Fever is a medical condition that describes a temporary increase in the body's thermoregulatory set point (usually by about 1°C-2°C). Fever differs from hyperthermia in that during a fever, the body's thermoregulatory set point itself is elevated, whereas in hyperthermia, body temperature rises above the set point. The increase in thermoregulatory set point in the hypothalamus during a fever can be attributed to the activity of the cytokines IL-1, IL-6, and TNF-α. It is important to note that fever is not a medical condition itself, but rather a symptom of other pathology.

CASE CORRELATES

- See Cases 28-36 (regulatory systems).

COMPREHENSION QUESTIONS

Refer to the following case scenario to answer questions 30.1-30.2:

A 28-year-old man is brought into the emergency department by EMS after having been found lying in the snow on the side of the road. He apparently had been walking along the road the night before and was struck by a car, which did not stop, and lay in the snow until someone noticed him this morning. His fingers and toes are blue and cool to the touch, and he is shivering uncontrollably.

30.1 Stimulation of which receptors are primarily responsible for triggering these responses to cold?

A. Thermoreceptors in the skin

B. Thermoreceptors in the posterior hypothalamus

C. Thermoreceptors in the anterior hypothalamus

D. Thermoreceptors in the lateral hypothalamus

30.2 What part of the hypothalamus receives input from cold-sensitive thermoreceptors and then generates signals that lead to conservation and generation of body heat?

A. Anterior hypothalamus

B. Posterior hypothalamus

C. Lateral hypothalamus

D. Medial hypothalamus

30.3 A 44-year-old manual laborer is brought to the emergency department after collapsing on the job. He had been working outside, paving a city street, when, according to witnesses, he said he needed to sit down and then collapsed. He did not lose consciousness at that time. His skin all over his body is red with very rapid capillary refill, and he is sweating profusely. He is correctly diagnosed with heat exhaustion, and treatment is begun. Which hypothalamic nucleus is involved in generating heat-dissipating responses?

A. Medial hypothalamus

B. Lateral hypothalamus

C. Posterior hypothalamus

D. Anterior hypothalamus

ANSWERS

30.1 **A. Thermoreceptors in the skin** are responsible for responses to cold. There are two areas in the body in which thermoreceptors are located: the skin and the anterior hypothalamus, specifically the preoptic area. These areas do not behave the same, however. There are significantly more cold receptors than warm in the skin and more warm receptors than cold in the hypothalamus. This means that skin-based thermoreceptors are more important for detecting cold conditions and hypothermia, while hypothalamic receptors are more important for detecting warm conditions and hyperthermia. Of course in an extreme example such as this, both the skin and the hypothalamic thermoreceptors would be generating signals to conserve or generate body heat.

30.2 **B.** The **posterior hypothalamus** generates the signals that cause an increase in heat conserving and producing behaviors. The posterior hypothalamus is also involved in sympathetic outflow, which is important in heat conservation. Sympathetic outflow is responsible for peripheral vasoconstriction and piloerection, both of which are heat conserving. Shivering, which is a heat-generating mechanism, is not mediated by the sympathetic system but rather by the shivering center, found in the dorsomedial posterior hypothalamus. Posterior hypothalamic projections to the cortex are involved in the more complex behavioral responses to cold, such as putting on more clothing, pacing, and going inside. The anterior hypothalamus, while being the location of hypothalamic thermoreceptors, is more sensitive to heat than cold and is involved in heat-dissipating activities.

30.3 **D.** The **anterior hypothalamus** is the area involved in heat-dissipating activities. It receives warm signals primarily from the preoptic hypothalamus but also from warm receptors located in the skin. Heat-dissipating activities include cutaneous vasodilation, achieved by sympathetic inhibition, which causes the skin to appear very red, and sweating. All heat-conserving activities are inhibited as well. There are also cortical responses to overheating, including fanning, removing clothing, and drinking cold drinks.

NEUROSCIENCE PEARLS

▶ Thermoregulation is controlled primarily by neural feedback mechanisms operating through the hypothalamus, namely the preoptic area.

▶ Involuntary responses for heat conservation are initiated by the hypothalamus and include vasoconstriction, piloerection, and thermogenesis (ie, sympathetic excitation and shivering).

▶ Fever differs from hyperthermia in that during a fever, the body's thermoregulatory set point itself is elevated, whereas in hyperthermia, body temperature rises above the set point.

REFERENCES

Bear MF, Connors BW, Paradiso MA, eds. *Neuroscience: Exploring the Brain*. 3rd ed. Baltimore, MD: Lippincott Williams & Wilkins; 2006.

Kandel ER, Schwarz JH, Jessell TM, Siegelbaum SA, Hudspeth AJ, eds. *Principles of Neural Science*. 5th ed. New York, NY: McGraw-Hill; 2012.

Squire LR, Berg D, Bloom FE, du Lac S, eds. *Fundamental Neuroscience*. 4th ed. San Diego, CA: Academic Press; 2012.

A 66-year-old Hispanic man presents to the emergency department with right-sided chest pain, weakness, and hemoptysis. He has a medical history of hypertension, COPD, recurrent episodes of pneumonia, and a 75-pack-year history for cigarettes. A chest X-ray in the emergency department demonstrates hyperinflated lungs, emphysema, and a large right-sided mass. Internal medicine admits the patient to the hospital for further evaluation. A neurologist is consulted to evaluate his new-onset weakness. His right upper extremity is diffusely weak, but there is profound weakness in his intrinsic hand muscles. He is also noted to have right ptosis and anisocoria; his right pupil is 2 mm and reactive, and his left pupil is 5 mm and reactive. It is soon found that he has a Pancoast tumor causing Horner syndrome.

- In what part of the lung are Pancoast tumors found?
- What explains the anisocoria?
- What explains the ptosis?

ANSWERS TO CASE 31:

The Sympathetic Nervous System

Summary: A 66-year-old man with frank pulmonary pathology and a heavy smoking history, presents with a chest mass, anisocoria, and muscle weakness.

- **Location:** Pancoast tumors are found in the apex of the lung.

- **Anisocoria:** Tumor invasion of the sympathetic chain, resulting in ipsilateral unopposed parasympathetic tone to the papillary constrictor muscle causing miosis.

- **Ptosis:** Also caused by tumor invasion of the sympathetic network. In this case, loss of innervation to Müller muscle causes ptosis.

CLINICAL CORRELATION

This is a classic presentation of a **Pancoast tumor** that invades the apex of the lung. These are often caused by squamous cell carcinoma of the lung. Patients often have ipsilateral shoulder and medial arm pain. These tumors often invade the brachial plexus but can also compromise the lower cervical or upper thoracic spinal nerve roots. Therefore, there is often sensory loss in the C8-T1 distributions. This results in sensory loss of the ipsilateral dorsomedial forearm and into the fifth and medial half of the fourth digits. There is often compromise of sympathetic chain ganglia and the stellate ganglion. This results in disruption of the sympathetic tone that returns to the head causing ptosis and miosis. Additionally, if the tumor invades the sympathetic fibers at or above the level of the carotid bifurcation, one would expect to find anhydrosis of the entire ipsilateral face. If the tumor is distal to this point, one would expect to find anhydrosis on the ipsilateral medial forehead and medial ipsilateral nose. It is important to remember that disruption of the sympathetic fibers at any point from their origin in the hypothalamus to their ascension along the carotid artery can result in Horner syndrome.

APPROACH TO:

The Sympathetic Nervous System

OBJECTIVES

1. Understand the anatomy of the sympathetic nervous system.
2. Know the neurotransmitters used by the sympathetic nervous system at the various synapses.
3. Be able to discuss the effects of the sympathetic stimulation of major end organs.

DEFINITIONS

THORACOLUMBAR NERVOUS SYSTEM: The sympathetic nervous system as defined with respect to the spinal cord levels containing the cell bodies of preganglionic sympathetic nerves.

PARAVERTEBRAL GANGLION: The sympathetic trunk containing the terminal fibers of white communicating rami and the cell bodies of gray communicating rami. This is a bilateral structure.

SUPERIOR CERVICAL GANGLION: Analogous to the paravertebral ganglion but has cell bodies of gray communicating rami that will eventually terminate in the head.

PREVERTEBRAL GANGLIA: The celiac, superior mesenteric, and inferior mesenteric ganglia, where white communicating rami terminate on the cell bodies of gray communicating rami that terminate onto most abdominal and retroperitoneal viscera.

DISCUSSION

The sympathetic nervous system is one of the two subdivisions of the autonomic nervous system. Together, these subdivisions govern the activity of **cardiac muscle, smooth muscle,** and **glands.** The autonomic nervous system is regulated by combined efforts from the **hypothalamus, cerebral cortex, amygdala,** and **reticular formation.** As with many other portions of the nervous system, it is often easier to facilitate an anatomical discussion by dividing the pertinent pathways and projections into their **central** and **peripheral** components.

The central organization of the sympathetic nervous system, which regulates the "fight or flight" responses, is best understood by focusing on the **hypothalamus.** At this level, both afferent and efferent information is processed and the tone of the sympathetic nervous system adjusted. **Afferent visceral** information about blood pressure, respiratory drive, and gastrointestinal status are carried by the glossopharyngeal and vagus nerves to the **solitary tract nucleus** in the brainstem. This nucleus accepts and redirects nerve impulses to various areas of the brain including the hypothalamus, insular cortex, amygdala, and adjacent respiratory centers in the medulla and pons. The insular cortex is involved with cardiac function. The **highest levels** of sympathetic control and largely efferent output are located in the **prefrontal, cingulate,** and **hippocampal cortices.** These areas both receive and project to other regions to achieve a maximal attenuation of sympathetic drive. By projecting to the hypothalamus, these cortical areas are able to manifest their output by altering the tone of the sympathetic nervous system. Most sympathetic fibers originate in the lateral and posterior regions of the hypothalamus. From here they descend without crossing into the **lateral** tegmentum of the **midbrain, pons, medulla,** and into the **spinal cord.** These axons terminate in the **interomediolateral cell column** of the spinal cord from **C8-L2.** Because the sympathetic nervous system is confined to this region of termination, it often carries the designation of **thoracolumbar.** The **peripheral sympathetic nervous** system begins with cell bodies located in the interomediolateral cell column of the gray matter of the spinal cord. These cell bodies project their **preganglionic myelinated axons,** or **white** communicating rami, through the ventral root exit zone of the spinal cord, through the ventral rootlet, and into the spinal nerve at each level. The white communicating rami will terminate in one of two structures: the **paravertebral ganglion** running alongside the vertebral column or the **prevertebral ganglia** in the posterior abdominal cavity (note that the paravertebral trunks are paired structures on both sides of the vertebral column). The three separate prevertebral ganglia include the **celiac ganglion**, the **superior mesenteric ganglion**, and the **inferior mesenteric ganglion**. By far, the majority of the sympathetic fibers traverse the **T5-L2** spinal nerves to their respective targets. The prevertebral and paravertebral ganglia accept the preganglionic sympathetic fibers and contain the cell bodies of **postganglionic unmyelinated sympathetic fibers**, or **gray communicating rami**. The axons of these cell bodies will terminate upon their end organ targets including sweat glands, intestinal glands, and cardiac and pulmonary tissue among others. Of note, the adrenal glands receive innervation by white communicating rami without synapses. The postganglionic sympathetic component of the adrenal gland is considered to occur when the gland releases its chemical product, epinephrine, into the blood stream.

In comparison to parasympathetic nervous system fibers, sympathetic preganglionic fibers are short and the postganglionic fibers are long. One preganglionic fiber innervates many postganglionic fibers, and it is this divergence that amplifies sympathetic outputs and coordinates sympathetic activation across different spinal levels (Figure 31-1).

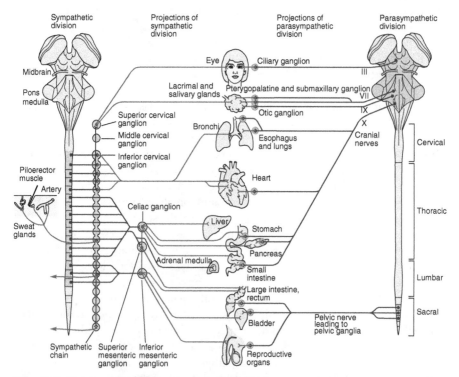

Figure 31-1. The sympathetic (thoracolumbar) division of the autonomic nervous system is noted on the *left. (With permission from Kandel ER, Schwartz JH, Jessell TM. Principles of Neural Science. 4th ed. New York, NY: McGraw-Hill; 2000:964.)*

The **head** receives its sympathetic network from **C8-T2** cord levels and terminates in the **superior cervical ganglion**. Postganglionic sympathetic fibers course along the **internal carotid arteries** to innervate the **salivary** and **lacrimal** glands, and the **sweat glands, smooth muscle,** and **blood vessels** of the head. The arms receive their postganglionic sympathetic innervation from the lower cervical and upper thoracic ganglia. The **heart** receives its sympathetic tone from the **upper thoracic ganglion**, while the abdominal ganglia receive their input from T5-T9/T10. The pelvis, legs, and descending colon are supplied by the upper lumbar ganglia.

It is important to remember that the sympathetic nervous system is not solely a visceral motor system. As mentioned earlier, there are **visceral afferent** pathways too. The cell bodies of these sensory axons are located in the dorsal sensory ganglion within the thecal sac and their afferent projections can terminate locally on the intermediolateral cell column to affect **local reflexes** or onto the dorsal horn of the spinal gray resulting in **central communication** of visceral sensations, such as a full bladder.

The neurotransmitter profile of the sympathetic nervous system is crucial for understanding the pharmacological interventions employed to treat many medical illnesses. The initial synapse of pre- and postganglionic neurons uses acetylcholine as a neurotransmitter, while the postganglionic end organ synapse uses norepinephrine. One exception is the **adrenal medulla**, where the synapse of the preganglionic sympathetic with the adrenal gland uses acetylcholine to stimulate release of **epinephrine** into the bloodstream. Two other exceptions that use **acetylcholine** release in the postganglionic sympathetic synapse at the target organ are the **sweat glands** and the **blood vessels**.

Sympathetic nervous system neurotransmitter receptors are classified into alpha and beta, with subgroupings of **alpha-1, alpha-2, beta-1,** and **beta-2.** Stimulation of alpha-1 receptors causes vasoconstriction, decreased gastrointestinal motility, and pupillary dilation. Alpha-2 receptors are located on the presynaptic terminal and upon stimulation cause attenuation of neurotransmitter release into the synaptic cleft. Beta-1 receptors increase heart rate and heart contractility. Beta-2 receptors cause vasodilation and bronchodilation.

CASE CORRELATES

- See Cases 28-36 (regulatory systems).

COMPREHENSION QUESTIONS

Refer to the following case scenario to answer questions 31.1-31.3:

A 73-year-old man comes into your office complaining of a droopy right eyelid ever since he had surgery to "clean out his neck artery" several weeks ago. He also notes that he does not think the right half of his face sweats as much as the left. On examination, you also note that his right pupil is several millimeters smaller than his left but that both are reactive to light. You suspect that this man has Horner syndrome as a complication of his recent carotid endarterectomy.

31.1 Where are the cell bodies of the damaged nerves located in this man?

    A. Paravertebral ganglia

    B. Prevertebral ganglia

    C. Intermediolateral column of the thoracic spinal cord

    D. Superior cervical ganglion

31.2 What neurotransmitters are normally released onto these damaged postganglionic neurons by the preganglionic neurons?

    A. Acetylcholine

    B. Norepinephrine

    C. Dopamine

    D. Epinephrine

31.3 What neurotransmitter is normally released by these damaged postganglionic neurons onto glandular and vascular targets?

    A. Acetylcholine

    B. Norepinephrine

    C. Dopamine

    D. Epinephrine

ANSWERS

31.1 **D. In the face, the ganglion involved is the superior cervical ganglion.** This man has Horner syndrome as a complication of his carotid endarterectomy. The sympathetic nerves that innervate the face run along the carotid artery and can be damaged (although rarely) with excess dissection around the artery in surgery. One feature of the sympathetic nervous system is that it has relatively short preganglionic axons, which synapse in ganglia near the spinal cord, and long postganglionic axons that then project from these ganglia to the end organs. Axons from cell bodies in the superior cervical ganglion ascend along the carotid artery and then follow its branches to their targets in the face and head. The first-order neurons that synapse in the superior cervical ganglion have their cell bodies in the intermediolateral column of the thoracic cord.

31.2 **A.** All preganglionic autonomic neurons (sympathetic and parasympathetic) release **acetylcholine** at autonomic ganglia. This acetylcholine activates ganglionic nicotinic acetylcholine receptors located on postganglionic neurons, triggering action potentials.

31.3 **B.** For the most part, postganglionic sympathetic neurons release **norepinephrine** onto their targets, including glands and blood vessels. The only targets that do not get norepinephrine neurotransmission are sweat glands. Sympathetic neurons innervating sweat glands release acetylcholine onto muscarinic receptors on the sweat glands.

NEUROSCIENCE PEARLS

▶ Most lesions of the sympathetic nervous system above the thoracic spine will cause an ipsilateral Horner syndrome.

▶ The various adrenergic receptors form the basis of pharmacological intervention for heart rate control, blood pressure control, and airway control.

▶ The hypothalamus is the principal nucleus and origin of the sympathetic nervous system.

REFERENCES

Bear MF, Connors B, Paradiso M, eds. *Neuroscience: Exploring the Brain.* 3rd ed. Baltimore, MD: Lippincott Williams & Wilkins; 2006.

Squire LR, Berg D, Bloom FE, du Lac S, eds. *Fundamental Neuroscience.* 4th ed. San Diego, CA: Academic Press; 2012.

Purves D, Augustine GJ, Fitzpatrick D, Hall WC, eds. *Neuroscience.* 5th ed. Sunderland, MA: Sinauer Associates Inc; 2011.

CASE 32

A 21-year-old previously healthy man presents to the emergency department with right-sided eye pain for the past 2 days. He has never experienced this symptom before. He states that he was watching his favorite movie when he noticed that colors seen by his right eye seemed faded and less intense than those in his left eye. Around this time he also developed worsening right eye pain. Upon further questioning, he states that he has had episodes of worsening urinary incontinence for the past month. He was embarrassed about this so he did not seek medical attention. He describes a strong urge to urinate but has difficulty initiating a stream. However, he adds that he often experiences urinary discharge when he senses no urgency at all. His cousin and sister both suffer from systemic lupus erythematosus. Eventually, he is diagnosed with multiple sclerosis (MS).

▶ Why is he incontinent of urine?
▶ What function does myelin serve?
▶ What cells are responsible for myelination in the central nervous system (CNS)?

ANSWERS TO CASE 32:

The Parasympathetic Nervous System

Summary: A 21-year-old man has urinary incontinence for the past month and eye pain with decreased vision and color acuity.

- **Mechanism of incontinence:** Compromise of the parasympathetic tone to the detrusor muscle and external urethral sphincter resulting in urinary incontinence secondary to a MS plaque.

- **Myelin:** Myelin serves to protect and insulate axons, increasing conduction velocity along them.

- **Myelinating cells in CNS:** Oligodendrocytes myelinate axons in the CNS. Schwann cells serve the same purpose in the peripheral nervous system.

CLINICAL CORRELATION

Multiple sclerosis is a demyelinating disease that can affect both the brain and spinal cord. Normally, this autoimmune disorder affects patients in their twenties, females more so than males. Twenty-five percent of patients will manifest MS as optic neuritis. In this disease there is often orbital pain worsened with eye movement, monocular vision loss, decreased color acuity, and a decreased afferent papillary response in the affected eye.

For the purposes of this discussion, we will focus on the bladder dysfunction. The bladder is composed of three muscle groups: the detrusor muscle, the internal urethral sphincter, and the external urethral sphincter. The internal sphincter is smooth muscle and receives innervation from the lower thoracic interomediolateral segments. The preganglionic sympathetic fibers traverse into the posterior abdominal cavity, where they synapse in the inferior mesenteric ganglion. Postganglionic sympathetic fibers innervate the smooth muscle composing the internal urethral sphincter, causing contraction and urinary retention. The external urethral sphincter is striated muscle and is under voluntary control. Cell bodies in the anterolateral horns of S2, S3, and S4 (Onuf's nucleus) send their axons along the inferior pudendal nerve to terminate onto the external sphincter, causing contraction. The detrusor muscle itself is innervated by both the sympathetic and parasympathetic nervous systems. Sympathetic fibers travel along the same pathways as the sympathetic fibers that terminate on the internal sphincter. When they terminate on the detrusor, they cause relaxation. Preganglionic parasympathetic fibers course through the inferior mesenteric ganglion and terminate onto postganglionic parasympathetic cell bodies located on the bladder wall. Firing of the postganglionic parasympathetic fibers causes detrusor contraction.

There is a cerebral micturition center located on the medial frontal cortex inferior to the accessory motor cortex. This structure is responsible for conscious continence management. There is a local reflex arc that stimulates bladder emptying based on afferent sensory impulses from the detrusor muscle. However, the micturition center projects fibers onto both Onuf nucleus and the cell bodies of parasympathetic preganglionic neurons to inhibit their firing. The sequence of micturition is as follows: voluntary relaxation of the perineum, flexion of abdominal wall muscles, detrusor contraction, and opening of the internal and then external sphincters.

In the case of MS, these pathological changes take place over time. The most common bladder manifestation is neurogenic bladder, resulting from spinal cord white matter lesions above T12. This results in a spastic bladder and loss of voluntary control of the external sphincter. A spastic bladder initiates an emptying reflex at lower bladder volumes and is analogous to muscle spasticity and hyper-reflexivity seen in upper motor neuron disease. This combination leads to urgency (depending on the degree of sensation that is still intact) and incontinence.

MS should be suspected when a patient has multiple neurological deficits that cannot be accounted for by a single lesion. **Loss of myelin sheaths and depletion of oligodendrocytes are seen within the plaques.**

APPROACH TO:
The Parasympathetic Nervous System

OBJECTIVES

1. Understand the anatomy of the parasympathetic nervous system (Figure 32-1).

2. Know the neurotransmitters used by the parasympathetic nervous system at the various synapses.

3. Be able to discuss the effects of the parasympathetic stimulation of major end organs.

DEFINITIONS

CRANIOSACRAL: Naming schema denoting the regions of preganglionic parasympathetic nuclei. The head distal to the splenic flexure is supplied by the brainstem nuclei, while the descending colon and pelvic organs are supplied by the sacral component.

NICOTINIC: The type of receptor found at autonomic preganglionic to postganglionic synapses as well as the receptor found at neuromuscular junctions.

MUSCARINIC: The receptor type found at postganglionic parasympathetic to end organ synapses.

DISCUSSION

The parasympathetic nervous system, which controls the "rest and digest" functions, is one of the two subdivisions of the autonomic nervous system. Together, these subdivisions govern the activity of **cardiac muscle, smooth muscle,** and **glands.** The autonomic nervous system is regulated by combined efforts from the **hypothalamus, cerebral cortex, amygdala,** and **reticular formation.** As with many other portions

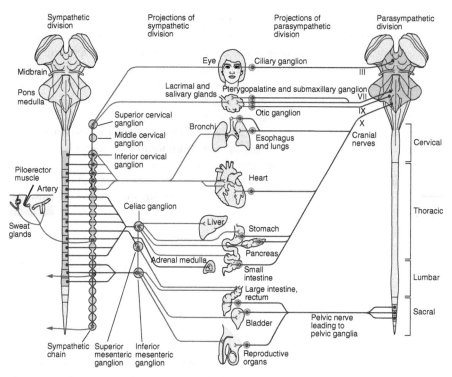

Figure 32-1. The parasympathetic (craniosacral) division of the autonomic nervous system is noted on the *right*. (*With permission from Kandel ER, Schwartz JH, Jessell TM.* Principles of Neural Science. *4th ed. New York, NY: McGraw-Hill; 2000:964.*)

of the nervous system, it is often easier to facilitate an anatomical discussion by dividing the pertinent pathways and projections into their **central** and **peripheral** components.

The central organization of the parasympathetic nervous system is best understood by focusing on the **hypothalamus**. At this level both afferent and efferent information is processed and the tone of the parasympathetic nervous system adjusted. **Afferent visceral** information about blood pressure, respiratory drive, and gastrointestinal status are carried by the glossopharyngeal and vagus nerves to the **solitary tract nucleus** in the brainstem. This nucleus accepts and redirects nerve impulses to various areas of the brain including the hypothalamus, insular cortex, amygdala, and adjacent respiratory centers in the medulla and pons. The insular cortex is involved with cardiac function.

The highest levels of parasympathetic control, largely achieved by efferent output, is located in the **prefrontal, cingulate,** and **hippocampal cortices**. These areas both receive and project to other regions to achieve a maximal attenuation of parasympathetic drive. By projecting to the hypothalamus, these cortical areas are able to manifest their output by altering the tone of the parasympathetic nervous system. Most parasympathetic fibers originate in the anterior regions of the hypothalamus. From here they descend into the midbrain, pons and medulla, and down the spinal cord. These axons terminate in the **interomediolateral cell column** of the spinal cord from **S2-S4**. Because the parasympathetic nervous system is confined to these two regions of termination, it often carries the designation **craniosacral**.

The brainstem nuclei that distribute preganglionic parasympathetic outflow include the **Edinger-Westphal nucleus** of the oculomotor nerve, the **superior salivatory nucleus** that contributes to the facial nerve, the **inferior salivatory nucleus** that contributes to the glossopharyngeal nerve, and the **dorsal motor nucleus** of the vagal nerve. Together, these four cranial nerves carry all preganglionic parasympathetic tone to the glands and blood vessels of the head (see Figure 32-1). The **Edinger-Westphal nucleus** projects preganglionic parasympathetic axons along the periphery of the oculomotor nerve as it exits the midbrain and enters the orbit through the **superior orbital fissure**. Here the preganglionic parasympathetic fibers travel along the **inferior division** of the oculomotor nerve and then branch off toward the **ciliary ganglion**. They then terminate, and postganglionic parasympathetic fibers innervate the **sphincter muscle** of the pupil and **ciliary muscles** allowing for **miosis** and **convergence**, respectively. The **superior salivatory nucleus** projects preganglionic parasympathetic fibers along the **facial nerve** as it exits the brainstem along the pontomedullary junction. These fibers branch off the facial nerve within the **greater petrosal nerve** just distal to the **geniculate ganglion**. These fibers reenter the cranium through the hiatus for the greater petrosal nerve and run anteriorly along the skull base. They exit the skull and travel to the **pterygopalatine fossa**, where they terminate in the **pterygopalatine ganglion**. Here postganglionic parasympathetic fibers travel superiorly through the **inferior orbital fissure** and terminate onto the lacrimal gland causing lacrimation. Other preganglionic parasympathetic fibers originating in the **superior salivatory nucleus** pass through the **geniculate ganglion**, travel with **chorda tympani**, and then follow the **lingual** nerve. These fibers terminate in the **submandibular ganglion**. Cell bodies of postganglionic parasympathetic fibers will then travel a short distance to innervate the **submandibular** and **sublingual** glands. The **inferior salivatory nucleus** projects preganglionic parasympathetic fibers to the **otic** ganglion. Fibers terminate here and postganglionic parasympathetic fibers travel to the **parotid gland** to provide innervation. The **dorsal vagal nucleus** located in the medulla projects preganglionic parasympathetic fibers along the vagus nerve. This nerve exits the skull through the **jugular foramen** and provides widespread parasympathetic innervation of the major organ systems of the chest and abdomen. Of note, the vagus nerve only supplies parasympathetic innervation to the ascending and transverse colon. The **descending colon** and **rectum** are innervated by the parasympathetic elements of the sacral spinal cord. Increased vagal tone results in bronchoconstriction, decreased heart rate, and increased gastrointestinal motility. Unlike within the head, the vagus nerve does not have discrete ganglia prior to reaching the target end organs. Rather, the preganglionic parasympathetic axons terminate in the **walls** of their respective targets, and the resulting postganglionic parasympathetic axons have a very short distance to travel. This is one of the principal structural differences between the sympathetic and parasympathetic nervous systems. That is, the postganglionic sympathetic axon traverses a **large distance** compared to the postganglionic parasympathetic axon. The principal functions of the sacral parasympathetic system are control of the bladder, descending colon, rectum, and pelvic organs.

The neurotransmitter used at both preganglionic parasympathetic–postganglionic parasympathetic and postganglionic parasympathetic–end organ synapses is **acetylcholine**. The ganglionic synapse receives acetylcholine at **nicotinic receptors**, while the end-organ synapse receives acetylcholine at **muscarinic receptors**.

CASE CORRELATES

- See Cases 28-36 (regulatory systems), and Case 4 (multiple sclerosis).

COMPREHENSION QUESTIONS

Refer to the following case scenario to answer questions 32.1-32.2:

A 31-year-old man comes into your office for a routine checkup that includes blood work. In order to expedite the process, you decide to draw the man's blood while you talk with him about his medical history. You successfully cannulate his vein, but as soon as he sees the blood filling up the tube his eyes roll back in his head and he collapses backward on the examination table. You elevate his legs and he rapidly regains consciousness.

32.1 In this case of vasovagal syncope, from which nucleus did the preganglionic parasympathetic fibers causing cardiodepression originate?

A. Edinger-Westphal nucleus

B. Intermediolateral neurons in the sacral spinal cord

C. Dorsal motor nucleus of the vagus nerve

D. Inferior salivatory nucleus

32.2 Where are the cell bodies of the postsynaptic nerves innervating this man's heart located?

A. Paravertebral ganglia

B. The musculature of the heart

C. Geniculate ganglion

D. Pterygopalatine ganglion

32.3 What neurotransmitter is released by both pre- and postganglionic neurons in the parasympathetic nervous system?

A. Norepinephrine

B. Epinephrine

C. Dopamine

D. Acetylcholine

ANSWERS

32.1 **C.** These fibers originate from the **dorsal motor nucleus of the vagus nerve**. Vasovagal syncope is a complicated reflex that results from a number of stimuli, one of which can be the sight of blood. The reflex involves increased parasympathetic outflow and decreased sympathetic outflow, which combine to cause hypotension sufficient to cause a loss of consciousness. The increased parasympathetic outflow causes a decrease in heart rate and myocardial contractility and is mediated by the vagus nerve. Presynaptic parasympathetic neurons in vagus nerve originate in the dorsal motor nucleus and supply the organs of the thorax and most of the abdominal viscera. The descending colon, rectum, and pelvic organs, however, are supplied by the sacral parasympathetic plexus, which originates in the sacral spinal cord. The Edinger-Westphal nucleus and the inferior salivatory nucleus supply parasympathetic innervation to organs in the head via the oculomotor and glossopharyngeal nerves.

32.2 **B.** In the thorax and abdomen, the preganglionic neurons synapse on postganglionic neurons that reside in the walls of the organs that they innervate, in this case in **the musculature of the heart**. In general, the parasympathetic nervous system has preganglionic neurons that extend all the way or nearly all the way to the target organs, unlike the sympathetic nervous system, which has relatively short preganglionic neurons and long postganglionic neurons. In the head and neck, the parasympathetic preganglionic neurons synapse on postganglionic neurons in discreet ganglia close to the target organs such as the geniculate and pterygopalatine ganglia.

32.3 **D. Acetylcholine** is released by all neurons in the parasympathetic nervous system, both pre- and postganglionic. The preganglionic neurons release acetylcholine onto ganglionic nicotinic receptors on the postganglionic neurons, which in turn release acetylcholine onto muscarinic receptors in the target organs.

NEUROSCIENCE PEARLS

► The parasympathetic nervous system is organized into brainstem and sacral components.

► Parasympathetic outflow generally opposes sympathetic outflow.

► The vagus nerve supplies the majority of major organ systems within the chest and abdomen.

► Multiple sclerosis is an autoimmune demyelinating disease that affects the white matter of the brain.

► MS is associated with loss of myelin sheaths and depletion of oligodendrocytes within plaques.

► MRI findings often correlate to the MS patient's clinical course.

REFERENCES

Bear MF, Connors B, Paradiso M, eds. *Neuroscience: Exploring the Brain*. 3rd ed. Baltimore, MD: Lippincott Williams & Wilkins; 2006.

Kandel ER, Schwarz JH, Jessell TM, Siegelbaum SA, Hudspeth AJ, eds. *Principles of Neural Science*. 5th ed. New York, NY: McGraw-Hill; 2012.

Squire LR, Berg D, Bloom FE, du Lac S, eds. *Fundamental Neuroscience*. 4th ed. San Diego, CA: Academic Press; 2012.

An 18-year-old Japanese woman presents to a medical clinic complaining of excessive sleepiness. Although she sleeps approximately 8 hours a night, she is constantly fatigued and sleepy during the day. Sleeping more during the night does not help. She also reports of waking up to find she has completed tasks which she has no recollection of doing. Additionally, she has been experiencing sudden bouts of muscular weakness, including slurred speech, weakened eyesight, and weakness at the knees. She notes that these experiences occur especially after laughing. After taking a polysomnogram to confirm the diagnosis, she is diagnosed with narcolepsy.

► What are common symptoms of narcolepsy?
► On what chromosome is the human leukocyte antigen (HLA) complex found?
► Which three major brain structures make up the limbic system?

ANSWERS TO CASE 33:
Sleep and the Limbic System

Summary: An 18-year-old Japanese woman complaining of excessive sleepiness during the day even after sufficient nighttime sleep, and muscular weakness after laughing.

- **Symptoms:** Excessive daytime sleepiness (EDS), cataplexy, slurred speech, impaired vision, weakness of facial and limb muscles, hypnagogic hallucinations, and sleep paralysis are common symptoms of patients with narcolepsy.

- **Chromosome:** The HLA complex is found on chromosome 6.

- **Limbic system:** The limbic system is composed of the hippocampus, amygdala, and hypothalamus among numerous other brain structures.

CLINICAL CORRELATION

Narcolepsy is a neurological condition characterized by EDS, often with episodes of muscular weakness known as cataplexy. Persons with narcolepsy tend to feel sleepy whenever they are awake and are often unable to keep themselves awake for long periods of time, even after adequate nighttime sleep. Cataplexy, which has been found to occur in nearly 75% of narcoleptic patients, is described as temporary muscular weakness or paralysis without loss of consciousness evoked by sudden emotional reactions, especially laughter. Other evoking emotions include anger, joy, fear, and surprise. Usually speech is slurred and vision is impaired, along with weakness of the face muscles and/or limbs. Narcoleptics also experience automatic behavior, meaning that they continue to function during sleep episodes but wake with no recollection of what they have done. Approximately 40% of narcoleptics experience this symptom. Other common symptoms include sleep paralysis (temporary inability to move or talk when waking up) and hypnagogic hallucinations (vivid, often frightening dreamlike episodes that occur when falling asleep and/or waking up).

Narcolepsy can be diagnosed with the help of a polysomnogram or a multiple sleep latency test. A polysomnogram involves continuous recording of brain waves and nerve and muscle functions during sleep. Narcoleptics fall asleep and enter rapid eye movements (REM) sleep quickly and may wake frequently during the night. In a multiple sleep latency test, the patient is allowed to sleep every 2 hours while observations of sleepiness and amount of time needed to reach various sleep stages are continuously taken. While true prevalence of narcolepsy is significantly underreported, the reported prevalence ranges from about 5 in 10,000 for North American and European populations to 16 in 10,000 in Japan. While there seems to be no significant difference between men and women, there does seem to be a strong genetic link. One factor which may predispose an individual to narcolepsy is the HLA complex on chromosome 6. Though not specifically defined, there appears to be a correlation between certain variations in HLA genes and narcolepsy. Recently, it has been discovered that narcolepsy is specifically related to the hypothalamus. **Brain cells which contain the neurotransmitter hypocretin and originate in the hypothalamus are**

found to be reduced by 85%-95% in people with narcolepsy. Scar tissue is found in hypothalamic regions where hypocretin-producing brain cells used to be, indicating the cells were present at birth but later died. It is thought that certain variations in the HLA complex increase the risk of an autoimmune response to hypocretin.

In terms of treatment, the common symptoms of EDS and cataplexy must be treated separately. EDS can be treated with amphetamine-like stimulants such as dextroamphetamine, or with modafinil, another type of CNS stimulant known as a "wake promoter." Cataplexy, sleep paralysis, and hypnagogic hallucinations can be treated using antidepressant drugs.

APPROACH TO:
Sleep and the Limbic System

OBJECTIVES

1. Know the signs and symptoms of narcolepsy.

2. Know the methods of treatment for EDS and cataplexy.

DEFINITIONS

VEGETATIVE FUNCTIONS: Body regulatory functions.

THERMORECEPTOR: A sensory receptor that responds to heat and cold.

OSMORECEPTOR: A receptor sensitive to plasma osmolality that exists in the brain to regulate water balance in the body by controlling thirst and the release of vasopressin.

ANTIDIURETIC HORMONE (ADH): Hormone secreted by the posterior pituitary gland and also by nerve endings in the hypothalamus; affects blood pressure by stimulating capillary muscles and reduces urine flow by affecting reabsorption of water by kidney tubules.

OXYTOCIN: A short polypeptide hormone released from the posterior lobe of the pituitary gland, which stimulates the contraction of smooth muscle of the uterus during labor and facilitates ejection of milk from the breast during nursing.

DISCUSSION

One of the oldest systems in the brain, the limbic system is involved in emotion and memory as well as numerous body regulatory functions collectively known as **vegetative functions**. While this system is composed of many structures, there are three main ones: the **hippocampus, amygdala, and hypothalamus.**

The Hippocampus

The hippocampi are found bilaterally deep in the medial temporal lobe and play a role in long-term memory and spatial navigation. Because the hippocampus is involved in transferring short-term memory to long-term memory, damage to this structure most commonly results in the inability to lay down new memories (antero-grade amnesia).

The Amygdala

Above the hippocampi lie the two almond-shaped neuron masses known as the amygdalae. The amygdala is involved in monitoring the environment for survival management and responds specifically to the stimulus of fear. The amygdala sends output to several structures involved in the response to fear, including (1) the hypo-thalamus to activate the sympathetic nervous system, (2) the reticular nucleus to increase reflexes, (3) the nuclei of trigeminal and facial nerves to modulate facial expressions, and (4) to various areas (ie, ventral tegmental area and locus ceruleus), which result in increased release of dopamine, norepinephrine, and epinephrine. The amygdala also contains receptors for estrogen and androgens, suggesting a role in sexuality. A bilateral lesion of the amygdalae results in the **Klüver-Bucy syndrome**, characterized by loss of fear, hypersexuality, hyperorality, and emotional blunting.

The Hypothalamus

Though extremely small, the hypothalamus is perhaps the most important structure of the limbic system. Its key central position in relation to the other limbic structures facilitates its two-way communication with all levels of the limbic system. Specifi-cally, the hypothalamus sends signals in three directions: up toward the higher areas of the diencephalon and cerebrum, down to the brainstem, and into the hypotha-lamic infundibulum to control hormonal secretion of the anterior and posterior pitu-itaries. With the help of the rest of the limbic system, the hypothalamus controls numerous aspects of emotional behavior and most of the bodily vegetative functions.

Cardiovascular Regulation   Stimulation of different regions of the hypothalamus is known to have different neurogenic effects on the cardiovascular system. For example, stimulation of the **posterior and lateral hypothalamus** (which is involved in sympathetic stimulation) increases heart rate and arterial pressure. On the other hand, stimulation of the **preoptic area** found in the anterior hypothalamus (which is involved in parasympathetic stimulation) causes a decrease in arterial pressure and heart rate. The specific cardiovascular control centers are found in the reticular regions of the brainstem, particularly the medulla and pons.

Regulation of Body Temperature   As discussed earlier in this book, the hypothal-amus is involved in thermoregulation. The thermostatic center is located in the pre-optic area, and **thermoreceptors** in this region are involved in monitoring the rise and fall of blood temperature. Increased blood temperature increases the activity of these thermoreceptors, while decreased blood temperature decreases their activity. The activity and signals from these receptors control a variety of thermoregulatory mechanisms, which are discussed in further detail in the case on thermoregulation.

Regulation of Body Water   The hypothalamus is also involved with the regula-tion of body water. Water homeostasis in controlled by two hypothalamic mecha-nisms: the **osmoreceptor-ADH** feedback system, and the **thirst mechanism**. The osmoreceptor-ADH feedback system regulates the amount of water absorbed in the kidney and excreted in the urine, while the thirst mechanism induces water drinking behavior whenever necessary. Special receptors known as osmoreceptors located in the supraoptic nucleus region are stimulated by changes in plasma sodium concentration. In turn, these osmoreceptors stimulate the secretion of ADH at nerve endings in the posterior pituitary. ADH acts on the collecting ducts of the kidney, causing increased water absorption. The thirst mechanism is controlled by the **thirst center**, composed of the Anteroventral third ventricle (AV3V) region and a specialized area near the preoptic nucleus. When blood osmolarity rises above normal levels, this center stimulates a thirst sensation which greatly increases an organism's desire to drink water.

Regulation of Uterine Contractility and of Milk Ejection from the Breasts
Stimulation of the **paraventricular nucleus** in the anterior hypothalamus induces secretion of **oxytocin**. In females, oxytocin increases contractility of the uterus and the myoepithelial cells surrounding the breast alveoli. It is released mainly after distension of the cervix and vagina during labor and after stimulation of the nipples, thereby stimulating childbirth and breastfeeding, respectively. A baby's sucking on the mother's nipple stimulates a reflex signal from the nipple to the posterior hypothalamus, inducing release of oxytocin, which then causes contraction of the breast milk ductules. Oxytocin is also known to play a role in circadian homeostasis, helping regulate body activity level and wakefulness.

Gastrointestinal and Feeding Regulation Hunger and appetite stimulation are controlled by several areas of the hypothalamus. The area most involved with stimulation of the hunger sensation, large appetite, and food-seeking behavior is the **lateral hypothalamic area**. Damage to this area can lead to starvation. On the other hand, the **satiety center**, which is located in the ventromedial nucleus, is involved in opposing the desire for food. When stimulated, this center causes an organism to lose interest in its food and stop eating. Damage to the satiety center can result in overeating and consequently to obesity. Important hormones involved include leptin, which signals satiety; ghrelin, which signals hunger; and insulin, which signals satiety and subsequent glucose storage. Additionally, the **mammillary bodies** in the posterior hypothalamus are involved in controlling feeding reflexes, such as licking the lips and swallowing.

Behavioral Functions Stimulation of different parts of the hypothalamus also affects behavior. For example, stimulation of the lateral hypothalamus increases the organism's general activity level, in addition to causing thirst and hunger as explained in the previous section. Conversely, stimulation in the ventromedial nucleus leads to satiety and tranquility. Stimulation of a thin area of the periventricular nuclei near the third ventricle can lead to fear and punishment reactions. Additionally, several areas, especially those found in the most posterior and anterior regions of the hypothalamus are involved with stimulating sexual drive.

CASE CORRELATES

- See Cases 28-36 (regulatory systems).

COMPREHENSION QUESTIONS

33.1 A 55-year-old man comes into the clinic for a routine checkup, the first he has had in many years. You perform the usual history and physical examination and note that his blood pressure (BP) is 165/94. In order to better assess this abnormality, you have him return several times over the next few weeks for BP checks, and each time it is a similar value. You diagnose him with hypertension and begin treatment. If this man's hypertension is because of hyperactivity of a hypothalamic nucleus, which nucleus is most likely affected?

A. Anterior nucleus

B. Arcuate nucleus

C. Posterior nucleus

D. Ventromedial nucleus

33.2 A 33-year-old woman is involved in an automobile accident and suffers a head injury. When she regains consciousness after 2 weeks in the ICU, she is not interested in eating at all. She says that she simply isn't hungry or interested in food in the least. If this lack of appetite is caused by a hypothalamic lesion, where would it most likely be?

A. Anterior hypothalamus

B. Lateral hypothalamus

C. Ventromedial hypothalamus

D. Dorsomedial hypothalamus

33.3 A 27-year-old woman is in labor with her first child, and you are supervising her care. After several hours of labor, it is apparent that she is not progressing appropriately. You decide to augment her labor by starting her on pitocin, a synthetic form of oxytocin. From what hypothalamic nucleus is oxytocin normally secreted?

A. Paraventricular nucleus

B. Arcuate nucleus

C. Preoptic nucleus

D. Suprachiasmatic nucleus

ANSWERS

33.1 **C.** The **posterior and lateral nuclei** of the hypothalamus have been associated with increased heart rate and arterial blood pressure. The anterior nucleus has been associated with decreased heart rate and blood pressure, the arcuate nucleus is involved in endocrine regulation, and the ventromedial nucleus is thought to be the satiety center.

33.2 **B.** The **lateral hypothalamus** has been associated with hunger, appetite, and food-seeking behavior. Bilateral lesions to this area in experimental animals have caused them to starve to death despite the presence of food. While it is highly unlikely that a person could suffer isolated bilateral lesions to a single hypothalamic nucleus, if it were to occur, this would likely be the result. The ventromedial nucleus is the satiety center, and a lesion to it would cause the opposite symptoms: overeating and obesity.

33.3 **A.** The **paraventricular hypothalamic nucleus** produces oxytocin, which is then transported viaits axons down the pituitary stalk to the posterior pituitary where it is secreted into the bloodstream when the appropriate stimulus is present. The arcuate nucleus is involved in anterior pituitary endocrine control, the preoptic nucleus is involved in temperature regulation, and the suprachiasmatic nucleus helps regulate circadian rhythm.

NEUROSCIENCE PEARLS

▶ Narcolepsy is the second most common cause of excessive daytime sleepiness (after obstructive sleep apnea), and is thought to be due to lack of the neurotransmitter orexin (hypocretin).

▶ The limbic system is involved in emotion and memory and consists of the hippocampus, amygdala, and hypothalamus.

▶ The hippocampus plays role in long-term memory and spatial navigation.

▶ A bilateral lesion of the amygdalae results in the Klüver-Bucy syndrome, characterized by loss of fear, hypersexuality, hyperorality, and emotional blunting.

▶ The hypothalamus sends signals in three directions: up toward the higher areas of the diencephalon and cerebrum, down to the brainstem, and into the hypothalamic infundibulum to regulate the anterior and posterior pituitaries.

▶ In females, oxytocin increases uterine contractions and secretion of milk from breast ductules.

▶ Damage to the lateral hypothalamic area leads to decreased appetite (*lesion of lateral = less eating*), while damage to the ventromedial hypothalamic area leads to increased appetite (*lesion of medial = more eating*).

▶ Leptin signals satiety, while ghrelin signals hunger.

REFERENCES

Bear MF, Connors B, Paradiso M, eds. *Neuroscience: Exploring the Brain.* 3rd ed. Baltimore, MD: Lippincott Williams & Wilkins; 2006.

Purves D, Augustine GJ, Fitzpatrick D, Hall WC, eds. *Neuroscience.* 5th ed. Sunderland, MA: Sinauer Associates Inc; 2011.

Squire LR, Berg D, Bloom FE, du Lac S, eds. *Fundamental Neuroscience.* 4th ed. San Diego, CA: Academic Press; 2012.

A 36-year-old woman presents to the clinic complaining of prolonged and extreme tiredness. She states that she has been suffering these symptoms for the past 6 months, and she feels just as tired after a full night's sleep as she did when she went to bed. When she exerts herself physically, she is often overcome with a debilitating fatigue that leaves her bedridden for several days. A physical examination reveals that the woman has tender lymph nodes. The woman states that she was suffering from the flu just prior to the onset of the symptoms she described. She never fully recovered, and her flu symptoms eventually degenerated into her current illness. Based on this presentation, the patient is diagnosed with chronic fatigue syndrome (CFS).

► Damage to what structure in the brain is associated with sleepiness?
► Where is the affected brain structure located?

ANSWERS TO CASE 34:

The Reticular-Activating System

Summary: A 36-year-old woman presents to clinic complaining of debilitating fatigue. She states that she had the flu just prior to onset of symptoms.

- **Structure damaged:** It is believed that damage to the ascending reticular-activating system (RAS), an area in the brain that extends upward from the reticular formation and is associated with sleep function, may contribute to onset of CFS.

- **Location of structure:** Brainstem.

CLINICAL CORRELATION

Chronic fatigue syndrome is a disorder marked by severe chronic mental and physical exhaustion, among other symptoms, arising in a previously healthy and active person. It is a highly debilitating disorder of uncertain etiology. Most cases of CFS begin immediately following a period of stress. While some cases start gradually, the majority start suddenly, often triggered by a viral illness. In acute-onset cases, patients report a sudden, drastic start to their illness, some being able to specify the date or even hour of onset. Many people report having the flu, exposure to an allergen, or an infection such as bronchitis, from which they are never able to fully recover and which evolves into CFS. In some instances, patients claim that vaccination, particularly vaccinations against hepatitis B, is the cause of acute-onset CFS. Other patients suffer from Lyme disease before sinking into CFS.

Patients who experience a gradual onset of CFS may not realize there is anything wrong for quite some time. These patients usually do not seek treatment until the condition is truly debilitating. Although CFS can affect people of any gender, age, race, or socioeconomic group, most patients diagnosed with CFS are 25-45 years old and female. Estimates of how many people are afflicted with CFS vary because of the similarity of CFS symptoms to other diseases and the difficulty in identifying it. The Centers for Disease Control and Prevention (CDC) has estimated that 4-10 people per 100,000 in the United States have CFS. According to the National CFIDS Foundation, about 500,000 adults in the United States (0.3% of the population) have CFS. This probably is a low estimate since these figures do not include children and are based on the CDC definition of CFS used for research purposes, which is very strict. While there is no known cause for CFS, many causes have been proposed. There is evidence that CFS may involve distinct neurological abnormalities, supporting many researchers' classification of CFS as a neurological illness. It is believed that damage to the ascending **RAS**, an area in the brain that extends upward from the reticular formation and is associated with sleep function, may contribute to onset of CFS. Imaging studies of the brain of CFS patients have shown metabolic abnormalities in the RAS, so it seems likely that damage to this area may be responsible for at least some cases of CFS. This damage may be caused by bacterial or viral damage, or an autoimmune attack on the region. Because there is no single identifiable cause for CFS, there is also no single treatment protocol. Frequently, medications such as antidepressants, hormones, and autonomic nervous system stimulants are administered to treat the symptoms of CFS.

APPROACH TO:
The Reticular-Activating System

OBJECTIVES

1. Describe the anatomy of the RAS.
2. Describe the interaction between the RAS and the cortex.
3. Describe the function played by the RAS in REM sleep.

DEFINITIONS

RETICULAR-ACTIVATING SYSTEM (RAS): The area of the brain that plays a major role in arousal and attention. Located within the brainstem, lesions affecting this system lead to impairments in consciousness.

PONTINE TEGMENTUM: Area of the brainstem located in the posterior aspect of the pons. This area houses the fibers and other structures of the reticular-activating system.

DISCUSSION

The **RAS**, composed of the reticular formation and its connections, is often called the attention center of the brain (Figure 34-1). The reticular formation is one of the oldest systems in the nervous system. The brains of primitive vertebrates are almost exclusively made up of a reticular formation. Humans retained this formation over the course of evolution as more organized components of the nervous system appeared. The RAS plays several roles, including nonspecific arousal, cortical activation and tone, and regulating sleep and wakefulness. Injury to the reticular system can cause a change in the level of consciousness, ranging from sleepiness to coma. The medullary levels of the RAS control vital respiratory and cardiovascular centers. Defects in these areas can impair respiratory rate, heart rate, and blood pressure.

The RAS has a diffuse arrangement of both ascending and descending neurons that form a system of networks. It is connected at its base to the spinal cord, where it receives information projected from the ascending sensory tracts, and runs all the way up to the midbrain. As a result, the RAS is a very complex collection of neurons that serve as a point of convergence for signals from the external world and from the interior environment.

The RAS is capable of generating dynamic effects on the activity of the cortex, including the frontal lobes, and the motor activity centers of the brain. The RAS acts as an information filter, managing what data to pass onto the cortex and what data to block. This function is vital, as there is too much competing information at

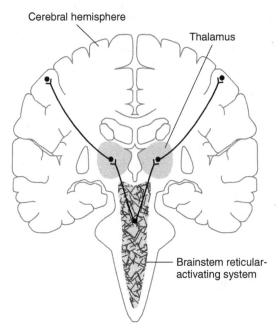

Figure 34-1. Brainstem RAS and its ascending projections to the thalamus and cerebral hemispheres. (*With permission from* Aminoff's Clinical Neurology. *New York: McGraw-Hill; 2005. 6th ed. Chapter 7, figure 7-6.*).

any given time for the brain to process all at once. The RAS also plays a major role in mediating and filtering ascending and descending sensory and motor information. It is the center of balance for the other systems involved in learning, self-control or inhibition, and motivation. When functioning normally, it provides the neural connections that are needed for the processing and learning of information, and the ability to pay attention to the correct task.

Researchers have hypothesized that the RAS also plays a role in anticipatory responding. It signals the cortex, alerting it to prepare to receive stimuli, allowing it to get into a heightened state of readiness. If the RAS fails to excite the neurons of the cortex as much as it should, the underaroused cortex will result in difficulty learning, poor memory, little self-control, and so on. If the RAS fails completely to stimulate the cortex, the result would be a lack of consciousness. Injury of the RAS is directly associated with coma. In contrast, an overstimulated RAS would arouse the cortex or other systems too much and cause restlessness and hyperactivity.

It is also evident that the RAS is the primary mechanism for turning REM sleep on and off. During wakefulness, the RAS maintains cortical arousal. The high activity in the ascending RAS stimulates the brain through projections into different neurological systems in the cortex. However, the midbrain reticular formation becomes activated immediately prior to REM sleep. At this time, the **medullary reticular formation** spurs a postsynaptic inhibition. This results in a loss of muscle tone.

The reticular formation in general decreases sensory input and reduces motor output during REM sleep. REM sleep-on cells, located in the **pontine tegmentum**, are particularly active during REM sleep, and are probably responsible for its occurrence.

The release of the monoamine neurotransmitters (norepinephrine, serotonin, and histamine) is completely inhibited during REM. This causes **REM atonia**, a state in which the motor neurons are not stimulated and thus the body's muscles do not move. Lack of such REM atonia causes REM behavior disorder; sufferers act out the movements occurring in their dreams.

CASE CORRELATES

- See Cases 28-36 (regulatory systems).

COMPREHENSION QUESTIONS

Refer to the following case scenario to answer questions 34.1-34.3:

A 35-year-old man is brought into the hospital following an industrial accident, where a large metal pipe fell and hit him in the back of the head. He is stabilized and placed in the ICU. Following treatment of his obvious traumatic injuries, his heart rate, respiratory drive, and blood pressure are all stable without medical intervention, but he remains unresponsive. An EEG demonstrates continuous slow-wave sleep.

34.1 Based on these findings, where would the physician expect to find a lesion in this man?

    A. Medullary tegmentum

    B. Midbrain tegmentum

    C. Cerebral cortex

    D. Cerebellum

34.2 In addition to maintenance of consciousness and arousal, in what other process does the ascending reticulating activating system (ARAS) participate in a normal brain?

    A. Motor control

    B. Sensory interpretation

    C. Information filtering

    D. Emotional regulation

34.3 Along with its role in consciousness and attention, what role does the ARAS play in sleep?

    A. Initiation of stage 1 sleep

    B. Initiation of REM sleep

    C. Inhibition of muscle tone during stage 4 sleep

    D. Inhibition of thermoregulatory mechanisms during REM sleep

ANSWERS

34.1 **B.** Damage to the **midbrain tegmentum** will sever all connection between the cortex and the ARAS, resulting in a coma as is seen here. The ascending RAS is located in the tegmentum of the brainstem, from medulla to midbrain. Projections from the ARAS to the cortex are necessary for the maintenance of consciousness and awareness. This man has a defect in this system as evidenced by his persistent sleeplike state with slow-wave sleep waves on EEG. Damage to the medullary tegmentum, in addition to causing problems with arousal, would likely damage some part of the heart rate, blood pressure, or respiratory control centers that are part of the medullary ARAS. Since the patient is stable, this does not seem like the likely place for the lesion.

34.2 **C.** The correct answer is **information filtering**. The ARAS sits at a very important point between the cortex, which perceives and analyzes information, and the brainstem, where a huge amount of sensory information, both about the external world and the internal body, is received. There is far too much information sensed at any one time for the cortex to be able to interpret, so this information must be filtered somehow. The ARAS is responsible to filtering this information, and depending on what information gets through, it plays a role in directing selected attention of the cortex.

34.3 **B.** Several areas of the pontine tegmentum that are also part of the ARAS are thought to be involved in the **initiation of REM sleep**. This REM-on system utilizes excitatory acetylcholine projections to the thalamus and cortex to generate REM sleep. Other areas of the ARAS seem to be involved in the inhibition of muscle tone during REM sleep but not during stage 4 sleep.

NEUROSCIENCE PEARLS

▶ RAS is known as the attention center of the brain; injury to the RAS can cause a change in the level of consciousness, ranging from sleepiness to coma.

▶ RAS is the primary mechanism for turning REM sleep on and off.

▶ Release of monoamine neurotransmitters (norepinephrine, serotonin, and histamine) is completely shut down during REM, resulting in REM atonia.

REFERENCES

Bear MF, Connors B, Paradiso M, eds. *Neuroscience: Exploring the Brain*. 3rd ed. Baltimore, MD: Lippincott Williams & Wilkins; 2006.

Purves D, Augustine GJ, Fitzpatrick D, Hall WC, eds. *Neuroscience*. 5th ed. Sunderland, MA: Sinauer Associates Inc; 2011.

Squire LR, Berg D, Bloom FE, du Lac S, eds. *Fundamental Neuroscience*. 4th ed. San Diego, CA: Academic Press; 2012.

A 77-year-old Caucasian man has a history of hypertension, atrial fibrillation, and severe alcoholism. He presents to the emergency department after his wife found him unconscious in their yard. Emergency medical services arrived at their home to find a nonresponsive man with an intact papillary reflex and spontaneous respirations. After obtaining this brief history, you examine the patient. He is still unresponsive, and you notice that his left pupil is significantly larger than his right. His respirations have become noticeably deeper, and he is breathing at a rate of 35 respirations per minute. A CT scan of his head without contrast shows an acute 14-mm left-sided subdural hematoma with midline shift. Based on the presentation, the patient is diagnosed with an altered respiratory drive secondary to an uncal herniation.

- Can other intracranial processes have similar effects on respiratory patterns?
- What is the difference between cortical and brainstem respiratory drives?

ANSWERS TO CASE 35:

Neural Control of Respiration

Summary: A 77-year-old man is noted to have an acute intracranial hemorrhage with increased intracranial pressure and uncal herniation.

- **Other intracranial processes:** Any intracranial process that causes an increase in intracranial pressure that results in herniation can cause uncal herniation (or any other of the herniation syndromes). Hydrocephalus, intracranial neoplasms, foreign bodies, and hemorrhage are all examples of lesions that can raise the intracranial pressure resulting in herniation.

- **Cortical and brainstem respiratory drives:** In general, cortical respiratory drives are involved with voluntary breathing patterns, whereas the brainstem drives control involuntary breathing patterns.

CLINICAL CORRELATION

In this case, the patient has a history of alcoholism and atrial fibrillation. The combination of these factors increases the risk of subdural hematoma. Alcoholism results in cerebral atrophy, which exposes bridging veins to sheering forces during cranial acceleration and deceleration. Atrial fibrillation is usually treated with anticoagulation, such as Coumadin, to prevent intraluminal cardiac thrombus formation. However, in this case, appropriate anticoagulation in the setting of a lacerated bridging vein can result in poor clot formation and further intracranial hemorrhage. The subdural hematoma in this case kept bleeding and resulted in increasing the intracranial pressure until the ipsilateral uncus was forced around and underneath the tentorium, resulting in transtentorial uncal herniation. The herniated uncus exerted unwanted pressure on the brainstem, leading to changes in the respiratory drive.

In addition to this clinical picture the physician would expect to find an increased pH, decreased pCO_2, and normal or increased pO_2 on arterial blood gas measurement.

Any insult to the brainstem, notably the pons and medulla, can result in alterations of respiratory function. Other brainstem lesions that could have similar outcomes include tumors, aneurysms, stroke, hemorrhage, and trauma.

APPROACH TO:

Neural Control of Respiration

OBJECTIVES

1. Know the three brainstem nuclear centers that are involved in respiration and understand their relative functions.

2. Realize there is a difference between automatic and voluntary respiration and how these are structurally represented in the spinal cord.

3. Be able to recognize the different types of pathological respirations.

DEFINITIONS

DORSAL MEDULLARY RESPIRATORY GROUP (DRG): A medullary nucleus containing mostly inspiratory neurons and that is one of the subnuclei of the solitary tract nucleus.

VENTRAL RESPIRATORY GROUP (VRG): A nucleus that is anterolateral to the nucleus ambiguous in the medulla. In its caudal portion, it contains the cell bodies of neurons that fire primarily during expiration and its rostral portion contains cell bodies of neurons that are synchronous with expiration.

PONTINE PAIR OF NUCLEI (PRG): Two nuclei located adjacent to each other in the pons. One fires during the transition from inspiration to expiration, and the other fires during the transition from expiration to inspiration.

PRE-BÖTZINGER COMPLEX: Located in the rostral portion of the VRG. Although its mechanism of action is not entirely clear, it appears to play some role in setting the automaticity of respiration.

DISCUSSION

Respiration regulation is one of the most fundamental homeostatic drives within the nervous system. There are both chemical and mechanical input pathways that influence respiratory patterns, an automated brainstem drive and the voluntary control mechanisms that begin in the **premotor cortices.**

Inspiratory neurons are concentrated in the **DRG** and **rostral VRG.** The fibers controlling automatic respiration course down the white matter tracts of the spinal cord and descend **lateral** to **anterior horn cells** of the **first three cervical** spinal cord sections to terminate on the anterior horn cells of C3-C5. The premotor cortex in the frontal lobe also gives rise to neurons that terminate onto the same anterior horn cells. These tracts containing the fibers controlling voluntary respiration course more dorsally in the cervical cord. If the ventral tracts are damaged, then automatic respirations are lost while voluntary are preserved. The **third, fourth, and fifth** cervical (C3-C5) segments project fibers that will ultimately become the **phrenic nerve** and will innervate the **diaphragm.** Although normal expiration is a passive process, there are clusters of expiratory neurons that provide upper motor innervation of accessory respiratory muscles as well as creating an **inhibitory force** on the **inspiratory neurons.** The caudal portion of the VRG and the rostral portion of the DRG contain these expiratory nerve cell bodies. There is some evidence suggesting the **PRG** serve as **binary switches** that control the transition between inspiration and respiration.

The precise mechanism accounting for the brainstem's ability to generate adequate respiration is unknown. There is a region in the VRG, the **pre-Bötzinger complex,** which appears to be involved with **autonomous rhythmicity.**

The **carotid sinus** is sensitive to hypoxia and changes in pH. Afferent fibers merge into the **glossopharyngeal nerve** and terminate in the **solitary tract nucleus**. There are also medullary chemoreceptors that detect pH changes in the extracellular fluid. J-type receptors detect material in the interstitial fluid of the lungs and can stimulate increased respiration.

When structural or metabolic factors divorce the brainstem respiratory centers from the cerebrum, **Cheyne-Stokes** respirations may result. This pattern of breathing is **alternating hyperpnea with hypopnea** that ends in **apnea** and then repeats itself. Bilateral hemispheric lesions, large unilateral hemispheric lesions, or metabolic encephalopathies can cause Cheyne-Stokes respiration. Because of the separation of communication between the brainstem centers and cerebral function, **carbon dioxide** accumulates until it triggers chemoreceptors to stimulate inspiration. This results in hyperpnea. As carbon dioxide is gradually removed from the body, the chemoreceptors fire less frequently until apnea occurs.

Central neurogenic breathing usually occurs with **midbrain-pontine** lesions. In this case, the minute ventilation is increased because both tidal volume and respiratory rate are increased. Thus, the pCO_2 drops and hyperventilation persists. This type of breathing is usually seen in **transtentorial uncal herniation**, as in the example in Case 2.

Pontine lesions often result in **apneustic breathing**, which consists of a pause between inspiration and expiration.

Medullary lesions result in completely disordered, or **ataxic**, breathing. It is thought that these pathological forms of respiration are interrelated and most patients will progress through various stages before complete respiratory failure ensues.

CASE CORRELATES

- See Cases 28-36 (regulatory systems).

COMPREHENSION QUESTIONS

Refer to the following case scenario to answer questions 35.1-35.3:

A 43-year-old woman with known insulin-dependent diabetes presents to the emergency department complaining of nausea, vomiting, generalized weakness, and increased urinary frequency and amount. She states that she has not taken any insulin for 4 or 5 days because she does not have the money to pay for it. On examination, she has a fruity odor to her breath and has a respiratory rate of 35 breaths per minute. A fingerstick blood glucose level of 573 mg/dL and an arterial blood gas test shows her pH to be 7.12. The physician correctly diagnoses her with diabetic ketoacidosis and begins appropriate therapy.

35.1 Her increased respiratory rate is at least in part because of afferent signals coming from which peripheral receptor that measure blood pH?

A. Carotid body

B. The J-type receptor in the lung

C. Medullary chemoreceptor

D. Carotid sinus

35.2 The increased signal from the peripheral chemoreceptors stimulates which CNS site to increase respiratory rate?

A. DRG

B. PRG

C. Premotor cortex

D. Pre-Bötzinger complex

35.3 Efferent signals from the central respiratory centers travel by what path to the diaphragm in this patient?

A. Via ventral respiratory spinal tracts to C3-C5

B. Via dorsal respiratory spinal tracts to C3-C5

C. Via ventral respiratory spinal tracts to C2-C4

D. Via dorsal respiratory spinal tracts to C2-C4

ANSWERS

35.1 **D. The carotid sinus**, located at the bifurcation of the internal and external carotid arteries and innervated by the glossopharyngeal nerve, measures arterial blood pH. It responds to increased concentration of hydrogen ions (decreased pH) by increasing its rate of firing, which stimulates central respiratory centers to increase respiratory rate. This increased respiratory rate will "blow off" excess carbon dioxide, thereby partially compensating for the acidemia. Medullary receptors also respond to decreased pH, but they are centrally located and do not directly measure blood pH but rather the pH of the extracellular fluid. The J-type receptor responds to changes in the interstitial fluid of the lungs.

35.2 **D. Pre-Bötzinger complex** is involved in automaticity. The dorsal and rostral ventral medullary respiratory groups are the primary sites responsible for inspiratory drive. They receive afferent connections from the carotid sinus and the medullary chemoreceptors, and other sites in the body converge in the nucleus of the solitary tract and from there project to the breathing centers, primarily the dorsal respiratory group. The PRG is involved in cycling from inspiration to expiration, the premotor complex is involved in voluntary breathing, and the pre-Bötzinger complex is involved in automaticity.

35.3 **A.** Because the ventral respiratory tracts carry signals related to involuntary respiration, and the diaphragm is innervated by the phrenic nerve, which carries fibers from spinal levels C3-C5, the correct answer is **ventral respiratory spinal tracts to C3-C5**. The anterior horn cells projecting fibers C3-C5 receive signals from both dorsal and ventral respiratory tracts in the spinal cord, but the ventral tracts, located lateral to the anterior horn, carry signals related to involuntary respiration, while the more dorsally located respiratory tract carries signals related to voluntary respiration.

NEUROSCIENCE PEARLS

▶ Inspiratory neurons are concentrated in the DRG and rostral VRG.

▶ The C3-C4 segments become the phrenic nerve, which innervates the diaphragm.

▶ The carotid sinus is sensitive to hypoxia andchanges in pH.

▶ When the connection between the respiratory centers and cerebrum is completely destroyed, Cheyne-Stokes respirations may result.

▶ Pontine lesions result in apneustic breathing, while medullary lesions result in ataxic breathing.

REFERENCES

Bear MF, Connors B, Paradiso M, eds. *Neuroscience: Exploring the Brain*. 3rd ed. Baltimore, MD: Lippincott Williams & Wilkins; 2006.

Kandel ER, Schwarz JH, Jessell TM, Siegelbaum SA, Hudspeth AJ, eds. *Principles of Neural Science*. 5th ed. New York, NY: McGraw-Hill; 2012.

Purves D, Augustine GJ, Fitzpatrick D, Hall WC, eds. *Neuroscience*. 5th ed. Sunderland, MA: Sinauer Associates Inc; 2011.

A 30-year-old man presents to the emergency room (ER) with a main complaint of painful priapism. He estimates the time of onset to be about 8 hours prior to admission to the ER. The patient also complains of severe muscle cramps in his arms and legs and seems anxious and irritable.

On physical examination, no signs of trauma to the spinal cord are found. Patient denies having injected any sort of substance into his penis, but visible track marks can be found on both his arms. He admits to sustained heroin use but claims to have been clean for several days. Based on this history, you inform the patient that these symptoms are most likely secondary to heroin withdrawal.

▶ What receptors does heroin act on?
▶ What would be an appropriate treatment for the patient's symptoms?

ANSWERS TO CASE 36:
Addiction

Summary: A 30-year-old man presents with priapism and severe muscle cramps in his limbs. Patient has a history of sustained heroin use but claims to have ceased usage.

- **Receptors:** Opioid receptors.

- **Treatment:** Because the elapsed time since onset has exceeded 6 hours, treatment of patient's priapism through medication is no longer an option. Surgical shunt and aspiration of the penis are the most viable procedures. Buprenorphine is recommended as a substitute opioid to help ease other symptoms of heroin withdrawal.

CLINICAL CORRELATION

Heroin withdrawal syndrome may manifest in a patient within 6-24 hours of discontinuation of sustained use of the drug. This time may vary depending on the degree of tolerance of the patient, as well as the amount of heroin in the last dose. Symptoms of heroin withdrawal include sweating, malaise, anxiety, depression, priapism in men, hypersensitivity of the genitals in females, a general feeling of heaviness, cramplike pains, yawning, insomnia, cold sweats, chills, severe muscle and bone aches, nausea and vomiting, diarrhea, goose bumps, and fever.

Many symptoms of opioid withdrawal are caused by rebound hyperactivity of the **sympathetic nervous system,** which can be suppressed using clonidine, a centrally acting α2 agonist primarily used to treat hypertension. Baclofen, a muscle relaxant, is often used to treat leg twitches, another symptom of withdrawal. Diarrhea can be treated with the peripherally active opioid drug loperamide. One of the most widely used opioid substitutes in the treatment of heroin withdrawal is **buprenorphine,** a partial opioid agonist/antagonist. Buprenorphine develops a lower grade of tolerance than heroin and results in less severe withdrawal symptoms when discontinued abruptly. Buprenorphine acts as a κ-opioid receptor antagonist, while simultaneously acting as a partial agonist at the same α-receptor where opioids like heroin exhibit their action. Because of the effects of buprenorphine on this receptor, patients with high tolerances are unable to achieve any euphoric effects from other opioids while using buprenorphine. There are three known **opioid antagonists** currently being used in the treatment of opioid addiction: naloxone and the longer-acting naltrexone and nalmefene. These medications act by blocking the effects of heroin and other opioids at the receptor sites.

APPROACH TO:

Addiction

OBJECTIVES

1. Describe the development of addiction.

2. Identify the mesolimbic dopaminergic system and its role in reward-related learning.

3. Describe the nature of drug tolerance and its implications in the withdrawal symptoms that occur on discontinuation of the drug.

DEFINITIONS

BUPRENORPHINE: A semisynthetic opioid analgesic used for the relief of moderate-to-severe pain.

OPIOID ANTAGONIST: A receptor antagonist that acts on opioid receptors, blocking the effects of opioids.

DISCUSSION

The development of **addiction** is thought to be a simultaneous process of increased focus on and engagement in a particular behavior and the simultaneous attenuation or shutting down of other behaviors. For instance, in certain experimental circumstances, test animals are allowed the ability to self-administer certain psychoactive drugs. Given an unlimited supply of the drug, the animals will show an extremely strong preference for it, forgoing food, sleep, and sexual intercourse in order to maintain access to the drug. From a neuroanatomical standpoint, it can be argued that the mechanisms involved in driving goal-directed behavior become gradually more selective for certain stimuli and rewards, exceeding the point at which the mechanisms involved in behavior inhibition can effectively preclude the action. In this case, the limbic system is thought to be the major driving force, and the orbitofrontal cortex is the substrate of the top-down inhibition.

The pleasure-reward processing of the brain is located in the limbic system. The **mesolimbic dopaminergic system** is the precise portion of the limbic system, which translates to motor behavior-learning and reward-related learning. This system is comprised of the **ventral tegmental area (VTA)**, the **nucleus accumbens**, and the bundle of dopamine-containing fibers that connect them. Located on the **anterior cingulated circuit**, the portion of the frontal lobe that incorporates much of the brain's motivational pathways, is the **ventral striatum**. This locus is of importance because it is where the nucleus accumbens are situated and where the release of dopamine takes place, a process that is believed to be a critical mediator of the reinforcing effects of stimuli, including drugs of abuse. This system is commonly implicated in the seeking out and consumption of rewarding stimuli or events, such as sweet-tasting foods or sexual interaction. However, its importance to addiction research goes beyond its role in "natural" motivation: while the specific site or mechanism of action may differ, all known drugs of abuse have the common effect in that they elevate the level of dopamine in the nucleus accumbens. This may happen directly, via the blockade of the dopamine reuptake mechanism, as is the case with cocaine use. It may also happen indirectly, such as through stimulation of the dopamine-containing neurons of the VTA that synapse with neurons in the accumbens, which occurs during opiate use. The euphoric effects of drugs of abuse are a direct result of the acute increase in accumbal dopamine.

The central nervous system, like the rest of the human body, has a natural tendency to maintain an internal equilibrium, or **homeostasis**. Prolonged elevated levels of dopamine will spur a decrease in the number of dopamine receptors. This process, known as **downregulation**, causes a change in postsynaptic cell membrane permeability. This in turn makes the postsynaptic neuron less excitable and less responsive to chemical signaling with an electrical impulse, or **action potential**. The resultant unresponsiveness of the brain's reward pathways contributes to an inability to feel pleasure, known as anhedonia, a phenomenon often observed in addicts. After the onset of **anhedonia**, a greater amount of dopamine is required to maintain the same electrical activity. This is the basis of physiological tolerance of a drug and the withdrawal syndrome associated with addiction.

Contrary to popular belief, drug overdoses are generally not the result of a user taking a higher dose than is typical but rather administering the same dose in a new environment. If a behavior occurs repeatedly and consistently in the same environment or contingently with a particular cue, the brain will adjust to the presence of these cues by decreasing the number of available receptors in the absence of said behavior.

Withdrawal symptoms occur on the absence of substances that the body has become physically dependent on. These substances include depressants of the central nervous system such as opioids, barbiturates, and alcohol. Withdrawal from alcohol or sedatives like barbiturates or benzodiazepines can cause seizures and may result in death. However, withdrawal from opioids, while still extremely uncomfortable, is rarely life threatening. In situations of particularly severe anhedonia, the body is so accustomed to high concentrations of a substance that it no longer produces its own natural versions. Instead, it produces opposing chemicals. When delivery of the substance is halted, the effects of the opposing chemicals can be devastating. For example, in instances of chronic sedative use, the body counters by producing chronic levels of stimulating neurotransmitters such as glutamate. High concentrations of glutamate can be toxic to nerve cells. This scenario is called **excitatory neurotoxicity.**

CASE CORRELATES

- See Case 28 (neurogenic diabetes insipidus), Case 29 (neuroendocrine axis), Case 30 (hyperthermia), Case 31 (sympathetic nervous system), Case 32 (parasympathetic nervous system), Case 33 (sleep and the limbic system), Case 34 (reticular-activating system), and Case 35 (breathing, neural control).

COMPREHENSION QUESTIONS

Refer to the following case scenario to answer questions 36.1-36.3:

You are working in a rehabilitation clinic, performing counseling for patients who are trying to stop using drugs of abuse, particularly heroin. A 33-year-old man is telling his story, about how his use of heroin cost him his job, his wife, his kids, his house, essentially everything he had. He says he knew it was destroying his life, but he just could not stop. He describes the amazing euphoric feeling he got whenever he used, and says that without the drug, he just cannot get that feeling anywhere.

36.1 Release of what neurotransmitter into the nucleus accumbens is most commonly associated with the euphoric effect common to almost all drugs of abuse?

A. Norepinephrine

B. Acetylcholine

C. Serotonin

D. Dopamine

36.2 What neural tract that connects the VTA to the nucleus accumbens is commonly thought of as the "pleasure center" of the brain?

A. Medial longitudinal fasciculus

B. Medial forebrain bundle (MFB)

C. Mammillothalamic tract

D. Spinothalamic tract

36.3 What is the molecular mechanism that causes more and more drug to be needed to achieve the same euphoric effect each time the drug is used and that also accounts at least in part for the withdrawal symptoms following cessation of a drug?

A. Receptor downregulation

B. Decreased receptor sensitivity

C. Exhaustion of dopamine stores

D. Depletion of acetylcholine

ANSWERS

36.1 **D.** The neurotransmitter most frequently associated with euphoria from drugs is **dopamine**. In a normal brain, dopamine release in the nucleus accumbens is associated with reward and helps to drive behavior. It can be naturally released through things like sexual activity and sweet foods.

36.2 **B.** The **MFB** is a tract of dopaminergic axons that project from the VTA to the nucleus accumbens. When stimulated, this tract releases dopamine onto the nucleus accumbens, which results in a euphoric feeling. In experimental animals, it has been shown that animals will autostimulate the MFB with an electrode to the detriment of everything to the point that they actually starve to death. This dopaminergic tract plays a very important role in reward and motivational drive, as well as addiction.

36.3 **A.** The phenomenon in which more and more drug is needed each time to achieve the same effect is known as tolerance, and the molecular mechanism responsible for this phenomenon is **downregulation of postsynaptic dopamine receptors** in the nucleus accumbens. Excessive stimulation by dopamine causes the postsynaptic cells to decrease the amount of dopamine receptors they express, resulting in less effect for the same amount of dopamine release. The individual receptors are just as sensitive, there are simply fewer of them.

NEUROSCIENCE PEARLS

▶ The portion of the limbic system responsible for the translation of motivation to motor behavior-related and reward-related learning is the mesolimbic dopaminergic system.

▶ The euphoric effects of drugs of abuse are a direct result of the acute increase in dopamine in the nucleus accumbens.

▶ Prolonged elevated levels of dopamine will spur a decrease in the number of dopamine receptors, a process known as downregulation.

REFERENCES

Bear MF, Connors B, Paradiso M, eds. *Neuroscience: Exploring the Brain.* 3rd ed. Baltimore, MD: Lippincott Williams & Wilkins; 2006.

Kandel ER, Schwarz JH, Jessell TM, Siegelbaum SA, Hudspeth AJ, eds. *Principles of Neural Science.* 5th ed. New York, NY: McGraw-Hill; 2012.

Squire LR, Berg D, Bloom FE, du Lac S, eds. *Fundamental Neuroscience.* 4th ed. San Diego, CA: Academic Press; 2012.

A 26-year-old right-handed Asian woman presents to the clinic with frequent burning and numbness in the palm of her right hand. The symptoms first began at night when the patient would feel the need to "shake out" her right hand, but now her hand frequently feels tingly and weak during the day. She works as an assembler in a manufacturing factory, and the symptoms have begun to interfere with her job. On neurological examination, the Tinel sign (applying pressure to the palmar aspect of her wrist) tested positive—the patient noted a shock-like sensation to her fingers. There was no indication of neurological deficit in any other area of the body. The patient also has a history of diabetes and was ultimately diagnosed with carpel tunnel syndrome.

- ▶ What nerve is likely affected?
- ▶ What is the pathophysiological mechanism of her symptoms?

ANSWERS TO CASE 37:
Axonal Injury

Summary: A 26-year-old Asian woman complains of progressive pain in the palm of her right hand.

- **Nerve affected:** Median nerve.

- **Mechanism of symptoms:** Carpal tunnel is the result of increased pressure and compression of the median nerve and tendons in the carpal tunnel. The syndrome is usually owing to having a smaller carpal tunnel, trauma and injury to the wrist, work stress, or other mechanical problems of the wrist joint.

CLINICAL CORRELATION

Carpal tunnel is more prevalent among populations of women, diabetics, and assembly-line workers because of a smaller carpal tunnel, related nerve effects that increase susceptibility to compression, and increased wrist trauma, respectively. This is a peripheral nerve disorder caused by a less obvious manifestation of axonal injury. The **compression of the median nerve** may have been traumatic enough to induce intrinsic mechanisms of axon cell death discussed below and in some cases can be severe enough to sever the axon but leave the basal lamina sheath of nerve intact to guide regeneration. Carpal tunnel represents the most common of the entrapment neuropathologies that are caused by the chronic compression of peripheral nerves, in this casethe median nerve, resulting in pain or loss of function. Affected patients complain of paresthesias of the thumb, index and middle finger. If more extensive, there can be motor impairment, such as weakness or atrophy of the thenar muscle (causing weakness of thumb opposition).

Entrapment syndromes result from chronic injury to a nerve as it travels through an osseoligamentous tunnel; the compression usually is between ligamentous and bony surfaces. Pathophysiology includes microvascular (ischemic) changes, edema, dislocation of the nodes of Ranvier, and structural alterations in membranes at the organelle level in both the myelin sheath (ie, focal segmental demyelination) and the axon. Severe cases of entrapment can result in Wallerian degeneration of the axons and permanent fibrotic changes in the neuromuscular junction that prevent reinnervation. If symptoms persist for 6 months or longer following nonsurgical treatment, this may be an indication of the continued physical compression of the nerve by the band of tissue surrounding the wrist that inhibits axon regeneration. This band is cut during surgery to reduce pressure on the median nerve.

APPROACH TO:

Axonal Injury

OBJECTIVES

1. Be able to understand disorders resulting from the chronic compression of peripheral nerves.

2. Know how to alleviate neuronal death and atrophy by experimental application of various trophic factors such as NGF.

DEFINITIONS

L-TYPE CALCIUM CHANNEL BLOCKERS: These blockers inhibit L-type voltage-dependent calcium channels. L-type channels have large sustained conductance, inactivate slowly, are responsible for the plateau phase of the action potential, and may trigger internal release of calcium ions.

CALPAINS: These are the calcium ion-dependent proteolytic enzymes that modulate cellular function. Cytoskeleton molecules are the main substrate for calpains; thus, calpain activation causes the degradation of the axonal cytoskeleton. Intuitively, inhibition of calpains can also prevent axonal degeneration in vitro.

TARGET-DERIVED TROPHIC FACTORS: Growth factors derived from the target of intended neuronal growth; for instance, a target neuron from which another neuron couldsynapse can release target-derived trophic factors to encourage neural growth.

RETROGRADE CELL DEATH: Following axonal injury retrograde cell death is the death of the neuron associated with the injured axon, usually following the retraction of the axon away from the severance site.

ANTEROGRADE CELL DEATH: Following axonal injury, anterograde cell death is the death of the neuron that synapses with the injured axon.

NERVE GROWTH FACTOR (NGF): A small protein secreted from target cells that causes differentiation, survival, and maintenance of sympathetic and sensory neurons.

CHROMATOLYSIS: The degradation of a chromophil substance, such as chromatin, within the nerve cell body. Typically occurs after peripheral cell damage or cell exhaustion.

DISCUSSION

Most nervous system damage will result in some sort of axonal injury. They are uniquely vulnerable, given their elongated physiology and how far removed they are from the cell body, from which they derive proteins for homeostasis and function (see Figure 37-1). When an axon is cut, it is necessarily removed from its source of protein synthesis compromising both the axon and its surrounding myelin. After injury,

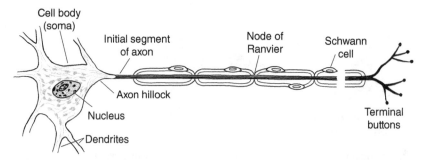

Figure 37-1. Motor neuron with myelinated axon. (*With permission from* Ganong's Review of Medical Physiology. *22nd ed. New York: McGraw-Hill; 2005. Page 48, figure 2-2.*).

there is an active inherent mechanism that causes axon death that is not dependent on intervention from outside cells for degeneration. This mechanism of axonal death harkens back to basic neuronal function of synaptic transmission.

There are two phases of calcium entry into the axon after injury. A diffusion of calcium ions from the extracellular pool occurs immediately, and is interrupted once the axonal membrane reseals. The second movement of calcium ions into the cell precedes axonal degeneration and takes placethrough the voltage-gated ion channels (L-type channels), as well as the sodium-calcium exchange pump. Entry of calcium through the first route is a consequence of axon depolarization; that is, a result of the axon losing its ability to maintain a normal resting membrane potential. Degeneration of the sensory axons that have been damaged has been blunted by **L-type calcium channel blockers**, which reduces the entry of extracellular calcium into the axon.

The sodium-calcium exchange pump that drives the second route of calcium entry is actually running in reverse following axonal injury. Normally, allowing the movement of sodium ions down concentration and potential gradients across the membrane drives ATP production that powers the sodium-calcium exchange pump to remove calcium from the cell. However, if the membrane becomes depolarized because of action potential activity or exchange pump failure, and the sodium gradient diminishes then the high concentration of extracellular calcium can drive the sodium-calcium counter-transport pump in reverse. The end result is a massive increase in calcium concentration within the axon exceeding the 200-μm threshold for calcium-activated enzymes, **calpains**, and possibly other mechanisms as well. Cytoskeletal molecules are the main enzymatic substrates of calpains; thus, calpain activation causes the degradation of the axonal cytoskeleton. Intuitively, inhibition of calpains can also prevent axonal degeneration in vitro.

The fallout from axonal injury can also extend beyond the axon distal to the cut to the neuron itself. Neuronal death can occur by apoptosis or necrosis. Death via apoptosis is usually caused by the loss of **target-derived trophic factors** that provide support to the neuron, or because of the intense influx of calcium mentioned earlier. Rapid necrotic cell death mainly occurs when the axon is damaged very close to the cell body, usually explained by a combination of physical trauma to the membrane and cytoskeleton of the neuron, and, again, the calcium influx that results from axonal membrane disruption. Both neurons in the vicinity of and some distance away from the lesion can be lost via apoptosis days after the injury. Adding on to the **retrograde cell death** of neurons whose axons have been cut, there is also some **anterograde cell death** of the neurons to which those damaged axons connect by apoptosis. Neuronal death in both these fashions is at least partly because of loss of access to trophic factors.

Trophic support for a neuron can be derived from several sources, including target neurons from which it synapses; neurons that synapse from it; Schwann cells, astrocytes, and oligodendrocytes that contact it at different points; and microglia that cluster around damaged neurons. Following axonal injury, a neuron clearly loses contact with its target cells and also loses their accompanying trophic support. Synapses on the cell soma have also been shown to retract, replaced first by microglia (the astrocytes). A new source of trophic support from the glial cells surrounding the area of damage replaces the support lost from the damaged neuron's efferent and afferent connections. Astrocytes become reactive in damaged regions of the CNS. Microglia also migrate to these regions where they undergo mitosis and become activated as well, producing a variety of neurotrophic and toxic **cytokines**. In the PNS, Schwann cells are activated following axonal injury which leads to a rapid accumulation of macrophages that secrete toxic molecules. Furthermore, neuronal death and atrophy can be alleviated by the experimental application of various trophic factors such as **NGF**. Thus, the overriding consensus is that loss of trophic support results from axonal damage since many neurons following axonal injury die or become atrophic.

Neurons that survive axonal injury undergo a predictable and well-described set of changes. Many of these changes are associated with the reinitiation of protein synthesis for axon growth and regeneration. Anatomically, these major changes in the pattern and quantity of protein synthesis are manifested in the neuron as **chromatolysis**, the dispersal of the large granular condensations of the rough endoplasmic reticulum accompanied by changes in the appearance of the nucleus. Many of the genes that underlie these changes are upregulated or downregulated after axonal injury and have been identified in the following categories: transcription factors, growth-associated proteins, cytoskeletal proteins, growth factor receptors and growth factors, and cytokines.

CASE CORRELATES

- See Cases 37-40 (nervous system damage and repair).

COMPREHENSION QUESTIONS

Refer to the following case scenario to answer questions 37.1-37.3:

A 17-year-old adolescent boy is dropped off outside the emergency department by some "friends" after having been shot in the arm. He will not discuss any specifics of the events surrounding the injury and does not know the caliber or type of bullet he was shot with. On examination there is a through-and-through wound of the proximal forearm, just distal to the cubital fossa. There is some bleeding from the wound, but surprisingly little, considering the proximity of the wound to numerous relatively large arteries. On examination, however, the patient has no ability to flex any of his fingers, and weak flexion of his wrist. He also has sensory loss of his volar forearm and his palmar thumb, index, and middle finger. The physician suspects that he has a traumatic transection of his median nerve.

37.1   The entry of what ion into the axon distal to the transection will activate cellular digestive enzymes that will degrade the cytoskeleton?

   A.   Sodium

   B.   Chloride

   C.   Magnesium

   D.   Calcium

37.2   In addition to degeneration of the axon distal to the site of injury, the entire nerve proximal to the injury may die as well. The absence of what substance, normally transported to the cell body via retrograde transport down the axon, can account for this cellular demise?

   A.   Neurotransmitter vesicle proteins

   B.   Unused peptide neurotransmitters

   C.   ATP

   D.   Neurotrophic factors

37.3   Following nerve injury in the peripheral nervous system, what glial cell type surrounding damaged neurons becomes active?

   A.   Schwann cells

   B.   Astrocytes

   C.   Microglia

   D.   Macrophages

ANSWERS

37.1 **D. Calcium** enters into the axonal cytoplasm following injury by two mechanisms. Initially, it enters through the disruption in the cellular membrane by simple diffusion, but this method is stopped as the membrane reseals. The second method is through voltage-gated calcium channels that open as the neuron depolarizes when it is cut off from the cell body and via the sodium-calcium exchange pump that runs in reverse in this pathological condition. The calcium in the cell quickly reaches threshold for activation of calpains, which begin to degrade cellular proteins (cytoskeletal elements in particular).

37.2 **D.** This cellular demise is because of absence of **neurotrophic factors** necessary for neuron survival. Normally, a neuron can derive trophic support froma variety of sources, including the cell it innervates. Trophic support is the supply of small molecules, like NGF, that are necessary for the survival of the cell. When a neuron is cut off from its supply of trophic factors, as is the case with axonal injury, apoptotic pathways within the cell become activated, resulting in cell death.

37.3 **A. Schwann cells,** the primary supporting glial cell of the PNS, become activated following axonal damage. When activated, they secrete a variety of cytokines, some of which can serve as neurotrophic factors which can prevent apoptotic neuronal death. They also secrete cytokines that are chemotactic for macrophages, which help to remove the damaged axon but are not themselves glial cells. Astrocytes and microglia are both active around damaged neurons in the CNS.

NEUROSCIENCE PEARLS

▶ Carpal tunnel syndrome is caused by compression of the median nerve at the wrist. Symptoms are tingling and pain of the thumb, index, and middle finger.

▶ Carpal tunnel syndrome is associated with repetitive hand movements, hypothyroidism, diabetes, dialysis associated amyloidosis, and pregnancy.

▶ Death of neurons via apoptosis is usually because of the loss of target-derived trophic factors or to the intense influx of calcium especially after axonal membrane disruption.

▶ Both neurons in the vicinity of and some distance away from the lesion can be lost via apoptosis days after the injury.

▶ Neuronal death and atrophy can be alleviated by the experimental application of various trophic factors such as NGF.

REFERENCES

Kandel ER, Schwarz JH, Jessell TM, Siegelbaum SA, Hudspeth AJ, eds. *Principles of Neural Science*. 5th ed. New York, NY: McGraw-Hill; 2012.

Purves D, Augustine GJ, Fitzpatrick D, Hall WC, eds. *Neuroscience*. 5th ed. Sunderland, MA: Sinauer Associates Inc; 2011.

Squire LR, Berg D, Bloom FE, du Lac S, eds. *Fundamental Neuroscience*. 4th ed. San Diego, CA: Academic Press; 2012.

CASE 38

A 40-year-old Caucasian man presents from the psychiatry unit exhibiting mood swings, increased irrationality, excessive weight loss, and with a recent history of attempted suicide. Changes in behavior were first noticed by alarmed family members who urged the patient to visit a psychiatrist. No psychological basis has been found for these sudden changes except a somewhat vague history in the patient's father of similar midlife crisis before passing away years later of pneumonia at home, and a neurological deficit is suspected. Neurological examination revealed uncontrolled dyskinesias of the arms and legs, and the patient remarked about having difficulties eating that have been getting worse. Imaging showed progressive degeneration of the striatum, with enlarged lateral ventricles and widened intercaudate distance. Genetic analysis revealed abnormality on chromosome 4 consisting of a polyglutamine repeat on a gene coding for a protein of unknown function.

▶ What is the most likely diagnosis?
▶ What is the molecular genetics behind the disorder?
▶ What is the cellular pathology and pathogenesis of this disease?
▶ What neurotrophins have been targeted for therapy?

ANSWERS TO CASE 38:

Nerve Growth Factors

Summary: A 40-year-old Caucasian man has physical and behavioral neurological deficits with an appreciable degeneration of the striatum and a genetic abnormality on chromosome 4.

- **Most likely diagnosis:** Huntington disease (HD).

- **Molecular genetics:** HD is an autosomal dominant disease inherited by 50% of the children of an affected parent, who will all express the disease and pass on the gene with the same probabilities. The critical gene codes for huntingtin, a protein of unknown function, and is located on chromosome 4. The gene contains a polyglutamine (polyQ) repeat that is normally 5-36 amino acids in length, but is lengthened in individuals with HD.

- **Cellular pathology and pathogenesis:** How the expanded polyQ repeat results in neuronal degeneration in the striatum is currently unknown. Abnormal inclusion bodies have been found in affected striatal neurons with heavy huntingtin staining, but whether this is a cause or consequence of cell death remains to be explored. The relationship between disturbed mitochondrial function observed in HD and pathology also continues to be delineated.

- **Neurotrophins:** Brain-derived neurotrophic factors are required for the correct activity of the corticostriatal synapse and the survival of the GABAergic striatal medium spiny neurons that die in HD. Neurotrophin 4/5 has been chosen for HD therapy because it increases the support of interneurons, so is protective for the more vulnerable medium spiny neurons. Also, FGF-2 has been targeted because it stimulates the development of new spiny neurons from stem cells already present in the brain.

CLINICAL CORRELATION

Huntington disease is an autosomal dominant neurodegenerative disorder that usually presents at the ages of 30-40 years. There can be aggressiveness, anxiety, and abnormal (choreiform) movements. The disease is marked by progressive degeneration of the striatum, usually identified by enlarged ventricles and widened intercaudate distance on MRI and CAT scans. Cell loss spreads in a dorsal medial to lateral ventral direction, and the GABA-containing neurons of the **caudate nucleus are usually affected earlier** and more extensively than those of the putamen. Medium spiny neurons are the most affected along with mild gliosis at the cellular level. As the disease progresses to advanced stages, other brain nuclei, particularly those of the basal ganglia, are affected. Cortical input atrophy and cell loss in the pallidal and substantia nigra pars reticulate outputs of the striatum are indications of advanced HD and translate into clinical signs of severe disability, and later death due to infection, pneumonia, or inanition years after diagnosis.

Currently there are no effective treatments available for HD, only drugs to suppress the motor and psychiatric symptoms. Neurotrophic protection in the striatum has been heavily investigated as a treatment for HD. Striatal neurons in models of the human disease can be rescued with application of a number of different nerve growth factors, including NGF, BDNF, NT-4/5, β-FGF, TGF-α, CNTF, and GDNF. The most important issue in developing a neurotrophin-based therapy for HD is the method of delivery of trophic molecules;either through direct infusion or through the use of genetically engineered cells.

APPROACH TO:
Nerve Growth Factors

OBJECTIVES

1. Describe the role of nerve growth factors in determining neuron survival.

2. Describe the typical brain pathological findings of a patient with HD.

3. Describe the speculated role of brain-derived neurotrophic factors in HD.

DEFINITIONS

ONTOGENETIC CELL DEATH: A cycle of massive neuronal cell death that occurs during development that serves a number of purposes for neural maturation. For instance, ontogenetic cell death occurs to eliminate redundant and multiplicative neural connections to the same target.

TRK RECEPTORS: Trk receptors belong to the tyrosine kinase class of receptors and are the receptor-binding sites on cells for many important neural growth factors.

CHOLINERGIC NEURONS: Neurons that synthesize and release acetylcholine, which functions as a neurotransmitter in the PNS and CNS during interneuronal communication.

SUBSTANTIA NIGRA: Large, pigmented, dopamine-producing cells that are layered in the mesencephalon. These cells have been implicated in Parkinson disease.

AUTOCRINE TROPHIC SUPPORT EFFECT: A motor neuron experiencing axonal injury is capable of achieving an autocrine trophic support effect of self-stimulation by producing its own tyrosine kinase B (TrkB) and BNDF. The extent to which this effect can delay cell death depends on the cell age and the position at which the axon is damaged.

ACETYLCHOLINESTERASE: Enzyme that removes acetylcholine from the synaptic cleft to allow repolarization to occur. It does this by converting acetylcholine into inactive choline and acetate through hydrolysis. Choline acetyltransferase synthesizes acetylcholine from choline and acetyl CoA.

DISCUSSION

Nerve growth or trophic factors have a well-characterized role in determining which neurons survived into adulthood, and which axonal processes connected to particular targets. Nerve growth factor or **NGF** was the first to be identified, and has served as a prototype for all that followed. NGF seems to have a role in keeping nerve cells alive (sensory and all sympathetic neurons) as well as inducing apoptosis during development. This factor is mainly secreted by the target tissues innervated by the sympathetic sensory neurons. When developing embryos are deprived of NGF, many of these neurons die before birth. It has been demonstrated in vitro that for neuronal survival it is essential that axons make successful connections to NGF-secreting tissue, and NGF applied to any part of the neuron, including the end of an axon, will promote survival. Aside from NGF, there are other trophic factors that have various roles in the adult CNS and PNS affecting neuronal and glial cell survival and plasticity; these trophic factors are active in both the normal adult brain as well as in regenerative responses to injury.

A few major families among the wide spectrum of molecules with neurotrophic effects account for most of the functional effects on neuroplasticity and regeneration. Factors NGF, BDNF, NT-3, and NT-4/5 constitute the first major family of centrally acting growth factors, and are structurally related to NGF. These factors bind to the **Trk receptors** in addition to a low-affinity NGF receptor, p75, which binds to NGF and BDNF. TrkA is found on some neuronal types, usually **cholinergic neurons**, while TrkB and TrkC are widely distributed throughout the nervous system, found on most CNS neurons. Ciliary neurotrophic factor (CNTF) receptors are also widely distributed in the CNS. The molecule itself is found in Schwann cells in peripheral nerves and is produced by astrocytes after CNS injury. Factor FGF-2 also has receptors widely distributed on neurons and glial cells, especially after injury. Astrocytes are the main sources of FGFs which promote survival and proliferation of most populations of neurons in the developing CNS. Factors FGF-2 and EGF have been specifically implicated in the promotion of division and proliferation of CNS progenitor/stem cells. Two other families of trophic factors are the TGF-β and the GDNF families. These two families have more cursory purposes in the nervous system. For example, GDNF and neurturin are of neuronal origin and persephin is of glial origin. Their receptors (GFR-α_1, GFR-α_2, RET) are found on neurons that respond to trophic factors, such as the dopaminergic neurons of the **substantia nigra**. Interestingly, GDNF appears to be the most potent yet identified for promoting survival and growth of dopamine neurons both in vitro and in vivo.

Peripheral nerve damage affects the axons of motor, sensory, and sympathetic neurons. These neurons generally survive axonal injury as long as it is some distance from the cell body, by mounting a massive regenerative response. Peripheral neurons lose contact with their targets following axonal injury and will consequently no longer receive target-derived trophic factors. However, support cells, such as Schwann cells, continue to supply trophic factors. NGF supports sympathetic axons, and CNTF released at the time of injury can have a protective effect on motor neurons. Trk receptors are found on almost all sensory axons: nociceptive and temperature-sensitive neurons carry TrkA and are supported by NGF, large proprioceptive fibers carry TrkC and are supported by NT-3, and fine touch and vibration-sensitive fibers carry TrkB and are supported by BDNF. Production of TrkB and BDNF in motor neurons following injury has an **autocrine trophic support effect**. The extent to which these receptors and complimentary trophic factors can prevent neuronal death is summarized in the following points:

- The likelihood of cell death in motor neurons following axonal injurydepends on the age of the cell and position at which the axons are damaged.

- Peripheral nerve damage in developing nervous systems leads to extensive motor neuron damage, but nerve damage does not cause motor neuron cell death.

- Motor neurons in adults lose choline acetyltransferase following axonal injury; this effect is absent in developing nervous systems.

- Trophic factors are effective in protecting against neuron death following axonal injury in both developing and fully matured nervous systems. Factors BDNF, GDNF, and CNTF are the most effective in preventing motor neuron death, and IGF, BDNF, and NT-3 are the most effective in the treatment of peripheral nerve damage.

Actions of trophic factors are not restricted to the CNS, nor are they limited to the promotion of neuronal survival. Nonspecific delivery of trophic factors causes unwanted side effects because of broad incorporation within the CNS. Thus, for a therapeutic benefit to be realized, factors need to be delivered very precisely to the neurons that need protection. For the time being, manipulation of factor-secreting cells seems to be the best option for such specialized delivery of nerve growth factors, although surgical approaches to the treatment of damaged peripheral nerves are still relatively effective.

CASE CORRELATES
- See Cases 37-40 (nervous system damage and repair).

COMPREHENSION QUESTIONS

38.1 In the developing nervous system, from which source do immature neurons receive the neurotrophic support that allows them to survive to become mature neurons?

A. The targets they innervate

B. Surrounding glial cells

C. Neighboring neurons

D. Themselves

38.2 During which developmental period do neurons, which are not receiving trophic support from properly innervated targets, undergo apoptosis and die?

A. Neurulation

B. Neurogenesis

C. Synaptogenesis

D. Ontogenetic cell death

38.3 A 42-year-old man sustains a crush injury to his left forearm while working on a construction site. On evaluation in the emergency department, his hand is well perfused, but he has no sensation in the distribution of his median nerve and severely limited flexion of his wrist and fingers. He is diagnosed with traumatic crush injury to his median nerve and scheduled for exploration and repair of the nerve. Which type of neurons in the nerve can produce their own trophic factors and thereby prevent cell death?

A. Sympathetic neurons

B. Motor neurons

C. Proprioceptive neurons

D. Nociceptive neurons

ANSWERS

38.1 **A.** Immature neurons receive neurotrophic support from **the target cells they innervate**. In the developing nervous system, there is typically redundancy in early development: multiple neurons will extend axons toward a given target. Once this happens, one of neurons (typically the one with the stronger connection) will be selected, and the others will degenerate. This selection process involves trophic factors secreted by the target cell that allow the neuron to persist. In the absence of these trophic factors, the nonselected neurons die. In mature neurons damaged during peripheral nerve injury, temporary trophic support is often received from surrounding glial cells, or produced internally, so that the cell bodies do not degenerate.

38.2 **D.** The period of **ontogenetic cell death** is a period in which a huge number of neurons, mostly neurons that are redundant or that have not reached the proper target, undergo apoptosis. During this period, neurons that have reached the appropriate target receive trophic factors and therefore do not undergo apoptosis. In the case of sensory neurons and neurons of the sympathetic nervous system, the trophic factor that prevents apoptosis is NGF.

38.3 **B.** Trophic factors are available from a number of other sources besides target cells, including Schwann cells, and in the case of **motor neurons**, themselves. Motor neurons can secrete their own BDNF, which provides autocrine trophic support.

NEUROSCIENCE PEARLS

▶ When developing embryos are deprived of NGF, many sensory and sympathetic neurons die before birth.

▶ Neurotrophic factors that are structurally related to NGF bind to Trk receptors.

▶ Neurons generally survive axonal injury as long as it is some distance from the cell body.

▶ Huntington disease (HD) is an autosomal dominant condition resulting from excessive CAG trinucleotide repeats in the Huntington gene on chromosome 4.

▶ HD is prone to a process called anticipation, in which subsequent generations are affected at earlier ages.

▶ The caudate nucleus is generally atrophied in HD, with loss of GABA-containing neurons.

REFERENCES

Bear MF, Connors B, Paradiso M, eds. *Neuroscience: Exploring the Brain.* 3rd ed. Baltimore, MD: Lippincott Williams & Wilkins; 2006.

Kandel ER, Schwarz JH, Jessell TM, Siegelbaum SA, Hudspeth AJ, eds. *Principles of Neural Science.* 5th ed. New York, NY: McGraw-Hill; 2012.

Squire LR, Berg D, Bloom FE, du Lac S, eds. *Fundamental Neuroscience.* 4th ed. San Diego, CA: Academic Press; 2012.

A 4-year-old boy is at the amusement park, where he complains of increasing headaches and begins to vomit. The parents take him to the children's hospital where he is evaluated. The family reports that the child had an uncomplicated delivery, received all his immunizations, and is doing well in school. The boy has been experiencing some headaches for the last 2 months, but they were attributed to excessive playing of video games. The radiological investigation reveals a pediatric brain tumor with findings of increased intracranial pressure. The physician tells the boy's mother that he most likely has a medulloblastoma.

▶ What part of the brain is most likely affected?
▶ What is the most likely cause of the increased intracranial pressure?

ANSWERS TO CASE 39:

Neural Stem Cells

Summary: A 4-year-old boy is evaluated for headaches for 2 months with increasing severity and frequency over the last 24 hours, along with multiple bouts of projectile emesis. The CT scan shows hydrocephalus and a dense mass lesion in the cerebellum. A contrast MRI of the brain is obtained and reveals an enhancing lesion, most likely a medulloblastoma.

- **Part of the brain affected:** Posterior fossa.

- **Most likely cause of increased intracranial pressure:** The obstruction of the cerebrospinal outflow via the fourth ventricle by the tumor led to obstructive hydrocephalus, causing increased intracranial pressure leading to the headaches and emesis. The trigger for emesis can be both from the increased pressure and from tumor compression on an area of the brainstem called the area postrema. The increasing headaches result when the brain's natural capacity to buffer the increased intracranial pressure is exhausted and small increases in pressure lead to dramatic increase in symptoms.

CLINICAL CORRELATION

Pediatric brain tumors are one of the most common malignancies in children and have a wide range of histopathologies. Some tumors can be removed surgically and with a complete resection, the chance of recurrence is minimal. Also, some tumors are successfully managed with chemotherapy and/or radiation, although treatment options are limited in children younger than 5 years, in whom radiotherapy leads to brain dysfunction. Yet, a significant proportion of pediatric brain tumors is of a highly undifferentiated aggressive histopathology and at the time of diagnosis often show cells outside of the primary tumor that have infiltrated the brain. These tumors continue to recur and are managed with repeated operations until the patient ultimately is no longer treatable. Examples of these tumors include medulloblastoma (Figure 39-1), anaplastic astrocytoma, and anaplastic ependymoma.

APPROACH TO:

Neural Stem Cells

OBJECTIVES

1. Know the definitions of the various types of stem cells.

2. Be able to describe the various sources of neural stem cells.

3. Be aware of the limitations and challenges that remain for neural stem cell–based therapy.

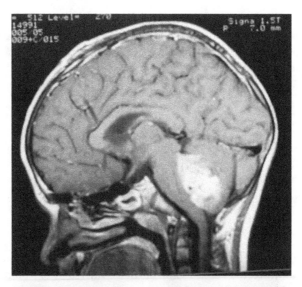

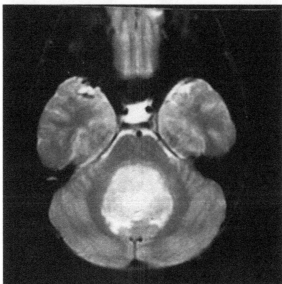

Figure 39-1. Medulloblastoma: MRI in the sagittal (*above*) and axial (*below*) planes, illustrating involvement of the cerebellar vermis and neoplastic obliteration of the fourth ventricle. (*With permission from* Adam and Victor's Principles of Neurology. *7th ed. New York: McGraw-Hill; 2000. Page 703, figure 31-11.*)

DEFINITIONS

TOTIPOTENT: Stem cells that can differentiate into embryonic and extraembryonic cell types. After the fusion of an egg and sperm cell, totipotent cells are produced by the first few divisions of the fertilized egg.

PLURIPOTENT: Stem cells which can differentiate into any cell type within the germ layers. A pluripotent cell can differentiate into any cell type of the mesoderm, endoderm, or ectoderm.

MULTIPOTENT: Stem cells which can differentiate into any cell type within its germ line lineage; for example, a neural stem cell exhibits self-renewal and can differentiate into astrocytes, oligodendrocytes, or neurons. It cannot, however, differentiate into a cardiomyocyte or intestinal cell.

UNIPOTENT: A progenitor cell, which can differentiate into only one cell type, such as astrocyte or neuron, but not both. These cells can self-renew and thus are considered stem cells.

EMBRYONIC STEM CELL (ESC): ESCs are cultures of cells derived from the inner cell mass of a blastocyst. A blastocyst is an early stage embryo approximately 4-5 days old in humans and consisting of 50-150 cells. ESCs are pluripotent.

ADULT (SOMATIC) STEM CELL: A stem cell that is derived from a fetus, child, or adult and can differentiate into the cells that populate one germ cell line, and are therefore considered to be multipotent.

EPIGENETICS: Nongenetic influences that influence cells. Typically this is the environment in which the cell is found, or the in vitro solution composition.

TRANSFECTION: The process of inserting a foreign gene into the host genome of a desired cell. Transfection leads to genetic modification that is stable or transient. Methods include biochemical, physical, and use of viruses.

DISCUSSION

Stem cells give rise to organs and maintain tissue integrity and homeostasis in the adult organism. There are different types of stem cells, including embryonic and somatic (fetal or adult-derived) from which new cells can be developed. A stem cell must have the following functional properties: (1) the ability to generate the cell types of the organ it was derived from, and (2) "self-renewal," that is, the ability to produce daughter cells with identical properties. The ability to populate a developing or injured region with appropriate cell types upon transplantation is another important stem cell feature that is well-established with hematopoietic stem cells and awaits standardization in other organ systems including the brain. There are two prototypical stem cells, the **embryonic stem cell (ESC)** and the **neural stem cell (NSC)**.

Embryonic stem cells have been derived from the inner cell mass of blastocysts of various species, including cells of human origin. They can be totipotent (be able to generate all cells types in an organism except the placenta), pluripotent (the ability to yield mature cell types from all different germ layers), or multipotent (be able to give rise to all cells within an organ). Currently, our understanding of human ESCs is increasing and knowledge is being accumulated on the improvement of cell culture conditions, long-term propagation, controlled differentiation, and transplantation into animal models of human disease. The list of various cell types differentiated from human ESCs (eg, neurons, cardiomyocytes, hepatocytes) is continuously increasing. The unlimited access to specific functional human cells is expected to play an important role not only in therapeutic cell replacement but also in disease modeling and drug screening.

In contrast to pluripotent ESCs, somatic stem cells are believed to be multipotent, or only capable of generating the major cell types found in their tissue of origin. Typically, the **NSC is capable of producing neurons, astrocytes, and oligo-dendrocytes**. Somatictissue-specific stem cells are the building blocks of organs during development and survive in specialized microenvironments ("stem cell niches") contributing to new cells throughout life.

Neural stem cells are multipotent, have the ability to populate a developing region and/or repopulate an ablated or degenerated region of the CNS with appropriate cell types, and undergo "self-renewal," that is, the ability to produce daughter cells with identical properties. They are highly abundant during embryogenesis, with a sharp decline shortly after birth. In the adult nervous system, NSCs are confined to the **subgranular zone (SGZ)** in the dentate gyrus of the hippocampus and the **subventricular zone (SVZ)** lining the lateral ventricles. Embryonic hippocampal neurons have been suggested to improve memory and address mood disorders such as stress and depression. In rodents, neuroblasts born in the SVZ migrate along the rostral migratory stream (RMS) to the olfactory bulbs where they differentiate into periglomerular and granule neurons. Isolation of cells from brain regions such as the amygdala, substantia nigra, and cortex has included cells with stem cell characteristics in vitro.

Morphologically, NSCs share properties with both astrocytes and radial glia. The main characteristic is a long process that extends radially. Although no definitive marker has been suggested for neural stem cells, a substantial amount of work shows that they are positive for nestin, an intermediary filament protein, and glial fibrillary acidic protein (GFAP), used traditionally to identify astrocytes.

Neural stem cell scan be generated from ESCs or directly isolated from the developing CNS (typically from the fetal brain) as well as from neurogenic regions of the adult brain (typically from a cadaveric brain specimen). Historically, the first established NSC lines exploited tumor viruses to achieve immortalization. However, NSC that have not been genetically modified can also be propagated in vitro for extended periods using high concentrations of mitogenic factors such as basis fibroblast growth factor (bFGF) and epidermal growth factor (EGF).

The challenge to treat brain tumors effectively pivots on the immense difficulty of attacking invading cells within the brain, as well as the delivery of chemotherapeutic modalities past the blood-brain barrier to tumors and tumor cells selectively. The unique ability of NSCs to home in on tumors has been demonstrated, even when transplanted at various sites outside of the tumor itself. This homing ability has been exploited to deliver therapeutics in various tumor models with remarkable efficacy in mice and may have promise for potential human therapy. More specifically, nude mice were inoculated with glioma cells and subsequently transplanted with human and murine NSCs at various locations (intratumoral, contralateral hemisphere, intraventricular, and tail vein) with clear demonstration of NSCs migrating to the tumor and distributing within the tumor. Interestingly, some NSCs appeared to track single cells invading brain parenchyma outside of the tumor mass. Subsequently, NSCs were transfected with a gene for cytosine deaminase (CD), a prodrug-converting enzyme that converts 5-FC to 5-FU, and transplanted some distance from the tumor. This technique led to an approximately 80% reduction in tumor burden.

Despite the promise of using NSCs to treat brain tumors, issues of patient safety must first be met before clinical trials can occur. Although NSCs are minimally immunogenic, it remains to be determined whether recipients of NSC transplants would need to be on immunosuppressive therapy. Also, it must be possible to follow the transplanted cells with imaging in the event they are migrating to unexpected areas outside of the brain. Lastly, the use of genetically modified cells is controversial, and concern exists over whether immortalized cells may grow uncontrollably and lead to tumor formation.

Treatment options for an intracranial mass begin with decompressing the hydrocephalus at the bedside with a ventricular catheter. Subsequently, an operation is performed to remove the lesion and obtain a histopathological diagnosis. Once tissue diagnosis is obtained, a postoperative plan for adjuvant therapy (radiation and/or chemotherapy) can be created. Even after resection and adjuvant therapy, most tumors with aggressive histology tend to recur. The treatment options for these recurrent brain tumors is limited and often includes more surgery to decrease the tumor burden, but fails to address the tumor cells that have infiltrated the brain and serve as foci for the seeding and spreading of the primary tumors. The horizon for experimental therapy for recurrent untreatable brain tumors may be the use of neural stem cells. Neural stem cells have demonstrated the ability to migrate toward tumors and tumor cells and in animal models some of these cells have beenshown to deliver chemotherapeutics and decrease tumor volume.

CASE CORRELATES

- See Cases 37-40 (nervous system damage and repair), Case 1 (glioblastoma), and Case 35 (uncal herniation).

COMPREHENSION QUESTIONS

39.1 Neural stems cells can be best described by which of the following terms?
  A. Totipotent
  B. Pluripotent
  C. Multipotent
  D. Unipotent

39.2 In the adult brain, in which of the following locations can neural stem cells be found?
  A. Ventricular zone
  B. Subgranular zone
  C. Deep white matter
  D. Midbrain

39.3 In which of the following locations can the physician find a totipotent stem cell?

    A. A blastocyst

    B. Embryonic mesoderm

    C. The subventricular zone in the human brain

    D. Skeletal muscle

ANSWERS

39.1 **C.** Neural stems cells are **multipotent** stem cells. This means that they are capable of dividing to form any of the cells that make up the brain and nervous system but not cells of any other organ system. The cell lines to which neural stem cells can give rise are neurons, astrocytes, and oligodendrocytes. Recall that other cells that inhabit the CNS and PNS (microglia and Schwann cells) are not derived from neuroectoderm, and therefore cannot be generated by neural stem cells.

39.2 **B.** In the adult brain, neural stem cells are located in two different places: the **subgranular zone** in the hippocampus and the subventricular zone lining the lateral ventricles. The subgranular zone generates neurons that enter into the dentate gyrus of the hippocampus and are thought to play a role in improving memory and affecting mood. The subventricular zone generates neurons that migrate to the olfactory bulbs in rodents. The ventricular zone is the location of neural stem cells in the developing nervous system, but they are not located there in the adult nervous system.

39.3 **A.** Embryonic stem cells are totipotent early in development, at the **blastocyst** phase, but their fate begins to be restricted soon after that. All the cells in the blastocyst are identical, and any one of them could give rise to any cell in the adult human. By the time a cell has differentiated to embryonic mesoderm, it no longer has the ability to become cells that derive from either endoderm or ectoderm. Neural stem cells in the subventricular zone are multipotent, because they can only become brain cells, and there are no stem cells in muscle tissue.

NEUROSCIENCE PEARLS

▶ Stem cells are able to self-renew indefinitely as well as differentiate into various cell types.

▶ Embryonic stem cells can differentiate into any cell and are therefore pluripotent.

▶ Neural stem cells can differentiate into astrocytes, oligodendrocytes, or neurons and are therefore multipotent.

▶ Stem cells can be genetically modified to carry various genes.

REFERENCES

Falk A, Frisen J. New neurons in old brains. *Ann Med.* 2005;37:480-486.

Lanza R, Gearhart J, Hogan B. *Essentials of Stem Cell Biology.* Oxford, UK: Elsevier Academic Press; 2006.

Lindvall OL, Kokaia Z, Martinez-Serrano A. Stem cell therapy for human disorders—how to make it work. *Nat Med.* 2004;10:S42-S50.

A 24-year-old African American man presents to the emergency room after a car accident resulting in paralysis of both legs. After respiratory stabilization, imaging and neurological assessment take place, paralysis of both legs is quickly appreciated. The patient has also lost bladder control. Imaging shows a narrowing disruption of the spinal cord at the T11-T12 region of the thoracic spine. A diagnosis of a spinal cord injury (SCI) is made.

▶ What areas are anatomically affected in spinal cord injury?
▶ What are the patient's treatment options?
▶ Is neural repair a likely outcome for this patient? Why, or why not?

ANSWERS TO CASE 40:

Neural Repair

Summary: A 24-year-old African American man presents with paralysis in both legs and a narrowing disruption of the spinal cord at the T11-T12 region of the thoracic spine.

- **Regions of spinal cord affected:** Spinal cord injury (SCI), also known as myelopathy, results in damage to the white matter tracts (myelinated fiber tracts) in the spinal cord that carry sensation and motor signals to and from the brain. It causes segmental losses of interneurons and motor neurons within the gray matter of the spinal cord.

- **Treatment options:** First and foremost, every precaution must be taken to stabilize the spine and not exacerbate the damage following injury. A significant amount of disability in patients with SCI results from insults sustained after the initial injury. Studies have shown that some neurological function can be recovered if methylprednisolone is administered immediately following injury, and this is routinely practiced in the United States. However, at present, there exists no real treatment capable of producing neural repair in the spinal cord. The majority of treatment is focused on rehabilitation and learning to function with disability.

- **Is neural repair likely:** While axons in the peripheral nervous system (PNS) experience spontaneous regeneration, central nervous system (CNS) axons do not, making natural recovery from axonal injury in the CNS, such as SCI, impossible. Axon growth and regeneration is a collaborative process that involves regenerative attempts made by the axon itself, and the entire environment surrounding the axon. Unfortunately, the CNS is naturally hostile to axon regeneration. It is riddled with astrocytes, oligodendrocyte progenitors, and oligodendrocytes that all have the ability to inhibit axon regeneration. Additionally, the axon itself does not retain the same vitality in regeneration as it has during developmental growth, and some mismatch between cell surface adhesion molecules on axons and those in the environment exists as well.

CLINICAL CORRELATION

Axon growth and regeneration is a collaborative process that involves regenerative attempts made by the axon itself and the entire environment surrounding the axon. When a CNS axon is severed, it retracts along with its myelin much like in the PNS. Similarly, CNS neurons with severed axons often die, along with their postsynaptic neurons. However, unlike the PNS, the remaining CNS neurons that do survive will atrophy, losing many of the enzymes associated with neurotransmitter production and cellular function. Many of the axons will also lose synaptic connections that link it to the cell body by microglia. Astrocytic processes quickly occupy these vacant connections. Furthermore, in the CNS there is little recruitment of macrophages to enter the axon and remove degenerated axonal and myelin debris except where blood cells and plasma have entered in the area of the lesion itself. Instead, degenerative debris is removed by the growth-inhibiting microglia, which are also present in much smaller numbers. Thus, debris persists in the CNS much longer, inhibiting regeneration. Following the microglia are oligodendrocyte progenitors, and then the lesion site fills up with reactive astrocytes and meningeal cells where the injury penetrates the meningeal surface. These last two types of cells work in tandem to produce a glial scar. All these cell types through these processes have regeneration-inhibiting properties.

APPROACH TO:
Neural Repair

OBJECTIVES

1. Know the effect of damage to CNS in mammals.

2. Be aware of the limitations of PNS repair.

3. Know the factors that can limit the regeneration.

DEFINITIONS

ADHESION MOLECULES: Molecules that regulate cell adhesion by interacting with the molecules on an opposing cell or surface. Adhesion molecules are sometimes referred to as receptors and their target molecules as ligands.

BASAL LAMINA SHEATH: An undamaged axon-Schwann cell unit is surrounded by a basal lamina sheath composed of collagen, laminin, and fibronectin.

BAND OF BUNGNER: The basal lamina sheath contains tubes of end-to-end Schwann cells that form the band of Bungner, which remains intact in the damaged nerve, and, barring mechanical disruption, spans the entire sequence of the nerve from the lesion to the area of axon termination.

PERINEURIAL FIBROBLASTS: A sheath of cells that give rise to connective tissue that disperse around a neural lesion.

GROWTH CONE: A dynamic actin-supported extension of the developing axon, the growth cone is composed of fine extensions known as filopodia made of actin that contain receptors important for axon guidance.

INTERLEUKIN-1 (IL1): A cytokine secreted by axon-invading macrophages during PNS repair that induces the NGF production by the Schwann cells; IL1 plays an important role in promoting the survival of a neuron following axonal injury.

SWOLLEN CLUB ENDING: The result of halted axonal growth because of fibroblastic scar tissue, also known as a neuroma.

NEURON-GLIAL ANTIGEN 2 (NG2): A proteoglycan that inhibits axonal growth, released by fibroblasts that make up the scar tissue around a neural lesion.

POLYNEURONAL INNERVATION WITHDRAWAL: A process of normal neural development in which an initial stage of multineuronal innervations of muscles occurs, followed by a withdrawal of such connections until each muscle fiber is innervated by only one axon. However, the persistence of this function after development creates positional breaks during regeneration following axonal injury.

DISCUSSION

In mammals, damage to the CNS usually leads to permanent incapacitation of the affected neurons and, in most cases, paralysis. However, the ability to regenerate axons in the PNS and regain much of the function that is lost after peripheral nerve damage has been retained in mammals. In general, axons regenerate at a rate of about 1 mm per day. PNS repair is seldom perfect; issues of axon guidance and other factors can limit regeneration.

When a peripheral nerve is damaged, the axons are disconnected from their cell body and begin to degenerate. Schwann cell processes surrounding the axons as myelin also begin to degenerate, and this axonal and myelin debris is removed by macrophages migrating to the degenerating nerve from the bloodstream. Schwann cells undergo several changes during this period, including secretion of nerve growth factors NGF and BDNF stimulated by cytokines released by invading macrophages; changes to the Schwann cell membrane surface where **adhesion molecules** L1, N-CAM, and N-cadherin are increased; and Schwann cell extracellular matrix transformations because of increases in tenascin and other proteoglycans. An undamaged axon-Schwann cell unit is surrounded by a **basal lamina sheath** composed of collagen, laminin, and fibronectin. This sheath contains tubes of end-to-end Schwann cells that form the **band of Bungner**, which remains intact in the damaged nerve, and, barring mechanical disruption, spans the entire sequence of the nerve from the lesion to the area of axon termination (Figure 40-1).

The fallout from nerve damage is dependent on the nature of the injury. A major disruption or severance of the nerve leads to complete Schwann cell annihilation, leaving just a strip of fibrotic tissue. Conversely, a localized crush injury is sufficient to kill the axon but leaves the basal lamina sheath intact. The preservation of the continuity of the Schwann cells seems to be essential for regeneration to occur. However, damage and repair also leads to fibroblastic scarring that can hinder regeneration. Therefore, neural repair occurs within this context of columns of demyelinated Schwann cells encased in a basal lamina sheath that have changed their surface conformations and have begun to secrete trophic factors, with **perineurial fibroblasts** dispersed around the lesion. It should be emphasized that nerves regenerate only through intact basal lamina sheaths, which ensures that the proper target is reinnervated.

Within an hour of injury, the damaged end of the axon seals off, and there is a formation of a **growth-cone** structure. This is the initial structural response to axonal injury, long before any molecules could have been exchanged between the site of damage and the cell body. Axons have an inherent ability to form motile growth cones without the production of new molecules . After a day or two, depending on how far from the cell body the axon was cut, major changes in gene expression and protein synthesis occur in the cell body, and new building block proteins such as

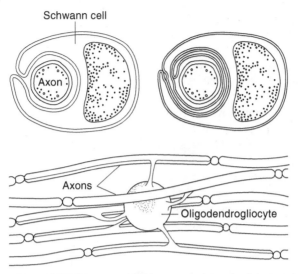

Figure 40-1. Top: Relation of Schwann cells to axons in peripheral nerves. Bottom: Myelination of axons in the central nervous system by oligodendrogliocytes. (*With permission from* Ganong's Review of Medical Physiology. *22nd ed. New York: McGraw-Hill; 2005. Page 49, figure 2-3.*).

tubulin are brought to the axon tip. There is a pattern of expression of cytoskeleton proteins (eg, tubulin) and microtubule-associated proteins in repairing neurons that both mimics and contrasts with neuronal development.

Regenerating axons are usually associated with bands of Bungner and grow in between the basal lamina sheath and the Schwann cell membrane. Half of the growth cone membrane maintains contact with the Schwann cell membrane, and half is connected to the basal lamina. Cell-to-cell interaction during regeneration must necessarily take place between these three elements. In vitro experiments have demonstrated that the basal lamina alone is not sufficient for axonal regeneration; it must be accompanied by Schwann cells. This is because Schwann cells provide the ideal substrate for regenerating axons and supply many of the vital trophic factors that boost nerve repair. Schwann cell division can be promoted by cytokine secretion by macrophages that enter the nerve in response to degeneration to remove axonal and myelin debris. Specifically, **interleukin 1** induces NGF production by Schwann cells. Similar changes in Schwann cells, resulting from a disconnect with the axons and spurred by macrophage interactions, make them a suitable substrate for regenerating axons.

Growth cones adhere to nearby surfaces to exert tension on growing axons through adhesion molecules on the growth cone surface and interactions between substrate adhesion molecules with ligands in the extracellular matrix. Adhesion molecules L1 and N-cadherin, via hemophilic interactions with the axonal surface molecules, and matrix molecules, via binding integrins and extracellular matrix, all promote axonal growth over Schwann cells. These Schwann cells, separated from the axon and stimulated by macrophage secretions, will produce many nerve growth factors that play a central role in enhancing neural repair. Other extracellular events, such as inflammation, which causes the release of triggering cytokines, also promote axon regeneration.

Some axons grow into the fibroblastic scar tissue that forms at regions of nerve damage. Axonal growth is halted, and a **swollen club ending** forms. These areas of scar tissue can become extremely sensitive to light touch, and can be quite painful. The more severe the damage, the greater the region of scar tissue formed, and the fewer the number of axons that can regenerate. Axons will not regenerate through regions of intense scarring unless the damaged region is surgically removed and replaced by a nerve graft. These regions of scar tissue are very effective at inhibiting axonal growth because of the lack of trophic support and the large amounts of highly inhibitory proteoglycan **NG2** released by fibroblasts.

Neural repair is a directed process because of the specificity of axonal connections required during regeneration to ensure the restoration of functionality. Regenerating motor axons are expected to find their target muscle, and sensory axons are expected to reconnect to target sensory structures. For this to occur there must be molecular recognition and guidance processes that allow repairing neurons to find their appropriate targets. Much of this specificity is owing to a molecular imprint left behind on the muscle end plate basal lamina. Molecules like agrin and s-laminin, which act as stop signals for axon growth and are found in synaptic junctions, and Schwann cells that envelope the end plate, allow repairing neurons to find these old denervated synaptic sites. However, axons still need to be guided by pathways formed by the damaged nerve and its bands of Bungner to the vicinity of the muscle before these effects will take place. An important caveat of muscle reinnervation is that because of lower numbers of regenerating axons, after **polyneuronal innervation withdrawal** takes place, each nerve fiber ends up innervating more muscle fibers than before, leading to much larger motor units. While this mechanism of reinnervation allows repairing neurons to find old sites of innervation, ensuring that particular axons continue to innervate the same exact sites as before is a much more imperfect science, and many positional errors are generally made by regenerating neurons because of disrupted bands of Bungner. Sensory reinnervation is less understood; regenerated nerve endings are rather different from those that existed before.

Neural repair in the CNS is a new frontier in experimental science since this process does not usually appear in nature. This is because of the nonpermissiveness of the CNS, which is in part due to the ability of astrocytes, oligodendrocytes, and progenitor cells to block axonal growth and also because of factors intrinsic to the axon itself: the reduced capacity for regenerative growth compared to developmental growth and the mismatch between cell surface adhesion molecules on axons and those present in the environment. Strategies to overcome these obstacles in neural repair include replacement of the hostile CNS environment with various grafts, removal of inhibitory cells and molecules, and treatments designed to increase the regenerative capacity of neurons.

CASE CORRELATES

- See Case 37 (axonal injury), Case 38 (nerve growth factors), and Case 39 (neural stem cells); and Cases 5 and 19 (spinal cord injury).

COMPREHENSION QUESTIONS

Refer to the following case scenario to answer questions 40.1-40.3:

A 42-year-old man presents to your office complaining of numbness on the dorsal aspect of his right hand and forearm and difficulty extending his wrist and fingers on that side. On further questioning, he admits that he got drunk and fell asleep with his arm draped at the back of a wooden chair several nights ago. Based on your suspicions, you perform additional tests and confirm that he has a radial neuropathy.

40.1 Given the nature of this man's nerve injury, what structure that should span from the distal end of the viable axon to the end of the nerve will allow for nerve regeneration?

A. Band of Bungner

B. Axon growth cone

C. Swollen club ending

D. Basal lamina sheath

40.2 At what rate will the peripheral nerve regenerate, given the appropriate structure to regenerate through?

A. 0.5 mm per day

B. 1 mm per day

C. 2 mm per day

D. 5 mm per day

40.3 Following peripheral motor nerve regeneration, which of the following best describes the reinnervation pattern of muscles compared to before nerve injury?

A. Each motor axon innervates more muscles.

B. Each motor axon innervates fewer muscles.

C. Each motor axon innervates the same number of muscles.

D. Motor axons do not undergo regeneration.

ANSWERS

40.1 **A.** The **band of Bungner** is a structure made up of Schwann cells and the basal lamina sheath of the endoneurium that used to surround the nerve and remains intact following injury to the axons of a peripheral nerve. As long as this structure is not otherwise damaged, it spans from the site of injury to the original endpoint of the nerve and serves as a hollow column down which the damaged axon can grow, allowing for at least partial regeneration.

40.2 **B.** Under optimal conditions, a damaged nerve will regenerate down the hollow center of a band of Bungner at a rate of approximately **1 mm per day** (or 1 inch per month).

40.3 **A.** When a peripheral nerve regenerates, it does so incompletely, resulting in a smaller number of axons reaching the target than were there originally. This translates into **each motor axon innervating a larger number of muscle fibers**. Since each motor unit is larger because of the decreased number of axons, the regenerated system has a lesser degree of fine movement control than did the original system.

NEUROSCIENCE PEARLS

▶ Nerves regenerate only through intact basal lamina sheaths, ensuring that the proper target is reinnervated.

▶ Regenerating axons are usually associated with bands of Bungner and grow in between the basal lamina sheath and the Schwann cell membrane.

▶ After a day or two, depending on how far from the cell body the axon was cut, major changes in gene expression and protein synthesis occur in the cell body, and new building-block proteins such as tubulin are brought to the axon tip.

▶ There are fewer axons regenerated after the injury, so each nerve fiber innervates more muscle fibers than before, leading to much larger motor units.

REFERENCES

Bear MF, Connors B, Paradiso M, eds. *Neuroscience: Exploring the Brain*. 3rd ed. Baltimore, MD: Lippincott Williams & Wilkins; 2006.

Purves D, Augustine GJ, Fitzpatrick D, Hall WC, eds. *Neuroscience*. 5th ed. Sunderland, MA: Sinauer Associates Inc; 2011.

Squire LR, Berg D, Bloom FE, du Lac S, eds. *Fundamental Neuroscience*. 4th ed. San Diego, CA: Academic Press; 2012.

A 15-year-old right-handed boy presents to the neurosurgery office for a preoperative evaluation. The patient describes a lifetime history of complex partial seizures. Initially, these seizures were controllable with medication, but over the last several years, they have become refractory to medical treatment. His seizures have increased to over six per day despite maximal dosage of medication and the implantation of a vagus nerve stimulator. He had been doing well at home and in school despite the seizures; however, the large doses of medication and increase in seizure frequency have begun to affect his cognitive and psychological development. He is diagnosed with medial temporal lobe epilepsy.

► What is the significance of the patient being right-handed?
► How will the surgeons decide which cerebral hemisphere is dominant?

ANSWERS TO CASE 41:
Brain Laterality

Summary: A 15-year-old right-handed boy undergoes epilepsy surgery for refractory tonic-clonic seizures. Through a combination of imaging and EEGs, he is diagnosed with medial temporal lobe epilepsy.

- **Significance of the patient being right handed:** The majority of both right-handed (>95%) and left-handed (70%) individuals are left-brain dominant. Fifteen percent of left-handed individuals are equally dominant in both hemispheres. This is important to know for surgical planning because it gives clues as to the possible function of the brain tissue that is involved in the epileptic activity.

- **Determination of cerebral hemisphere dominance:** One method of determining if the patient is right- or left-brain dominant is the Wada test. In the Wada test, the patient receives an anesthetic injection into either the right or the left carotid artery. This isolates the function of cerebral hemisphere not receiving the injection, allowing for evaluation of function. Preoperative or intraoperative mapping is also utilized.

CLINICAL CORRELATION

For this patient, the epileptic focus is removed surgically, and the patient recovers without complication. In consultation with the neurologist, the postoperative epilepsy medication is modified. Six months postoperatively, the seizure frequency has decreased dramatically and the patient has been able to significantly decrease his antiseizure medication. This has led to an improvement in school function and overall quality of life.

Through unknown mechanisms, complex partial seizures originate from a specific area of the brain and then spread to surrounding regions. This focus of abnormal neuronal activity can be surgically removed. Seizure-free rates after surgery range from 75% to 80%. This is not the case for seizures that begin diffusely throughout the cortex, as there would be too much morbidity associated with the removal of the affected brain tissue. An option for generalized epilepsy would be to sever the connections between the two hemispheres of the brain, known as a corpus callosotomy. Up to 63% of patients with seizures have decreased epileptic episodes with this intervention. Both of these surgical treatments for epilepsy accentuate the lateralized functions of the two cerebral hemispheres.

Determination of candidacy for epilepsy surgery: First, the patient must fail medical management. It is important to determine if the seizure focus is diffuse or focal. Focal lesions may be amenable to surgical removal. One must also consider the location of the abnormal firing. If it difficult to access, or associated with essential functions such as movement or language, surgical resection may carry unwanted morbidity. Anteromedial temporal resection for medial temporal lobe epilepsy has shown particularly good results.

APPROACH TO:
Brain Laterality

OBJECTIVES

1. To understand the differences in function between the two cerebral hemispheres.

2. Predict the manifestations of injury or disease depending upon laterality.

3. Conceptualize how each brain hemisphere perceives stimuli.

DEFINITIONS

IPSILATERAL: Affecting the same side.

CONTRALATERAL: Affecting the opposite side.

CORPUS CALLOSUM: The cerebral commissure of white matter connecting the left and right hemispheres.

PROSOPAGNOSIA: An inability to recognize a familiar person by looking at his or her face.

ANOSOGNOSIA: The denial of obvious illness or disability.

ANOSODIAPHORIA: An indifference or jocularity to a grave weakness or debilitation, distinct from denial.

NONBELONGING: The feeling that an affected limb does not belong to the individual.

AUDITORY AGNOSIA: An inability to interpret sounds.

AMUSIA: An impairment to music perception.

DISCUSSION

The concept that the brain's two hemispheres have distinct yet integrated roles was first described in the late 1800s in Paul Broca's work on the lateralized functions of speech. From there, physicians and scientists have continued to characterize differences in function between the right and left cerebral hemispheres. Generally speaking, the **nondominant hemisphere has been described as the "artistic" side responsible for holistic function, whereas the dominant side is the more "scientific" half, dealing with analytic attention to detail**. The dominant hemisphere separates complex stimuli into its discreet parts (when seeing a forest, focuses on the trees), whereas the nondominant hemisphere synthesizes the diverse parts into a whole (when seeing trees focuses on the forest). Both are required, and the communication between the cerebral hemispheres through the corpus callosum allows for seamless coordination of the separate functions. The lateralized functions of the brain can be categorized as motor function, language, music, visual/spatial perception, and executive capacity.

Descending motor information from one side of the cerebrum controls the opposite side of the body. Likewise, the majority of ascending sensory information crosses and maps out to the contralateral side of the brain. The coordination and memory of complex motor skills (eg, playing an instrument, dancing) appears to be controlled by the nondominant hemisphere.

Language, like mathematics, is a sequential activity. Language is lateralized to the left hemisphere in most individuals; however, 15% of left-handed individuals exhibit language control in both hemispheres. **The cerebral hemisphere in control of language is what determines "dominance."** The dominant hemisphere controls the nonemotional aspects of language. This includes speech, writing, reading, comprehension, naming, verbal memory, vocabulary, concept formation, and language structure (spelling, grammar, syntax). The nondominant hemisphere controls organization, prosody, abstract language, interpretation and expression of mood, and the subtleties of which include intonation, connotation, body language, and facial expression. **In a conversation, the dominant hemisphere concerns itself with the content of speech, while the nondominant hemisphere focuses on the general mood and affect of the interaction.**

While the nondominant hemisphere is often thought of as the "artistic" side, the dominant hemisphere controls the rhythm and language of music. The nondominant hemisphere contributes to musical understanding, and temporal lesions may cause auditory agnosia or amusia.

The dominant and nondominant hemispheres "see" the external world differently. This difference can be demonstrated in facial recognition. When the dominant hemisphere perceives a face, it sees the hairline, the mouth, the skin pigment, but cannot place all of the different pieces into one complete picture, that of a face. The nondominant hemisphere, on the other hand, compiles this information, determining not only that this is a face, but also that this is a familiar or unfamiliar face, and can consider the mood expressed by the face.

The nondominant hemisphere is largely responsible for the perception of space, which is why it is central to the creation and appreciation of art. It is responsible for conceptualizing the "whole picture," including spatial relations, pattern recognition, geometrical shapes and forms, and esthetics. These elements perceived by the nondominant hemisphere are all concurrently present, in comparison to the work of the dominant hemisphere, which deals with sequential tasks. In fact, spatial memory is stored in the nondominant hemisphere. The concept of personal space is controlled by the dominant hemisphere, whereas extrapersonal space resides in the nondominant hemisphere. For example, a lesion to the nondominant parietal lobe may lead to anosognosia. Patients deny concurrent debilitations, for example, believing there is nothing wrong with a paralyzed half of their body. Patients may even deny that a paralyzed limb belongs to them. Anosognosia carries significant morbidity as the patients are at risk of injury caused by the lack of recognition of their limitations.

Higher executive functions are also lateralized. The dominant hemisphere is responsible for abstract and rational thinking, analytical reasoning, initiative, attention, and linear thought processes. It also has the power of introspection and a sense of self. This provides the ability to recognize and adhere to social norms and behavior, allowing for integration into society. The nondominant hemisphere is more intuitive, giving a general gestalt to an interaction, situation, or perception. It is also responsible for the ability to daydream, and to have complex and rich dream imagery during sleep.

> ## CASE CORRELATES
> - See Cases 41-49 (cognition).

COMPREHENSION QUESTIONS

41.1 A 72-year-old man comes into the emergency department complaining of weakness of his left arm. He states that the weakness began a few days ago and is getting progressively worse. On examination, you note 3/5 strength in the left distal upper extremity, 4/5 strength in the left proximal upper extremity, and you also note some left-sided nasolabial fold flattening and asymmetric facial movements. On further questioning, the man reports that he fell getting out of the shower several weeks ago. Based on your suspicions, you perform a CT of his head, which shows a chronic subdural hematoma. Given the nature of his symptoms, which part of his brain do you expect to be affected?

A. Left precentral gyrus

B. Left postcentral gyrus

C. Right precentral gyrus

D. Right postcentral gyrus

41.2 A 67-year-old left-handed man is brought into the clinic by his family because they say he has been talking strangely for several days. They say he used to be a very animated speaker, with lots of gestures and changes in his voice, but for the past several days has been speaking in a near complete monotone with few gestures and few facial expressions. He does seem to understand body language and inflection, however. Based on this description of symptoms, where would the physician expect to find a neurologic lesion?

A. Pars triangularis and opercularis of the left inferior frontal gyrus

B. Pars triangularis and opercularis of the right inferior frontal gyrus

C. Posterior part of the left superior temporal gyrus

D. Posterior part of the right superior temporal gyrus

41.3 A 42-year-old right-handed architect comes into the clinic because he has been having a lot of trouble doing his job recently. In particular, he seems to be having a lot of trouble with "the big picture" of his projects. He can design individual parts or a project but can't seem to get them to go together in the right way like he used to be able to do. If there is a brain lesion responsible for these symptoms, where would the physician expect to find it?

   A. Right frontal lobe
   B. Left frontal lobe
   C. Right parietal lobe
   D. Left parietal lobe

ANSWERS

41.1 **C.** The most likely region to be affected is the **right precentral gyrus**. This man shows signs and symptoms of motor system dysfunction, and a lesion affecting his primary motor cortex could account for this. The primary motor cortex is located in the precentral gyrus and paracentral lobule on the contra-lateral side of the body. Since his dysfunction is left sided, involving the upper extremity and face, we would expect the lesion to be located on the opposite side of the body.

41.2 **B.** The most likely location for the lesion is the **pars triangularis and opercu-laris of the right inferior frontal gyrus**. This man presents with what appears to be a productive aprosody: he can understand intonation and body language but not produce it. Prosody is produced in the nondominant hemisphere in areas analogous to the speech areas in the dominant hemisphere. Since this is a productive aprosody, we would expect the lesion to be located where Broca's area is, only in the nondominant hemisphere. Even though this man is left handed, it is still likely that he is left-hemisphere dominant (70% of left-handed people are left dominant), so this lesion should be on the right side of his brain.

41.3 **C.** The most likely location of the lesion is **the right parietal lobe**. This man is having trouble with his visuospatial perception and with artistic/aesthetic design, both of which are nondominant hemisphere processes that are local-ized to the parietal or parietal-occipital areas of the brain. Since he is right handed, it is very likely that he is left-hemisphere dominant, so we would expect his lesion to be on the right side of his brain.

NEUROSCIENCE PEARLS

▶ The majority of the population is right-brain dominant, as determined by handedness and the location of language control.

▶ The dominant cerebral hemisphere is analytic, compared to the more holistic nondominant hemisphere.

▶ The corpus callosum facilitates communication and coordination between the two distinct hemispheres.

REFERENCE

Engel J Jr, Wiebe S, French J, et al. Practice parameter: temporal lobe and localized neocortical resections for epilepsy. *Epilepsia*. 2003 Jun;44(6):741-751.

A 24-year-old man presents to his family medicine clinic with concerns over recent sexual dysfunction, including decreased libido and impotence. He is otherwise healthy. A thorough review of symptoms reveals new-onset headaches for 2 months. On physical examination the physician notices several bruises on his forearms, which he says is from frequently bumping into and tripping over objects. Examination of the chest reveals that a milky white fluid can be expressed out of his nipples.

▶ What is the most likely cause of the patient's symptoms?
▶ What is the mechanism for the patient's changes in vision?

ANSWERS TO CASE 42:
Visual Perception

Summary: A 24-year-old man presents with galactorrhea, sexual dysfunction, decreased peripheral vision, and new-onset headaches.

- **Most likely cause of the patient's symptoms:** This patient's tumor is a prolactinoma, a benign prolactin-secreting neuroendocrine tumor of the anterior pituitary gland, which is likely a pituitary macroadenoma (>10 mm diameter). Under physiological conditions, prolactin secretion causes milk production and lactation; however, this is generally inhibited by dopamine. This negative-feedback mechanism is overwhelmed by the prolactin-secreting tumor. Other common symptoms include headache and visual disturbances.

- **The mechanism for the patient's changes in vision:** The pituitary gland sits in a bony depression at the base of the skull called the *sella turcica* or "Turkish saddle." The optic nerves and chiasm are closely associated with the superior portion of the pituitary gland. Enlargement of the pituitary compresses the optic nerves and/or chiasm. Involvement of the optic chiasm selectively destroys the crossing optic fibers carrying information from the nasal hemiretinas, leading to bitemporal hemianopia.

CLINICAL CORRELATION

For this patient, his general practitioner orders an MRI that reveals a pituitary mass compressing the optic chiasm. Medical treatment with dopamine agonists is attempted; however, the patient continues to suffer from visual and endocrinologic symptoms. Consequently, the patient undergoes a transsphenoidal resection of the tumor. Postoperatively, the levels of prolactin normalize and the patient develops transient diabetes insipidus. He is discharged from the hospital with endocrine and ophthalmology follow-up.

Pituitary tumors are just one type of tumor that may present with visual symptoms. Depending upon the location of the tumor, the visual symptoms will differ. In the above case, the crossing fibers of the optic chiasm were compressed, leading to a decrease in peripheral vision. Symptoms may improve with the removal of the offending mass. If tumor resection involves the resection of neural tissue responsible for the transmission of visual information, that visual function will be permanently lost. Prolactinomas in general have an excellent prognosis, and many patients recover their visual function.

APPROACH TO:
Visual Perception

OBJECTIVES

1. Understand the pathways of visual input.

2. Describe the basics of visual imprinting.

3. Predict the visual defect from a neurological lesion.

DEFINITIONS

HEMIANOPIA: A defect in half of the visual field.

BITEMPORAL HEMIANOPIA: The elimination of visual input from the temporal half of the visual field, secondary to involvement of the optic chiasm.

QUADRANTANOPIA: A defect in one quadrant of the visual field.

AMBLYOPIA: The improper development of a nonfavored eye in childhood leading to decreased ability of that eye to see details; a "lazy eye."

SCOTOMA: An area of diminished vision within the otherwise-seeing field. This may result from a lesion to the neurons or fibers in the retinocortical pathway.

BINOCULAR RIVALRY: When different optical images falling on corresponding points in the retinal map are seen by each eye, there is alternate displacement of the two images and occasionally the images are superimposed.

DIPLOPIA: "Double vision"; a single object is perceived as two images, secondary to the same optical image falling on noncorresponding points in the two eyes. This may occur as a result of oculomotor nerve or muscle dysfunction.

NEAR-RESPONSE COMPLEX: Accommodation, convergence, and pupillary constriction.

PERIMETRY: Visual field testing where a stationary eye perceives a target spot of various sizes and intensity, moved systematically through the visual field of each eye.

HOMONYMOUS HEMIANOPIA: A defect in the same side of the visual field in both eyes.

BINOCULAR FUSION: Images produced by left and right retina are fused into a single image by projecting to the same position in the visual cortex. This melding of monocular images enables inference of depth.

ASTIGMATISM: Warping of the curvature of the cornea or lens leading to a refractive error and resulting distortion of the image on the retina.

FOVEA: A depression in the macula of the retina with a high concentration of cones, producing the highest visual acuity and the greatest color discrimination.

INCONGRUOUS: Differences between the two eyes in terms of their visual field deficits.

DISCUSSION

The eye's perception of bilateral visual fields begins with the retinal photoreceptor cell's transduction of wavelengths of energy into electrical signals (cones—color, rods—black and white). These signals travel from the bipolar and ganglion cells to the optic nerve (cranial nerve II). At the optic chiasm, the nerve fibers carrying visual input from the nasal half of each retina cross. From the optic chiasm, the optic tracts carry information from the contralateral visual field (contralateral nasal and ipsilateral temporal hemiretina). The optic tracts synapse in the lateral geniculate nucleus (LGN) of the thalamus as a topographic map of the contralateral visual half field. From the LGN, the visual information leaves the thalamus via the optic radiations, part of the retrolenticular limb of the internal capsule. The optic radiations then synapse in the primary visual cortex (striate/calcarine cortex) surrounding the calcarine sulcus (area 17) of the occipital lobe. The separate perceptions of the world by each eye are melded into one single image (binocular fusion) because the optic radiation neuron projections, corresponding to specific retinal points, synapse at exactly the same point on the cortical map of the central visual field. Adjacent points in the visual field synapse at adjacent locations in the primary visual cortex map. The **fovea** projects to the posterior striate cortex, and as the fovea has higher acuity, there are more neurons committed to this area of the map. The information in the striate cortex is upside-down and backward to the external visual environment. Therefore, the striate cortex below the calcarine fissure responds to visual field input from the contralateral upper quadrant.

An example of cross-innervation can be seen in the pupillary light reflex. Following a unilateral stimulus, some of the optic nerve fibers travel to the pretectal nucleus. Here, some cross in the posterior commissure to synapse in bilateral **Edinger-Westphal nuclei**. The signal to constrict the pupil leaves the Edinger-Westphal nuclei through the pretecto-oculomotor tract , joining the oculomotor nerve (CN III). With CN III, these fibers synapse in the ciliary ganglion, finally transmitting the signal to the bilateral pupillary constrictor muscles.

Different areas of the visual cortex have different function in visual perception and integration. For example, the primary visual cortex reacts to lines, linear boundaries, and bars with specific rotational orientation. The endpoint of such a line or angle is perceived by the prestriate cortex. This area also responds to form, motion, and color. The neurons detecting multiple lines and patterns converge onto feature detector neurons, leading to complex image recognition, such as that of a familiar object or face. The inferior temporal gyrus allows for discrimination and understanding the significance of visual forms and colors. For example, the inferior temporal gyrus perceives hands and faces, and interprets and recalls visual memories. The posterior parietal lobe allows for appreciation of the spatial relationships between objects, providing localization and navigation of visual space.

This ability to recognize lines and angles can be limited by exposure to images during early brain development. For example, in children with strabismus (crossed eyes) or an astigmatism, the distorted angle of the eye will not lead to the stimulation of the appropriate area of the retina, leading to conflicting visual signals being transmitted to the cortex. This may lead to a condition known as amblyopia, where one eye is favored by the brain and the other eye is ignored in order to block out the conflicting information. Therefore, the visual cortex responding to the nonfavored eye does not develop appropriately. Animal models show that exposure to lines of only one orientation leads to cortical development only of cells responding to lines of that particular orientation.

The **superior colliculus** is also involved in tracking (responses to stimulus movement), orienting, and saccadic eye movement. It receives topographic input from the ipsilateral optic tract and visual cortex, projecting to the thalamus. In hydrocephalus, impingement on the superior colliculus is what leads to **Parinaud syndrome** ("setting sun" sign).

Understanding the visual pathways allows one to approximate the location of the lesion causing certain symptoms. For example, a lesion distal to the optic chiasm (eg, optic nerve, retina) would lead to symptoms only in the affected eye. As in this case, lesions to the optic chiasm lead to bitemporal hemianopia. After the crossing of fibers in the chiasm, each optic tract contains input only from the contralateral half visual field, perceived on the ipsilateral hemiretina of bilateral eyes. Therefore, a lesion of the optic tracts leads to an incongruous homonymous hemianopia. Optic radiation lesions lead to bilateral quadrantanopia. A lesion of the temporal optic radiation leads to a contralateral upper visual field deficit, whereas a lesion of the parietal lobe radiation leads to a contralateral lower visual field deficit. A similar result occurs with injury to the primary visual cortex. For example, unilateral damage to the primary visual cortex inferior to the calcarine fissure (input from the inferior contralateral hemiretina) leads to vision loss of the superior visual field. Lesions of both the primary visual cortex and the radiations generally spare the macular area. This area is damaged secondary to posterior lesions and leads to a homonymous hemianopia of the central visual field.

Treatment options for a prolactinoma: Because dopamine normally inhibits prolactin secretion, prolactinomas can be treated with dopamine agonists such as bromocriptine or cabergoline. Radiotherapy has a limited role in the treatment of prolactinoma. If medical therapy fails to restore normal pituitary function, is poorly tolerated, or if symptoms and tumor size progress or persist, the patient should consider a transsphenoidal pituitary adenomectomy. Unfortunately, a high number of these tumors do recur (20%-50%).

CASE CORRELATES

- See Cases 41-49 (cognition), Case 23 (sella tumor), and Case 29 (acromegaly).

COMPREHENSION QUESTIONS

42.1 A 27-year-old woman comes into the emergency department following a fall from a five-story building. She is unresponsive, and examination shows that while her left pupil responds appropriately to light in either eye, her right pupil is maximally dilated and does not react to light shined in either eye. Interruption of nerves with cell bodies in what brainstem structure is responsible for this phenomenon?

A. Superior colliculus

B. Trochlear nucleus

C. Oculomotor nucleus

D. Edinger-Westphal nucleus

42.2 A 57-year-old woman comes into your office for regular follow-up of her glaucoma. She has been using all her medications as prescribed and has not noticed any changes in her vision since the last visit. You check her ocular pressure, which is a little elevated, and note that on fundoscopic examination she appears to have additional damage to her optic nerve. The axons in this nerve, which originate from retinal ganglion cells, terminate on which thalamic nucleus?

A. Medial geniculate nucleus

B. LGN

C. Dorsomedial nucleus

D. Ventral posteromedial nucleus

42.3 You see a 2-month-old child in clinic with a hemangioma growing on her upper eyelid. At this time it is not bleeding and does not appear to be obstructing her vision, but the parents note that it has increased in size by about 50% in the last several weeks. You are concerned that if the tumor grows anymore it will begin to obstruct vision in that eye, so you immediately refer the patient to plastic surgery for evaluation and removal of the lesion. What irreversible ocular complication are you trying to avoid by promptly removing the tumor?

A. Amblyopia

B. Strabismus

C. Astigmatism

D. Myopia

ANSWERS

42.1 **D.** Interruption in the **Edinger-Westphal nucleus** would account for the symptoms. This patient has a "blown pupil," which is a sign of transtentorial brain herniation. This is likely from increased intracranial pressure from an intracranial bleed she suffered from her fall. The uncus herniates downward, compressing, among other structures, the oculomotor nerve, which is carrying the parasympathetic fibers from the Edinger-Westphal nucleus to the sphincter muscle of the iris. Interruption of these fibers results in unopposed sympathetic innervation to the dilator muscle, resulting in a maximally dilated pupil that does not react to light.

42.2 **B.** The **lateral geniculate nucleus (LGN)** is the thalamic relay for vision. Each nucleus receives information from the contralateral visual field (ipsilateral temporal and contralateral nasal fibers) in a somatotopically oriented manner. It engages in some processing of the image, and then projects to the striate cortex (primary visual cortex), retaining its somatotopic organization. The medial geniculate nucleus is the thalamic relay for sound.

42.3 **A.** The concern with this tumor is that if it continues to increase in size, it will block vision in the child's eye, which will result in deprivation **amblyopia**. When the developing visual cortex does not receive information from one of the eyes because of a defect in the visual pathway, it begins to ignore the input from that eye since it conflicts with what the other eye sees. In a young child like this, deprivation for as little as 1 week can result in this process being irreversible. Although structurally the visual system works, there will never be any conscious perception of vision out of the affected eye.

NEUROSCIENCE PEARLS

▶ The information in the striate cortex is upside-down and backward to the external visual environment.

▶ Lesions distal to the optic chiasm lead to symptoms in one eye only.

▶ Examples of cross-innervation include the pupillary light reflex and tracking.

REFERENCES

Bear MF, Connors B, Paradiso M, eds. *Neuroscience: Exploring the Brain*. 3rd ed. Baltimore, MD: Lippincott Williams & Wilkins; 2006.

Kandel ER, Schwarz JH, Jessell TM, Siegelbaum SA, Hudspeth AJ, eds. *Principles of Neural Science*. 5th ed. New York, NY: McGraw-Hill; 2012.

Squire LR, Berg D, Bloom FE, du Lac S, eds. *Fundamental Neuroscience*. 4th ed. San Diego, CA: Academic Press; 2012.

A 67-year-old right-handed man returns to the neurology clinic with his wife for follow-up after a right middle cerebral artery (MCA) stroke. His wife says that he has been recovering well; however, he has some behaviors which are concerning to her. For example, she must be standing on his right side in order for him to acknowledge her. When he eats, he only eats food of the right side of his plate. Looking at the patient, you realize he has only shaven the right side of his face. Physical examination is notable for a left facial droop, grade 2 strength of the left arm, and grade 3 strength of the left leg. When you call the patient's name while standing on his left, he turns his head to the right. The patient does not blink to threatening stimuli from the left side. The patient is asked to draw a clock and is only able to reproduce the right side.

▶ What is the reason for this patient's behavior?
▶ What other symptoms are frequently seen with a right MCA stroke?
▶ What other tests can be used to elicit these symptoms?

ANSWERS TO CASE 43:

Spatial Cognition

Summary: A 67-year-old right-handed man presents with left-sided neglect follow-ing a right MCA stroke.

- **Reason for the patient's behavior:** Hemineglect, which is more common in patients with cerebral lesions on the right (33%-85%) rather than the left side of the brain (0%-24%).

- **Other symptoms seen with a right MCA stroke:** The motor deficits including weakness of the face and arm more than the leg. As with 95% of the right-handed population, Broca and Wernicke areas appear to be on the left in this patient and are therefore not affected by the stroke. Other language deficits include problems with prosody and nonverbal cues (see Case 46 for language disorders), behavior and attention (eg, extinction), autonomic dysfunction, and hemianopia.

- **Tests to elicit the symptoms of neglect:** When patients are asked to bisect a line, they do so far to the right of center. If you show a photo or drawing to the patients and ask them to copy the image, they may copy only the right half of the image, or transfer the entire image to the right side of the paper. One may also ask the patient to verbally identify objects in the neglected field of vision. When asked to identify the right-sided limbs, patients may not be able to rec-ognize them as their own.

CLINICAL CORRELATION

Strokes provide important information about the function of the brain under normal as well as pathological circumstances. In this case, the stroke elucidated the impor-tance of the right brain in global spatial cognition. Overall, for the majority of patients the poststroke visuospatial dysfunction will resolve. For the remaining 10%-20% of patients who have difficulties lasting more than 3 months, rehabilitation and safety are the most concerning issues. The motor weakness secondary to stroke is compounded by the difficulties in rehab participation. For example, to have this patient practice strengthening his right arm muscles, he must first believe he has a right arm to move.

APPROACH TO:

Spatial Cognition

OBJECTIVES

1. Know the neuroanatomy involved in spatial cognition.

2. Predict how injuries to these areas will lead to challenges in spatial cognition.

3. Be familiar with the terminology and disorders of spatial cognition.

DEFINITIONS

EXTINCTION: Inattention to one stimulus when two stimuli are presented simultaneously.

VISUAL (HEMI)NEGLECT: Failure to give proper attention to the external environment.

HEMI-INATTENTION: Failure to give proper attention to one side of the personal space.

CONSTRUCTIONAL APRAXIA: Inability to synthesize and comprehend discreet parts as a whole.

DRESSING APRAXIA: Inability to manage spatial aspects of dressing oneself; a manifestation of hemi-inattention.

PROSOPAGNOSIA: Inability to identify a familiar face, without injury to visual nervous system.

AGNOSIA: Inability to recognize and identify objects or persons despite intact memory, knowledge of the objects or persons, and intact sensory function; generally limited to specific senses.

ALLESTHESIA: Consistently attributing of sensory stimulation on one side to stimulation of the other side.

IDEATIONAL APRAXIA: Improper sequencing of events (eg, drinking from a cup and then filling it).

DISCUSSION

In the majority of the population, **the nondominant hemisphere is responsible for the composition and perception of spatial relations.** This includes synthesizing individual parts of a visual image into a whole, as well as the perception of geometric designs and esthetic patterns. **Constructional apraxia describes the inability to synthesize discrete parts into a whole image.** A cortical or subcortical lesion of the forebrain may have this result.

The function of the nondominant hemisphere includes the interpretation of extrapersonal space. **Self-awareness in terms of personal space and the concept of one's own body lies in the dominant parietal lobe.** A lesion to this area leads to confusion of left vs right and an inability to identify body parts. This has to do with the posterior parietal lobe's function in the spatial localization of stimuli, which allows for the recognition of spatial relationships from the individual to an object, or between objects. Without this essential analysis of visuospatial information, the individual cannot orient herself or objects, nor can she navigate space appropriately, secondary to an inability to interpret spatial relationships. The superior colliculus is also involved in the perception of an object's location. Split brain studies suggest the function of the superior colliculus to be bilateral.

The **posterior parietal lobe fixes visual attention** on a particular stimulus of interest. Right-sided neglect, secondary to an assault to the left side of the brain, is generally accompanied by an aphasia and therefore may be difficult to detect.

Neglect, while generally thought to be secondary to a lesion of the parietal lobe, can also result from injury to the frontal lobe, thalamus, or caudate nucleus. The hemineglect syndrome embodied by the above example has multiple forms but can consist of hemineglect, hemi-inattention, visual extinction, allesthesia, anosognosia, anosodiaphoria, nonbelonging, visual-field defects, and gaze paresis.

Motor neglect refers to the under use of the side of the body contralateral to the cerebral insult. This may be mistaken for hemiparesis. With extraordinary care by the diagnostician, the patient may be encouraged to demonstrate function. Regardless, on examination the affected side shows decreased withdrawal and a lack of routine movements, such as repositioning a limb when in an uncomfortable position. Likewise, when the patient falls toward the affected side, there is no reflexive effort to protect against injury. Attempts at rehabilitation may be successful, and a recent study indicated that limb activation and cueing may lead to a reduction in unilateral visual neglect.

Bilateral occipital temporal lesions result in visual agnosia. As the primary visual cortex is intact, the patient has normal visual fields and acuity; however, when a known object is presented visually, it cannot be named. The object can be verbalized when presented via other sensory modalities such as tactile, auditory, or olfactory. Occipitotemporal lesions may also lead to environmental agnosia.

CASE CORRELATES
- See Cases 41-49 (cognition).

COMPREHENSION QUESTIONS

43.1 A 67-year-old woman is brought into the clinic because she has recently begun behaving very strangely. Most notably, she has become very clumsy recently, dropping objects when trying to set them down, and running into objects that have been in their location for years while walking around her house. On examination, you further note that when asked to perform an action with her left hand, she unpredictably uses her right hand instead, and the same occurs when asked to use her right. Given this conglomeration of symptoms, in what part of her brain would you expect to find a lesion?

A. Nondominant parietal lobe

B. Dominant parietal lobe

C. Nondominant temporal lobe

D. Dominant temporal lobe

43.2 A 57-year-old man is brought into the emergency room because of acute onset of weakness and sensory loss over the left side of his body. On examination, he can move the right side of his body without difficulty to command but cannot move anything on the left, and he denies feeling anything to touch or pain stimuli on the left. Additionally, when presented with his own left hand, he denies that it is his and also denies that there is anything wrong with him. He will also not respond to you unless you are standing on his right side. You suspect that he has had an acute stroke, which is confirmed by emergent neuroimaging. A lesion to what area of his brain most likely accounts for his inability to recognize his own hand?

A. Nondominant parietal lobe

B. Dominant parietal lobe

C. Nondominant frontal lobe

D. Dominant frontal lobe

43.3 A 54-year-old man presents to your clinic because he feels like he is "losing his mind." Recently, he has become less and less able to identify objects while looking at them, although he has no difficulty seeing and describing what objects look like. On examination you find that he is in fact not able to name objects based on visual stimulus alone, but when he is able to touch the object as well, he has no difficulty in naming the object. A lesion to what part of the brain would best account for these symptoms?

A. Dominant hemisphere occipital-temporal region

B. Nondominant hemisphere occipital-temporal region

C. Bilateral occipital-temporal region

D. Bilateral occipital lobe

ANSWERS

43.1 **B.** Deficits in visuospatial perception are attributed to lesions in the **dominant parietal lobe**. Lesions to the nondominant parietal lobe are more commonly associated with a contralateral hemineglect syndrome.

43.2 **A.** Hemineglect syndromes (when the patient cannot recognize nor respond to stimuli that are coming from either the right or the left half of his body or visual field) occur from lesions in the contralateral parietal lobe, and occur more commonly with lesions in the **nondominant parietal lobe** than in the dominant parietal lobe. This patient may additionally have some left-sided paresis and anesthesia, although it is very difficult to diagnose that in the face of a neglect syndrome.

43.3 **C.** This man is presenting with visual agnosia, an inability to recognize objects based on vision, which is caused by a lesion to the **bilateral occipital-temporal region** that interrupts the communication between the visual cortex and the association cortex where objects are identified. If the lesion is unilateral, the object can still be identified if it is viewed in the ipsilateral visual field (which projects to the contralateral brain, where there is no lesion). Vision remains intact, but no information is getting to the association cortex, so identifying the object becomes impossible. Communication between other sensory cortices and the association area are intact, however, so when presented with stimuli of a different modality, the object can be recognized.

NEUROSCIENCE PEARLS

▶ Spatial perception and composition is generally the complex work of the nondominant hemisphere.

▶ Disruptions in spatial processing leads to significant disability with varied manifestations.

▶ Attention is an important part of spatial cognition and is for the most part controlled by the posterior parietal lobe.

REFERENCES

Bailey MJ, Riddoch MJ, Crome P. Treatment of visual neglect in elderly patients with stroke: a single-subject series using either a scanning and cueing strategy or a left-limb activation strategy. *Phys Ther.* August 2002;82(8):782-797.

Mort DJ, Malhotra P, Mannan SK, et al. The anatomy of visual neglect. *Brain.* September 2003; 126(9);1986-1997.

A 72-year-old man presents to the emergency room following a sudden onset of headache, weakness, and inability to speak. His family says 2 hours ago he was out in the yard gardening when he complained of a severe headache. He began to experience weakness on the right side of his body and eventually could not respond verbally to his family. His daughter states that the patient has a history of high cholesterol and high blood pressure. The patient is right handed. On examination he has a left facial droop, his eyes look to the right, and he does not spontaneously move the left arm. The patient, while awake, does not follow commands and does not communicate with the physician or his family members. The physician diagnoses the patient as having a cerebrovascular accident (CVA).

▶ What is the most likely artery affected in this patient?
▶ What are the treatment options for this patient?

ANSWERS TO CASE 44:

Language Disorders

Summary: A 72-year-old right-handed man presents with sudden-onset hemiplegia and global aphasia and is diagnosed with a stroke.

- **Artery most likely affected:** As mentioned above, this patient has suffered an ischemic stroke of the **middle cerebral artery (MCA).** The occlusion of this vessel at the trunk leads to contralateral hemiparesis, global aphasia, and ipsilateral eye deviation (secondary to injury to the lateral gaze center). In 70%-90% of right-handed people, the left cerebral hemisphere controls language.

- **Treatment options for MCA stroke:** In patients with depressed levels of consciousness, one must first ensure the airway is protected. Blood pressure control is important to maintain perfusion to the brain, and the parameters depend upon if the patient is a candidate for tissue plasminogen activator (tPA). Treatment with tPA is most successful if given within 3 hours from the onset of symptoms. It is important to monitor for brain edema and resulting increases in intracerebral pressure. Treatment with mannitol or hyperventilation may be necessary.

CLINICAL CORRELATION

Death from stroke continues to be the third highest cause of mortality in the United States, with an incidence of 750,000 new strokes each year. For the nearly 80% of people who survive the initial insult, the morbidity is significant, making strokes the leading neurological cause of long-term disability. The most common vessel to be affected is the middle cerebral artery (MCA). The language centers of the brain are supplied by this vessel, the occlusion of which leads to difficulties in speech and/or interpretation of language. In fact, the most common cause of aphasia is stroke.

APPROACH TO:

Language Disorders

OBJECTIVES

1. Understand the neuroanatomy of language.

2. Be familiar with multiple language disorders.

3. Be familiar with the terminology of speech, language, and the resulting disorders.

DEFINITIONS

AFFECTIVE AGNOSIA: The inability to perceive or comprehend emotional intonation of speech, occurring as a result of injury to the right temporoparietal region.

AGRAPHIA: Acquired inability to write, or express oneself in written language; may occur without aphasia, but generally seen in association.

ALEXIA: Acquired complete inability to recognize or comprehend written language, seen in association with aphasia.

ANOMIA: Word-finding difficulty or inability to name an object. Affected patient may describe the function of the object rather than provide the name. Patient may recognize the name of the object when given.

ANOSOGNOSIA: Disabled patient unaware of his or her disability.

APHASIA: Acquired lack of comprehension or expression of language, which may include speech, writing, or signing.

APRAXIA: Normal muscle function but an inability to coordinate voluntary movements, such as in speech formation.

APROSODY: Lack of intonation or emotion in speech. Occurs as a result of injury to the right hemisphere (temporoparietal) and basal ganglia, or in Parkinson disease.

AUDITORY AGNOSIA: The loss of ability to interpret auditory stimuli, despite having knowledge of the characteristics of the stimuli. For example, one may know what a car is and what it sounds like, but when presented with the sound of a car, it cannot be identified.

BROCA AREA: Injury to this area in the inferior frontal lobe leads to an inability to formulate speech, with intact comprehension (expressive aphasia).

DYSARTHRIA: Weakness of the muscles required for speech production, or damage to the lower brain functions that control speech.

DYSLEXIA: Developmental disorder of perception of written language.

DYSPHAGIA: Difficulty swallowing.

DYSPRAXIA: Inability to organize movement.

JARGON APHASIA: Meaningless sentences filled with neologisms, incoherent arrangement of standard words, or a combination of both.

NEOLOGISTIC DISTORTIONS: A neologism is a recently created term or phrase. In neurologic disease, this may include words where the definition is only known by the patient.

PARAPHASIC ERRORS: Instead of the desired word, the patient produces the wrong word or wrong wordlike sounds (substitution or addition of inappropriate sounds, syllables, or words) in efforts to speak. Includes literal (phonological), neologistic, and semantic (verbal).

WERNICKE AREA: Injury to this area in the superior temporal lobe leads to an inability to comprehend language (fluent/receptive aphasia).

DISCUSSION

Interpersonal communication is essential for an independent existence. Language disorders limit a patient's ability to communicate and, therefore, carry significant morbidity. Language disorders occur as a result of mechanical or cognitive problems and can be either receptive (input) or expressive (output). The main areas of understanding and composing language are Wernicke and Broca areas, respectively. Deeper brain structures, including the thalamus, also appear to be heavily involved.

The right and left superior temporal lobes (primary auditory area) receive auditory input from the eighth cranial nerve (vestibulocochlear). Bilateral lesions to these areas lead to cortical deafness. Patients with cortical deafness are able to "hear" sound (appropriate signals arrive to the thalamus and amygdala), but these sounds cannot be processed, and therefore they cannot perceive or comprehend sounds or language. This auditory agnosia is because of a disconnection between the bilateral primary auditory areas and the auditory association areas. Patients with global auditory agnosia cannot transfer external auditory stimuli to Wernicke area for language comprehension, and as a result they are still able to speak but cannot perceive their own speech.

Examples of mechanical problems include damage to the anatomy required for speech formation and damage to the nerves innervating the musculature. For example, prolonged intubation can injure the vocal cords, leading to difficulty in phonation. Amyotrophic lateral sclerosis or Lou Gehrig disease leads to a selective attack of central nervous system motor neurons that control voluntary muscles. Therefore, many patients present with "thick speech" secondary to weakness of the muscles required to speak. This type of language disorder is referred to as dysarthria. Additional causes of dysarthria include brain injury to the motor cortex (stroke, traumatic, tumors, cerebral palsy), developmental disorders (dyspraxia, spasmodic dysphonia), and other neurodegenerative disorders (Parkinson disease, Huntington disease, multiple sclerosis).

A significant amount of knowledge about the function of certain regions of the brain has been derived from stroke patients. The limitations of a patient after a stroke can be correlated with the area of the brain affected by an ischemic incident. **An infarction of** the dominant inferior frontal lobe (insula and frontoparietal operculum, **Broca area**), **leads to an expressive or motor aphasia**. If the patient has a Broca aphasia secondary to stroke, frequently this will be accompanied by a motor weakness because of the fact that Broca area and the motor strip are adjacent. In Broca aphasia, the patient is able to understand language; however, he or she has great difficulty producing language. These patients speak in short phrases without intonation, using agrammatic or "telegraphic speech." For example, if a patient wanted to say "I would like to eat dinner out on Friday," she instead may say, "Dinner eat." The inability to put into context these two words makes it difficult to convey the meaning of the phrase. Patients with Broca aphasia can hear and understand their own speech impediments and therefore may become frustrated with their inability to communicate effectively. If the right frontal-temporal emotional-melodic speech areas are functional, the patient may retain the ability to swear and sing. Deep frontal or cingulate gyrus lesions lead to the loss of prosody.

Damage to Wernicke area (posterior temporal, inferior parietal, lateral temporo-occipital regions), leads to a **fluent or receptive aphasia**. This may result from the occlusion of the lower division of the MCA bifurcation or one of its branches. The patient creates speech easily and with inflection; however, the string of words lacks meaning. These strings of nonsensical gibberish may include neologistic and phonetic distortions, word and sound substitution, confusion of semantic and phonetic information, perseveration, and the loss of normal conversation pauses. Language structure (eg, grammar) remains intact. Rather than saying, "I would like to eat dinner out on Friday," patients with Wernicke aphasia would say, in a pressured sentence, "You see with seeds on the day to donsumit that I wanted you to know I like schwaddle for the eating of that porridge so what's up do you see." This can deteriorate into jargon aphasia. Unfortunately, patients with Wernicke aphasia cannot understand their own speech and as a result are unaware of their own limitations.

Information is transmitted from Wernicke to Broca areas via the arcuate fasciculus. Lesions involving the supramarginal gyrus, angular gyrus, insula, auditory cortex, or left temporal lobe may injure these axonal fibers connecting the two regions. This leads to conduction aphasia, where patients cannot repeat words or read out loud and have difficulty naming objects. The ability to comprehend language (both oral and written) would remain intact. These patients understand that they have some difficulty with language and if they make a mistake will attempt to self-correct.

Whereas in conduction aphasia the language areas are separated from each other, **in transcortical aphasia the speech areas are connected to each other but not to the rest of the cerebral axis**. This isolation of the speech area occurs because of ischemia of the surrounding cortex, disconnecting the process of language from any meaningful association or interpretation by higher brain functions. The patient has the ability to create and absorb language; however, because of the disconnect, he or she has nothing to say and cannot interpret what is said. The patient does have the ability to generate automatic responses of previously learned phrases, prayers, or songs.

As in the initial case example, injury to both Broca and Wernicke areas leads to a global aphasia. Global aphasia means the patient can neither understand nor formulate meaningful language. This is a devastating condition, as the patient has no means to communicate with the outside world, nor can the world communicate with the patient. Generally, patients with global aphasia also have a corresponding hemiplegia.

All of the above examples generally result from an insult to the dominant hemisphere for language (the left, for the majority of the population). Language disorders also occur with injury to the right or nondominant hemisphere. For example, patients with right language-area lesions may have difficulty organizing their speech, leading to difficulties in communication. Likewise, reasoning disorders lead to an inability to understand abstract language (parables, sarcasm, metaphors, humor) and pragmatics (intonation, body language, facial expression). Finally, social judgment may be impaired, leading the patient to say inappropriate things without realization that he or she has done so.

> ## CASE CORRELATES
> - See Cases 41-49 (cognition).

COMPREHENSION QUESTIONS

44.1 A 56-year-old right-handed woman presents to the clinic because of difficulty speaking. Over the past few months she has had increasing difficulty express-ing herself. She knows what she wants to say but can only express herself in short, agrammatic phrases. She is capable of understanding what others are saying to her and seems very frustrated with her inability to speak properly. A lesion to which area of the brain best accounts for these findings?

A. The dominant pars opercularis and triangularis of the inferior frontal gyrus

B. The nondominant pars opercularis and triangularis of the inferior frontal gyrus

C. The dominant posterior aspect of the superior temporal gyrus

D. The nondominant posterior aspect of the superior temporal gyrus

44.2 A 42-year-old left-handed man is brought to the clinic by his family because of bizarre speech. They state that 2 days ago, he suddenly began speaking in sentences that do not make any sense, and he does not appear to understand what they are telling him. On examination the man speaks easily, and his speech seems to have normal structure and intonation, but the combinations of words simply do not make sense. He is also completely unable to follow any commands that the physician tells him. Damage to what area of the brain is mostly likely responsible for these findings?

A. The right pars opercularis and triangularis of the inferior frontal gyrus

B. The left pars opercularis and triangularis of the inferior frontal gyrus

C. The right posterior aspect of the superior temporal gyrus

D. The left posterior aspect of the superior temporal gyrus

44.3 A 56-year-old woman comes into your clinic complaining of difficulties with speech. Most notably, she says, she cannot read out loud, and she also occasionally has difficultly naming objects. You do a full speech evaluation, and in addition to the complaints that she reports, you find that she cannot repeat words or phrases that you ask her to, and she also replaces words with incorrect words that sound similar. Based on these findings, a lesion in what location is most likely responsible for her symptoms?

A. The nondominant arcuate fasciculus

B. The dominant arcuate fasciculus

C. The dominant pars opercularis and triangularis of the inferior frontal gyrus

D. The dominant posterior aspect of the superior temporal gyrus

ANSWERS

44.1 **A.** The lesion is most likely in the **dominant pars opercularis and triangularis of the inferior frontal gyrus**. This woman is presenting with Broca aphasia, also known as a productive aphasia. She is able to understand speech but has difficulty in producing it. Because she can understand her own speech, she is able to recognize that it is deficient, which commonly causes a great deal of frustration in these patients. Broca area is located in the inferior frontal gyrus of the dominant hemisphere, adjacent to the motor cortex responsible for controlling the speech organs.

44.2 **D.** This man is presenting with Wernicke aphasia (fluent or receptive aphasia) caused by a lesion in Wernicke area, which is located in the **posterior aspect of the superior temporal gyrus** of the dominant hemisphere (the **left** in this patient). This area is responsible for comprehending speech and generating comprehensible speech. Damage to this area results in an aphasia where the patient has no problem creating speech (it is fluent); it simply does not make sense. Although this man is left handed, there is still a 70% likelihood that he is left-hemisphere dominant.

44.3 **B.** This patient is presenting with a conduction aphasia secondary to a lesion in the **dominant arcuate fasciculus**, the fiber tract connecting Wernicke and Broca areas. Because this disorder affects the communication between Wernicke and Broca areas, common findings include difficulties with reading aloud and the inability to repeat words and phrases.

NEUROSCIENCE PEARLS

▶ In 70%-90% of right-handed people, the left cerebral hemisphere controls language.

▶ Broca area (inferior frontal lobe) and Wernicke area (posterior temporal lobe) have different (output vs comprehension) yet integrated and essential functions.

▶ The nondominant hemisphere contributes to the nuances of language, such as intonation and emotion.

REFERENCES

Antonucci SM, Beeson PM, Rapcsak SZ. Anomia in patients with left inferior temporal lobe lesions. *Aphasiology*. 2004;18(5-7):543-554.

Carandang R, Seshadri S, Beiser A, et al. Trends in incidence, lifetime risk, severity, and 30-day mortality of stroke over the past 50 years. JAMA. 2006 Dec 27;296(24):2939-2946.

Delazer M, Semenza C, Reiner M, Hofer R, Benke T. Anomia for people names in DAT: evidence for semantic and post-semantic impairments. *Neuropsychologia*. 2003;41(12):1593-1598.

A 67-year-old woman is brought into the neurology clinic by her husband for increasing episodes of forgetfulness. The husband states that over the last several years she has had difficulty remembering the locations of her belongings, recalling friends' names, and has even gotten lost several times in their neighborhood where they have lived for the last two decades. The woman suffers from osteoarthritis but takes only ibuprofen and is otherwise healthy.

The patient is alert and oriented only to self, and partly to place (can name the state, but cannot recall the city). She is mildly disheveled. A thorough neurological examination reveals no focal deficits; however, the patient at times has difficulty following directions. She scores 12 out of 30 on the mini-mental state examination (MMSE). Her copy of the interlocking pentagons is as follows.

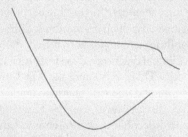

The remainder of her physical examination is normal. Routine laboratory tests show mild hyperlipidemia, but her complete blood count, cobalamin, liver enzyme levels, blood cortisol levels, thyroid-stimulating hormone, and rapid plasma reagent are all normal. A head CT reveals diffuse cerebral atrophy with dilation of the perihippocampal fissure. The patient is placed on a combination of medical and behavioral interventions.

► What is the most likely diagnosis?
► What are the cortical lesions associated with this disease?
► What are the treatment options for this condition?

ANSWERS TO CASE 45:
Memory

Summary: A 67-year-old otherwise healthy woman presents with several years of progressive memory loss.

- **Most likely diagnosis:** Alzheimer disease

- **Cortical lesions of Alzheimer disease:** These include neurofibrillary tangles and senile plaques, occurring mainly in the medial temporal lobe.

- **Treatment options:** Behavioral modification and psychotropic medication seek to mediate the clinical manifestations of Alzheimer disease. Cholinesterase inhibitors attempt to counter the pathophysiology of the cerebral decrease of acetylcholine. N-methyl-D-aspartate antagonists are utilized at the end stages of the disease. Antidepressants help significantly with corresponding mood disorders. Other investigational treatment modalities include estrogen receptor modifiers, anti-inflammatories, free-radical scavengers, and certain antibiotics.

CLINICAL CORRELATION

Alzheimer disease (AD) is a progressive neurodegenerative disorder mainly affecting elderly individuals. As the most common cause of dementia, it is a major cause of morbidity and mortality in the United States. It affects both men and women, the majority of whom are more than 60 years old. Initially, short-term memory is affected more than long-term memory. As the disease progresses, patients begin to experience behavioral changes, psychiatric symptoms, difficulty with higher cortical function, and decreased abilities to care for activities of daily living.

Alzheimer disease is a clinical diagnosis; therefore, it is important to rule out any treatable cause for dementia. There is no cure for AD. **Other causes of dementia include** vascular or multi-infarct dementia, pharmacological agents, HIV, delirium, Lewy body dementia, Pick disease, head trauma, Huntington disease, Parkinson disease, normal pressure hydrocephalus, frontotemporal dementia, Lyme disease, neurosyphilis, prion-related diseases, thyroid disease, and Wilson disease. Amyloid deposits form senile plaques in the brain, which is typical of AD.

Current treatments look to prevent or slow the progression of the disease, so far with minimal success.

> # APPROACH TO:
> ## Memory

OBJECTIVES

1. Know the terminology of memory.

2. Understand the cerebral circuitry of memory.

3. Understand the amnesic syndromes.

DEFINITIONS

IMMEDIATE MEMORY: The ability to repeat a short set of numbers or words; may last for seconds.

DECLARATIVE SHORT-TERM (RECENT) MEMORY: Contains seven or fewer pieces of information at one time and is easily disrupted with distraction; may last from seconds to minutes.

DECLARATIVE LONG-TERM (REMOTE) MEMORY: Contains explicit information about facts; for example, distant personal or public memories, reciting the helping verbs, completing a multiplication table.

PROCEDURAL LONG-TERM MEMORY: Involved in actions that improve with repetition; for example, riding a bicycle, playing a musical instrument, crocheting; nonconscious (ie, automatic).

ANTEROGRADE AMNESIA: The inability to retain new information through the conversion of short-term memories to long-term memories.

RETROGRADE AMNESIA: The inability to recall events occurring before the onset of amnesia; may be temporally graded (ie, distant events are easier to recall than more recent ones).

CONSOLIDATION: Refers to short-term memory conversion to long-term memory.

LONG-TERM POTENTIATION (LTP): The molecular process that strengthens groups of repeatedly used synapses.

HIPPOCAMPUS: Located in the medial temporal lobe; from the Greek word for seahorse; responsible for memory consolidation.

DISCUSSION

The neurocircuitry of memory includes the medial temporal system (right and left hippocampus, entorhinal cortex) and the medial diencephalic system (mediodorsal nuclei of the thalamus, mamillary bodies, mamillothalamic tracts). Working or short-term declarative memory is stored in the prefrontal cortex. Conversion into long-term declarative memory uses both the medial temporal and diencephalic systems. Long-term memories are stored diffusely throughout the cortex, most likely in the areas responsible for the perception of the initial stimulus. The hippocampus does not seem to play a role for short-term memory, procedural memory, or the end storage of memory. The hippocampus is required for declarative memory consolidation. The cerebellum integrates procedural memory, which is then transmitted to the basal ganglia for storage and coordination.

The physiology of the formation of memories most likely involves LTP. Long-term potentiation is only one way that memories are formed, but it appears to play a significant role. Studies have shown that hippocampal neurons, when regularly stimulated, convert this constant level of presynaptic stimulation into a larger postsynaptic output. The LTP of a particular neuron may vary depending upon anatomical location and age. The specific mechanisms detailing how LTP leads to the formation of memory have yet to be elucidated.

Declarative memory follows a particular pathway through the hippocampus. Important structures to note include the dentate gyrus, CA1, CA2, CA3, subiculum, and the entorhinal cortex (part of the parahippocampal gyrus). A memory begins as a stimulus perceived by the sensory cortex. This information is transmitted by neurons of the entorhinal cortex (hippocampus input), which traverse the subiculum and dentate gyrus (perforant path) to synapse in the dentate gyrus. The "memory" then travels to CA3 (mossy fibers), then CA1 (Schaffer collaterals), and ends up in the subiculum. The subiculum (hippocampus output) projects axons via the fornix to the hypothalamus and mamillary bodies or returns to the entorhinal cortex. At that point the entorhinal cortex relays the information back to the sensory cortex.

Disruption of this pathway leads to an inability to consolidate new memories, while distant memories remain intact. The most famous clinical example of hippocampal injury is from the patient HM. On August 25, 1953 HM underwent resection of his medial temporal lobes, thought to be the etiology of his intractable epilepsy. This included removal of his hippocampus formation, amygdala, and entorhinal and perirhinal cortices. From that date onward, HM was no longer able to store new declarative memories to his long-term memory cortex. He also suffered from moderate temporally graded retrograde amnesia. HM was able to acquire new motor skills (ie, to form new procedural memories); however, he could not recall that he had learned them. Therefore, HM demonstrates the result of bilateral lesions of the medial temporal or medial diencephalic system: anterograde amnesia with temporally graded retrograde amnesia, intact short-term and procedural memory, and intact personal identity.

The most common diseases of the hippocampus include AD and epilepsy. Alzheimer disease severely affects CA1, disrupting memory formation. The hippocampus is also particularly sensitive to global ischemia.

CASE CORRELATES
- See Cases 41-49 (cognition), and Cases 5 and 14 (Alzheimer disease).

COMPREHENSION QUESTIONS

Refer to the following case scenario to answer questions 45.1-45.2:

A 26-year-old man is brought into your clinic because of anterograde amnesia that he has been experiencing for the past several weeks since he was involved in a motor vehicle accident. During the accident, he was ejected from his vehicle and afterwards remained in the intensive care unit for several weeks. You perform an MRI, which indicates that he has bilateral damage to his medial temporal lobes in the area of the hippocampus.

45.1 Which of the following processes is most likely disrupted because of these lesions?

A. Short-term memory

B. Storage of long-term memories

C. Retrieval of long-term memories

D. Storage of procedural memory

45.2 Which neurophysiologic process is thought to be extremely important in storing new memories?

A. Short-term potentiation

B. Ionotropic neurotransmission

C. Long-term potentiation

D. Metabotropic neurotransmission

45.3 A 33-year-old woman comes into the physician's office for follow-up after a relatively severe head injury that she suffered on the job several months ago. Overall she seems to be doing well, but she says she has noticed "something kind of weird." Recently she has been trying to learn how to knit to help pass the time, since she is still not well enough to return to work, but she is having a lot of trouble doing it. She says she used to be very good with her hands, but that every time she picks up the needles it is like it is the first time she has ever used them. She wants to know if this could be related to her injury. Damage to what structure involved in storing procedural memory would be most likely to cause this problem?

A. Cerebellum

B. Hippocampus

C. Prefrontal cortex

D. Mamillary bodies

45.4   A 26-year-old woman is being evaluated for entry into the airforce. She has normal cognitive function and is found on testing to be homozygous for the apolipoprotein E4 allele. In the future, this individual is most likely to develop which of the following disease/disorder?

A.   Alzheimer disease

B.   Diabetes mellitus

C.   Familial hyperlipidemia

D.   Hypertrophic cardiomyopathy

E.   Premature menopause

ANSWERS

45.1   **B.** The function of the hippocampus as it relates to memory is in the role of consolidating **long-term memories,** or transition from short-term to long-term memory. Short-term memory involves the prefrontal cortex, and long-term memory seems to be stored diffusely throughout the cortex, but the hippocampus is integral in the transition between the two. Because of the location of the injury in this patient, he can recall long-term memories formed prior to the accident without difficulty, and has a perfectly functional short-term memory. His deficiency comes when he has to store his short-term memories as long-term memories, which he cannot do.

45.2   **C. Long-term potentiation,** in which a given stimulus causes a postsynaptic neuron to alter the way in which it responds to stimuli for a long time (or permanently), is thought to be very important in the storing of long-term memories. Long-term potentiation involves metabotropic neurotransmission, which alters gene expression in the target cell, but metabotropic neurotransmission is not sufficient alone. The mechanism by which long-term potentiation results in memory storage has not yet been specifically determined.

45.3   **A.** The **cerebellum** is thought to play a very important role in the learning of new procedural memory. The other structure that seems to be highly involved in structural memory formation and retrieval is the basal ganglia. The hippocampus and mamillary bodies are involved in consolidation of declarative memory, and the prefrontal cortex is involved in short-term memory.

45.4   **A. Alzheimer disease** has a strong familial predisposition, with about one-third of affected individuals having a family history. Onset is categorized as having an early or late manifestation. Early onset has been defined as clinically apparent prior to age 60 and is associated with mutations of the amyloid precursor protein (APP) gene, the presenilin-1 gene (chromosome 14), and the presenilin-2 gene (chromosome 1). Late-onset AD is associated with the E4 allele of the apolipoprotein gene, although the mechanism is unknown.

NEUROSCIENCE PEARLS

▶ The neurocircuitry of declarative memory includes the medial temporal and diencephalic systems.

▶ The cerebellum and basal ganglia integrate, store, and coordinate procedural memory.

▶ As distant memories are stored diffusely throughout the cortex, disruption of the hippocampal pathway leads to an inability to consolidate new memories, while distant memories remain intact.

▶ LTP plays a significant, but still undefined, role in memory formation.

REFERENCES

Bear MF, Connors B, Paradiso M, eds. *Neuroscience: Exploring the Brain.* 3rd ed. Baltimore, MD: Lippincott Williams & Wilkins; 2006.

Kandel ER, Schwarz JH, Jessell TM, Siegelbaum SA, Hudspeth AJ, eds. *Principles of Neural Science.* 5th ed. New York, NY: McGraw-Hill; 2012.

Squire LR, Berg D, Bloom FE, du Lac S, eds. *Fundamental Neuroscience.* 4th ed. San Diego, CA: Academic Press; 2012.

A 32-year-old right-handed man has a history of atonic seizures since childhood. He presents to the neurosurgery clinic for a preoperative appointment. His seizures have been managed unsuccessfully with various high doses of medication, and he continues to have multiple "drop attacks" each day. He is resigned to wearing a helmet at all times. He gives a history of multiple occasions when he was not wearing his helmet, suffered an atonic seizure, and severely injured his head and body. Currently, he requests a corpus callosotomy to attempt to decrease the frequency of his atonic seizures.

► What is the purpose of the corpus callosum?
► What complications may occur with corpus callosotomy?

372   CASE FILES: NEUROSCIENCE

ANSWERS TO CASE 46:

Disconnection Syndromes

Summary: A 32-year old right-handed man with medically refractory atonic seizures prepares for a corpus callosotomy.

- **Purpose of corpus callosum:** Communication between the right and left cerebral hemispheres.

- **Complications with corpus callosotomy: The acute disconnection symptoms and signs.** Immediately following a corpus callosotomy, patients may experience acute symptoms and signs from the lack of communication between the two hemispheres. For example, patients may suffer from mild akinesia or competitive movements between the two hands. In fact, some patients express surprise at the purposeful and independent actions of an "alien" left hand. On examination, bilateral Babinski responses indicate damage to the brain. Brain edema may occur secondary to retraction on the brain to reach the surgical site. In patients with speech and handedness stored in opposite hemispheres, complete callosotomy may lead to difficulties in the production of spontaneous speech.

CLINICAL CORRELATION

Corpus callosotomy is most effective for atonic seizures, tonic-clonic seizures, and tonic seizures. Response to the treatment is high, with up to an 80% decrease in seizure frequency after partial callosotomy. The abnormal focus origin of the seizure will remain; however, that abnormal electrical activity will not be able to generalize to the other half of the brain. This leads to a decrease in not only the frequency but also the severity of the seizures. In the long term, signs and symptoms of disconnection via corpus callosotomy are indistinguishable from individuals with intact circuitry, with the exception of certain memory problems and responses under special lateralizing testing situations. The lack of interhemispheric communication does not seem to affect everyday life. For example, hemispheric specialization and the inhibition of interhemispheric transfer can be demonstrated when a right-handed individual is given a familiar object in the left hand (graphesthesia) but cannot name the item. Apparently, split-brain patients are able to compensate for these deficits through tricks to trigger the opposite hemisphere. For example, as the right hemisphere may recognize individual words, a patient may say out loud the name of an object in order to facilitate retrieval with the left hand.

APPROACH TO:
Disconnection Syndromes

OBJECTIVES

1. Be able to identify different types of disconnection signs and syndromes.
2. Know the tests which can be performed to isolate the function of either hemisphere of the brain.

DEFINITIONS

AKINESIA: "Freezing" of body movements; difficulty in starting or maintaining a body motion.

AGENESIS OF THE CORPUS CALLOSUM: Congenital lack of interhemispheric connecting neurons.

BABINSKI RESPONSE: Firm stroking of the outer, dorsal heel to the pad of the foot leads to extension upward and spreading of the toes. Normal in infants less than 2 years of age. In adults is a sign of brain or spinal cord injury.

GRAPHESTHESIA: Tactile ability to recognize letters written on the skin.

VERBAL ANOSMIA: Inability to name smells presented only to the right nostril; however, the left hand can use tactile information to find a corresponding object.

DOUBLE HEMIANOPIA: Inability to indicate the side of a visual stimulus with the contralateral hand.

UNILATERAL ANOMIA: Inability to name an object with purely tactile stimuli. The object can be named when placed in the right hand.

UNILATERAL AGRAPHIA: Inability to write with the left hand, whereas the right hand writes well, or vice versa.

UNILATERAL APRAXIA: Inability to perform a verbal command with the left hand that is easily performed by the right hand, or vice versa.

UNILATERAL CONSTRUCTIONAL APRAXIA: Inability of the right hand to perform tasks that require the right hemisphere's abilities of spatial cognition. Examples include copying geometric forms and completing mathematics that requires writing down procedural steps.

DISCUSSION

Cortical disconnection signs and syndromes present secondary to subcortical white matter lesions connecting two brain regions. These can be classified as either intrahemispheric or interhemispheric.

Intrahemispheric disconnection syndromes involve damage to deep white matter connections between structures within the same hemisphere. In the dominant hemisphere, this can include structures involved in language. The arcuate fasciculus connects Wernicke and Broca speech areas. Disruption of this connection leads to retention of fluent speech (although dysphagic) and comprehension; however, patients have difficulty with repetition. This is referred to as a conduction aphasia, as the information received in **Wernicke area** cannot be transferred to **Broca area**. Injury to the white matter connecting the primary auditory cortex with the auditory association areas leads to a pure word deafness. The patient can "hear" (ie, the physical and neuronal mechanisms for conducing auditory stimuli from the ear to the primary auditory cortex are intact), and audiometry examinations will be normal. Yet the patient will not be able to perceive and process the information, making the patient functionally deaf. A patient with pure word deafness will exhibit impaired comprehension of speech but will be able to produce speech and comprehend written language normally.

Another disorder of language involves the motor apparatus used to produce speech. Apraxia of the buccal and lingual areas can occur secondary to disconnection between the association motor cortices in the subcortical region. The patient's signs include right brachiofacial weakness and apraxia of tongue, lip, and left limb movements.

Three main structures allow for communication between the cerebral hemispheres: the anterior commissure (olfactory and limbic), the hippocampal commissure (limbic), and the corpus callosum (hemispheric cortical processing). Interhemispheric disconnection results from a lesion of the corpus callosum. One example of a symptom that may be seen in the patient in this case during the acute postoperative period is left-hand apraxia to verbal commands. This occurs secondary to disconnection of the right motor strip from the left language area. Unlike unilateral lesions, the acute disconnection syndrome does not include symptoms such as hemineglect and aphasia. In fact, within months most patients appear normal, both in initial social interactions and on routine neurological examinations. Patients do appear to suffer from subtle memory, language, and personality deficits, yet split-brain patients provide an important example of brain plasticity. For example, initially left-hand apraxia to verbal commands is a manifestation of the right hemisphere's poor language comprehension. As there is a contralateral increase in language comprehension and ipsilateral increase in motor control, the left-hand apraxia improves.

Agenesis of the corpus callosum, when accompanied by other brain defects, carries severe symptoms (seizures, mental retardation, hydrocephalus, spasticity). There are, however, some individuals with isolated agenesis of the corpus callosum, who have normal intelligence and whose deficits are only elicited upon exercises requiring matching of visual patterns. Many of the minimally affected patients with callosal agenesis do not manifest the disconnection signs, whereas older individuals with a corpus callosotomy do.

Multiple tests can be performed to isolate the function of either hemisphere. For example, when a split-brain patient sees a chimeric portrait, one image is sent to the left side of the brain and a distinct image to the right. When asked to point to the whole picture, the patient will usually choose the image sent to the right hemisphere (facial recognition); however, when required to verbally choose the picture, the patient will say the image sent to the left hemisphere (speech dominance). Object location is also unffected, as the superior colliculus has bilateral inputs for spatial localization. Stereognostic input from one side is only perceived by the contralateral hemisphere. For a patient with an interhemispheric disconnection syndrome, an object held in the left hand can only be perceived by the right hemisphere and cannot be named. When different words are presented in either ear, the ipsilateral pathway from the left ear is suppressed by the dominant contralateral right ear, and the patient will repeat the word presented in the right ear.

> **CASE CORRELATES**
> • See Cases 41-49 (cognition).

COMPREHENSION QUESTIONS

46.1 A 13-year-old girl has severe intractable generalized epilepsy. She has daily tonic-clonic seizures that are unresponsive to treatment with therapeutic dosages of many different antiepileptic drugs. In an effort to control her seizures, her family is considering surgery to help manage her epilepsy. What structure that normally allows for interhemispheric cortical communication can be ablated to help control seizures in extreme cases like this?

   A. Anterior commissure
   B. Corpus callosum
   C. Hippocampal commissure
   D. Posterior commissure

46.2 A 62-year-old man is brought into the physician's office by his family because of the abrupt onset of strange behavior. According to his wife, one day before, he simply stopped responding when she spoke to him. On examination, he responds to sounds by turning toward them, can read and understand perfectly, and can generate comprehensible speech without any problem. The only deficit seems to be in understanding speech. What term best describes the deficit this patient is experiencing?

   A. Wernicke aphasia
   B. Global aphasia
   C. Pure word deafness
   D. Cortical deafness

46.3 A 57-year-old man presents to the clinic with some difficulties speaking, particularly when it comes to reading aloud and repeating words. The physician performs an MRI, which shows a lesion in the area of the arcuate fasciculus. A lesion in this area could result in what speech-related deficit?

A. Productive aphasia

B. Receptive aphasia

C. Transcortical aphasia

D. Conductive aphasia

ANSWERS

46.1 **B. The corpus callosum** is a bundle of white matter that in the normal brain serves as the major communication between the cerebral hemispheres and is involved in interhemispheric cortical processing. It can be severed to help control severe epilepsy, with fairly good results in patients with incapacitating seizures. The anterior commissure and hippocampal commissure are involved in the limbic system, and the posterior commissure is involved in the pupillary light reflex.

46.2 **C.** This patient is experiencing what is known as **pure word deafness**, which is caused by an insult to the white matter that connects the primary auditory cortex, located in the transverse temporal gyrus of Heschl, to Wernicke area, located on the posterior aspect of the superior temporal gyrus. This insult inhibits the arrival of information about heard speech at the comprehension center in Wernicke area, making the patient functionally deaf when it comes to speech. They can hear, however, and will turn to face noises, and they comprehend written words, because the connections between the visual cortex and Wernicke area are intact. In cortical deafness, lesions to the bilateral auditory cortex make patients deaf, although lower brain structures respond to sound, and aphasia patients have difficulty generating speech.

46.3 **D.** The arcuate fasciculus connects Wernicke and Broca areas, and a lesion to it results in what is known as a **conduction aphasia.** The primary manifestations of this are difficulties in repeating words and in reading aloud, as the comprehension of the words in Wernicke area cannot be communicated to Broca area for speech generation.

NEUROSCIENCE PEARLS

▶ Cortical disconnection signs and syndromes present secondary to subcortical white matter lesions connecting two brain regions.

▶ Interhemispheric disconnection results from a lesion of the corpus callosum.

▶ Patients with a corpus callosotomy appear normal in initial social interactions, and only specific neurological examinations can elicit the disconnection signs.

REFERENCES

Bear MF, Connors B, Paradiso M, eds. *Neuroscience: Exploring the Brain*. 3rd ed. Baltimore, MD: Lippincott Williams & Wilkins; 2006.

Kandel ER, Schwarz JH, Jessell TM, Siegelbaum SA, Hudspeth AJ, eds. *Principles of Neural Science*. 5th ed. New York, NY: McGraw-Hill; 2012.

Squire LR, Berg D, Bloom FE, du Lac S, eds. *Fundamental Neuroscience*. 4th ed. San Diego, CA: Academic Press; 2012.

CASE 47

A 23-year-old man is brought into the clinic by his mother. She states he has not been acting like himself ever since he received a head injury in the war. Previously a polite and easygoing individual, she says he is now ill-tempered and makes frequent inappropriate comments. As a result the patient has not been able to hold down a steady job and has alienated many of his friends. She says he has noticed simple tasks to be more frustrating lately but otherwise does not think he has changed since his injury.

On examination, the patient has a scar over his right forehead. His neurological examination is normal with the exception of perseveration on his cerebellar function testing.

▶ What area of the brain has most likely been injured, leading to this patient's symptoms?
▶ Why has the patient experienced personality changes?
▶ How will the patient respond when presented with a novel task?

ANSWERS TO CASE 47:
Executive Function

Summary: A 23-year-old posttraumatic brain injury patient presents with symptoms of disinhibition and personality change.

- **Region of the brain most likely injured:** The frontal lobes, the site of executive cognitive function, are the most likely area to have been damaged. The frontal lobes are commonly involved in traumatic brain injury (TBI), owing to location and size. Likewise, the sphenoid wing and orbit of the skull damage the brain when applied with force to the tissue.

- **Most likely cause of the patient's symptoms:** The frontal lobes are responsible for executive cognitive function and are considered the source of personality and emotional control. Patients with dysexecutive syndrome have difficulty in social settings requiring intricate thought processes and intuitive nuances.

- **When the patient is presented with a novel task:** The executive system facilitates learning, making actions and problem solving less effortful with repetition. The frontal lobes also function in conceptualization of abstract problems and following through with a complex, predetermined plan. Inability to adapt to novel situations and interactions combined with a disinhibition of emotional control and social norms may cause patients with frontal lobe damage to become easily enraged.

CLINICAL CORRELATION

The symptoms exhibited by this patient suggest that he has a frontal lobe lesion. Lesions in the frontal lobes are known to cause aggressiveness in patients (ie, Phineas Gage phenomenon) and hinder the patient's ability to carry out executive functions. These functions include behaviors such as planning multistep tasks, the capacity for quick switching to the appropriate mental mode, the ability to withstand distractions or internal urges, anticipation, logical analysis, working memory, multitasking, and decision making.

APPROACH TO:
Executive Function

OBJECTIVES

1. Be able to define the role of the frontal lobes in higher cognitive function.

2. Understand the common symptoms associated with frontal lobe lesions and traumatic brain injury.

3. Be familiar with the mechanisms for testing executive function.

DEFINITIONS

PERSEVERATION: Persistent repetition of an activity, word, phrase, or movement without any apparent stimulus for it.

COUP/CONTRECOUP: Contusions both at the site of impact (coup) and on the opposite side (contrecoup) of the brain.

THURSTONE (LETTER FLUENCY, WORD GENERATION) TEST: Ask the patient to generate as many words as possible beginning with the letter F in 1 minute. A normal score for a native English speaker with at least a high school education is 8 words or more.

SERIAL 7S: Counting down from 100 by 7s. Tests concentration and memory. Spelling the word *world* backward is commonly used as a substitute for patients who cannot perform calculations of 7s.

DIGIT SPAN: Also a measure of attention and concentration. A normal span is seven digits forward and five backward. An abnormal digit span is the most common neuropsychological deficit in patients with head injury.

DISCUSSION

The **frontal lobe** is an area in the brain of mammals located at the front of each cerebral hemisphere. The frontal lobes are positioned anterior to the parietal lobes, and the temporal lobes are located beneath and behind the frontal lobes. In the human brain, the precentral gyrus and the related cortical tissue that folds into the central sulcus comprise the primary motor cortex, which controls voluntary movements of specific body parts associated with areas of the gyrus. The frontal lobes have been found to play a part in impulse control, judgment, language production, working memory, motor function, sexual behavior, socialization, and spontaneity. The frontal lobes assist in planning, coordinating, controlling, and executing behavior. People who have damaged frontal lobes may experience problems with these aspects of cognitive function, being at times impulsive, impaired in their ability to plan and execute complex sequences of actions, and perhaps persisting with one course of action or pattern of behavior when a change would be appropriate.

The executive functions of the frontal lobe involve abstract thinking, rule acquisition, problem solving, planning, cognitive flexibility, selecting relevant sensory information, initiating behavior, monitoring behavior via judgment and impulse control, motor function, executive memory, language, sexual behavior, and decision making. It **allows for adaptation to new environments and the acquisition of new skills or manners of thinking.** This includes insight and error correction, enabling one to navigate dangerous or technically difficult situations. Important in social functioning, executive cognition **allows for the suppression of strong habitual responses or temptations.**

The executive system organizes and prioritizes cognitive resources in such a way as to allow for complex conceptualization of abstract problems and execution of detailed plans and actions. It is required for foresight or temporal planning, critical analysis of rules, limitations and abilities, focus for attainment of goals, and reflection for troubleshooting and adaptation. For previously encountered situations, a patient with dysexecutive syndrome can rely on learned, automatic physical and psychological responses. In novel situations, however, the patient does not have these automatic responses to fall back on or the abilities to organize his thoughts in order to deal with the new circumstances. Patients are unable to reach a particular goal secondary to the fact that they cannot plan and execute the smaller individual steps to achieve the desired end. The end result is thatthe patient is seen as apathetic and easily distractible. This can also be frustrating for a patient, which may lead to outbursts of anger.

Damage to the frontal lobes may occur through traumatic injury, mass lesions, cerebrovascular and degenerative disease, or infection. For example, the frontal lobes are affected by normal aging, and by pathological degenerative processes, such as multiple sclerosis, Huntington's disease, and AD. Cerebrovascular disease, such as the occlusion or hemorrhage of the anterior communicating artery, also may lead to damage of the frontal lobes. The dominant neurotransmitter in the frontal lobe is dopamine. Low levels of dopamine often present as problems in frontal lobe functioning, one example of such being schizophrenia.

The most commonly cited example of a dysexecutive patient is that of Phineas Gage. Mr Gage sustained injury to his prefrontal cortex (PFC) following a railroad construction accident in 1848, when a 1.1-m, 6-kg tamping iron went in under his left cheek bone and out through the top of his head. Mr Gage went from being an efficient foreman to an impulsive, profane individual with an inability to plan and carry out productive goals. **While intelligence is intact, patients with frontal lobe damage show difficulty with emotional control, social interaction, and memory.** This disinhibition is attributed to a disconnection between the PFC and the autonomic nervous system. Without a "warning" to alert the individual to a dangerous, undesirable, or sensitive situation, executive decision making is severely hampered. As such, executive cognition may not act so much to inform us of what behaviors are appropriate but more to deter us from inappropriate interactions. The difficulty interpreting environmental stimuli contributes to the previously described social dysfunction and impaired learning (feedback to guide behavior).

The impulse-control difficulties leading to social dysfunction also may manifest as utilization behavior. For example, without instruction or reason, patients may be compelled to use daily objects within their visual field. For example, if a patient were to see a toothbrush in front of him, he would begin to brush his teeth, and do so repeatedly, even if he had already brushed his teeth. When asked why they are using this object without purpose, patients tend to confabulate. Patients may also perseverate in speech and action.

Spontaneity appears to originate in the frontal lobes as well. Patients with frontal lobe damage have limited facial expression and can have either increased spontaneity of speech (right frontal in a right-handed individual) or decreased (left frontal). Patients may also have deficits of declarative memory or memory for temporal order of events. Working memory and attention are frequently impaired. This is in contrast to the declarative memory deficits found with hippocampal damage.

> **CASE CORRELATES**
> • See Cases 41-49 (cognition).

COMPREHENSION QUESTIONS

Refer to the following case scenario to answer questions 47.1-47.2:

A 29-year-old man is hospitalized following a motorcycle accident in which he was thrown from his bike unhelmeted and suffered a head injury. Following several weeks of sedation and intubation in the ICU, he physically recovers enough to be sent home with his family. They report, however, that his personality has changed from before the accident, and they want to know if his head injury could be responsible.

47.1 The patient's family report that prior to the accident he was very organized and knew what he wanted out of life. Now he just sits around, not really caring about anything, "lets people walk all over him," and is completely incapable of planning anything. Damage to what region of this man's brain could account for these symptoms?
   A. Dorsolateral frontal cortex
   B. Orbitofrontal cortex
   C. Parieto-occipital cortex
   D. Temporo-occipital cortex

47.2 The patient's family reports that prior to the accident he was a well-behaved, somewhat reserved person who was very reliable and dependable. Since the accident he has behaved very inappropriately, shirking responsibility, and bouncing from one activity to another. They say he doesn't seem to have any kind of internal filter anymore, and doesn't seem to care about social norms—he is doing whatever he wants, whenever he wants. Damage to what region of the brain could account for these symptoms?
   A. Dorsolateral frontal cortex
   B. Orbitofrontal cortex
   C. Parieto-occipital cortex
   D. Temporo-occipital cortex

47.3 The physician sees a patient in an inpatient psychiatric hospital who was "treated" for schizophrenia with a frontal lobotomy a number of years ago. He is very calm and shows very little emotion, and is relatively disinterested in what is going on around him. In addition to these findings, in which process is he most likely to have deficits?
   A. Visuospatial processing
   B. Language production
   C. Motor learning
   D. Solving novel problems

ANSWERS

47.1 **A.** The **dorsolateral frontal cortex** is associated with planning and strategy formation and lesions here usually cause apathy, poor planning ability, and excessively passive behavior, as seen in this patient. The frontal lobes in general are involved in executive function, which consists of, among other things, planning, problem solving, abstract thinking, judgment, and behavioral inhibition.

47.2 **B.** This man has a lesion in the **orbitofrontal cortex**, which is associated with behavioral inhibition. Patients with lesions in this area typically behave poorly in social situations and often engage in hypersexual behavior, abuse alcohol and drugs, and may engage in compulsive gambling.

47.3 **D.** This patient, who has had a frontal lobotomy, would be expected to have deficits in executive function such as **solving novel problems.** Patients with frontal lobe dysfunction can often work their way through problems that they have experienced before because they can rely on the learned behavior from prior encounters. When confronted with a novel situation, however, they have great difficulty in problem solving and planning what to do.

NEUROSCIENCE PEARLS

▶ The executive function of the frontal lobes coordinates adaptive and problem-solving abilities when presented with a new situation.

▶ The frontal lobes allow for warnings of social norms, allowing for the suppression of strong habitual responses or temptations.

▶ Traumatic brain injury involving the frontal lobe leads to problems with emotional control, social interaction, and memory, despite the fact that intelligence is unaffected.

REFERENCES

Adolphs R, Tranel D, Damasio AR. The human amygdala in social judgment. *Nature.* Jun 4 1998; 393(6684):470-474.

Saver JL, Damasio AR. Preserved access and processing of social knowledge in a patient with acquired sociopathy caused by ventromedial frontal damage. *Neuropsychologia.* 1991;29(12):1241-1249.

A 19-year-old man, who suffers an accident on his dirt bike is brought into the emergency department by paramedics. He cannot recall the crash but says he was not wearing a helmet. Currently he is alert and oriented, complaining of nausea, vomiting, and has a left-sided scalp contusion, laceration, and an underlying bony step-off. You send the patient for a CT scan of the head and call for an urgent neurosurgery consult. As you are waiting for the operating room, the patient becomes more difficult to arouse and confused. Based on the patient's presentation, you make a diagnosis of an epidural hematoma (EDH).

► Which artery is likely to be affected leading to EDH?
► What is the most likely explanation for the patient's symptoms?
► What is the CT scan of the patient's head likely to reveal?

ANSWERS TO CASE 48:

Consciousness

Summary: A 19-year-old man, after an unhelmeted motorcycle accident, presents with head trauma and decreasing level of consciousness. He is amnesic of the accident, has a lucid interval, then following CT scan becomes obtunded. The CT scan reveals an EDH.

- **Artery affected:** The **middle meningeal artery** runs on the surface of the dura, and a rupture of this artery leads to trapping of blood between the dura and the skull, also known as an EDH.

- **Most likely cause of this patient's symptoms:** Although not experienced by all patients with an EDH (<20%), he has a classic lucency window, where the patient is alert but then deteriorates rapidly.

- **The CT scan of the patient's head is likely to reveal:** An EDH, with a depressed open skull fracture. The classic finding of an EDH is an extra-axial, homogeneous, biconvex or lenticular lesion, generally confined by suture lines. This is in contrast to a subdural hematoma, which follows the convexity of the brain.

CLINICAL CORRELATION

Traumatic brain injury and the resulting increases in intracranial pressure (ICP) may lead to changes in a patient's level of consciousness. With EDH, as with other supratentorial space-occupying lesions, pressure is placed on the brain, leading to midline shift. Epidural hematoma commonly results from laceration of the middle meningeal artery. Should the ICP continue to increase, contralateral hemiparesis and ipsilateral pupil dilation (secondary to subfalcine herniation and compression of the third cranial nerve) occurs. Without intervention, this may progress to flexor and then extensor posturing, and patients may exhibit a Cushing response: hypertension, bradycardia, and widened pulse pressure. The mortality of EDHs can range from 5% to 50% depending upon the patient's presentation (Glasgow Coma Scale), progression, and other factors, including age, and location (eg, posterior fossa vs temporal) and size of the lesion. An excellent recovery is predicted for patients with a normal level of consciousness preoperatively, and the mortality increases for obtunded (10%) and comatose patients (20%).

APPROACH TO:
Consciousness

OBJECTIVES

1. Know the definition of consciousness.

2. Know how to evaluate consciousness.

3. Understand the neuroanatomical relationship between increased ICP and decreased level of consciousness.

DEFINITIONS

CONSCIOUSNESS: Awareness of self and environment (referred to as content of consciousness), and ease of arousal (referred to as level of consciousness).

OBTUNDED: Dulled or blunted.

STUPOR: Reduced sense or sensibility.

PERSISTENT VEGETATIVE STATE: Wakefulness without awareness, with intact sleep/wake cycles.

COMA: Unarousable, either by external stimuli or inner needs, absence of sleep/wake cycles.

LOCKED-IN SYNDROME: Paralysis of voluntary muscles except for ocular movements, no disturbance of awareness.

OCULOCEPHALIC REFLEX (DOLL'S EYE REFLEX): Conjugate eye movement opposite to head movement in order to maintain forward gaze during neck rotation. Brainstem lesions lead to absent or asymmetric eye movement.

CALORIC REFLEX: Ice water placed in the external auditory meatus leads to a slow, ipsilateral, conjugate eye deviation.

OCULAR BOBBING: Conjugate, bilateral fast downward jerk followed by slow return to mid-position.

DISCUSSION

Consciousness involves the ability to be awake, alert, and aware. Like awareness, arousal is not an all-or-nothing concept, but rather it ranges from inattentiveness to stupor and obtundation. Consciousness is a complex process, centered around the reticular-activating system (RAS). **The RAS is a portion of the rostral pons (paramedian tegmental zone), continuous caudally with the spinal cord and rostrally with the subthalamus, hypothalamus, and thalamic nuclei.** Consciousness depends on intact diencephalic connections with the RAS. The ascending acetylcholinergic neurons from the RAS that project to the thalamus act as an *on-off* switch, determining if ascending information arrives at the cortex, and if descending information is likewise transmitted. The acetylcholine pathways sensitize the thalamic neurons, allowing sensory input. This leads to an "awake" state.

Disturbances of consciousness have multiple etiologies, including traumatic injury, metabolic disturbances, psychiatric, infarction/hemorrhage, brainstem disorders, neoplastic, and toxins. Coma results from bilateral damage or lesions in the cerebral hemispheres, the thalamus, the hypothalamus, and/or the RAS. As described in this case, increases in ICP can lead to tentorial herniation and compression of the brainstem RAS. Comas are generally caused by lesions rostral to the level of the pons.

The **Glasgow Coma Scale (GCS) score** is a validated physical examination scoring system to evaluate levels of consciousness, shown to be predictive of certain outcomes following neurological injury. It is determined by three categories: Eye (E), verbal (V), and motor (M). Tests are scored as follows, with total scores ranging from 3 to15:

- **Eye**
  Opens spontaneously (4), opens to sound (3), opens to pain (2), does not open (1).

- **Verbal**
  Responsive and appropriate (5), confused (4), unintelligible (3), moans at pain (2), silent (1).

- **Motor**
  Follows commands (6), localizes pain (5), withdraws from pain (4), decorticate or flexor response (3), decerebrate or extensor response (2), no movement (1).

The GCS arbitrarily defines coma as a failure to open eyes in response to verbal command (E2), performing no better than weak flexion (M4), and uttering only unrecognizable sounds in response to pain (V2).

The eye examination provides important insight into the origin of the decrease in consciousness. In an insult to the bilateral hemispheres, the papillary examination and oculocephalic response is normal. As was mentioned in this case, supratentorial mass lesions lead to secondary brainstem compression and a blown pupil on the same side as the lesion. Brainstem lesions lead to abnormal oculocephalic responses. Patients with metabolic disturbances have normal pupils. An intact pontine reticular formation allows the patient to blink, either spontaneously or in reaction to stimuli. Roving eye movements occur with intact third nerve nuclei and connection, meaning the insult is likely toxic/metabolic or bihemispheric. Seizures lead to contralateral conjugate eye deviation. Acute pontine lesions lead to ocular bobbing. Absence of the oculocephalic response indicates progression to the caloric reflex evaluation.

The red nucleus is important for the localization of a lesion affecting consciousness. Output from the red nucleus reinforces antigravity flexion of the upper extremity. When this is lost, unregulated reticulospinal and vestibulospinal tract output reinforces the extension tone of the extremities. If the patient has an upper motor neuron lesion above the red nucleus, the patient will have flexor or decorticate posturing. Upon receipt of painful stimulus, the patient will have upper limb flexion with pronation of the forearm, and lower limb extension with foot inversion. With lesions below the level of the red nucleus but above the level of the vestibulospinal and reticulospinal nuclei, decerebrate posturing will occur. Decerebrate posturing is the extension and pronation of the upper extremities with extension of the lower extremities. With a lesion of the medulla, the descending corticospinal tract is disrupted, leading to acute flaccidity.

CASE CORRELATES

- See Cases 41-49 (cognition).

COMPREHENSION QUESTIONS

48.1 A 32-year-old woman is brought into the ED by the paramedics following a rollover motor vehicle accident in which she was trapped in the car and had to be released with the jaws of life. On arrival at the ED, she is noted to be completely unresponsive and has a GCS score of 4. What normal CNS structure(s) is/are required for the maintenance of consciousness?

A. Cerebral cortex

B. RAS

C. Spinal cord

D. Both the cerebral cortex and the RAS

48.2 You are examining a comatose patient in the ICU. Your first item of examination is the basic brainstem reflexes, and you start with the pupillary reflexes. Shining your light into either of the patient's eyes results in equal, direct, and consensual pupillary constriction. Based on this finding alone, where would you localize the lesion causing the coma?

A. Diffuse cortical damage

B. Midbrain damage

C. Pontine damage

D. Medullary damage

48.3 You are examining an unresponsive patient, brought into the ED by EMS. When presented with a painful stimulus, the patient responds by flexing all of his extremities (decorticate posturing, scoring 2 in the motor section of the GCS). This response indicates motor output directed by what level of the nervous system?

A. Cerebral cortex

B. Red nucleus

C. Pontine reticular formation

D. Vestibular nuclei

ANSWERS

48.1 **D.** Consciousness depends on the interaction between the intact **cerebral cortex and the RAS,** and damage to either or both of these structures or the communications between them can cause lack of consciousness, or coma. The RAS serves as a gate which determines whether or not sensory input is relayed to the cortex, so without it, no information reaches the cortex, making it unable to interact with the outside world. Diffuse injury to the cortex with an intact RAS can also cause coma, because although information can reach the cortex, it cannot be processed.

48.2 **A.** Because of the variety of this patient's symptoms, it appears there are multiple lesions accounting for **diffuse cortical damage.** Intact pupillary reflexes indicate intact and functioning midbrain and brainstem. The reflex depends on intact structures in the superior colliculus and the oculomotor complex, both of which reside in the midbrain. In this patient, diffuse cortical dysfunction is the likely cause, possibly from hypoxic-ischemic injury to the cortex or a metabolic encephalopathy.

48.3 **B.** Motor outflow from the **red nucleus** through the rubrospinal tract stimulates antigravity flexion of the extremities. Interruption of descending motor control between the cortex and red nucleus leaves it as the primary motor outflow, resulting in decorticate (flexor) posturing. Interruption of motor control below the level of the red nucleus results in motor outflow being driven by the reticulospinal and vestibulospinal tracts, which stimulate extensors, resulting in decerebrate (extensor) posturing.

NEUROSCIENCE PEARLS

▶ A portion of the rostral pons, the RAS plays a key role in consciousness, acting as an *on-off* switch between the spinal cord and thalamus.

▶ The Glasgow Coma Scale is a validated assessment of consciousness and neurological function in patients with brain injury.

▶ Lesions either rostral or caudal to the red nucleus lead to decerebrate or decorticate posturing.

REFERENCE

Bateman DE. Neurological assessment of coma. *J Neurol Neurosurg Psychiatry.* September 2001;71 (suppl 1):i13-i17.

A 47-year-old woman presents to the emergency room complaining of severe headache, neck stiffness, nausea, and vomiting for the last 6 hours. The patient's husband says that she was complaining of the worse headache that she has ever had for about 4 hours. The patient has had no medical problems previously. Currently, she is oriented to person and place only and is somewhat somnolent. Her blood pressure is 130/70 mm Hg, her heart rate is 85 beats per minute, and her respiratory rate is 16 breaths per minute.

▶ What is the most likely diagnosis?
▶ What is the most common cause for her condition?
▶ What are risk factors for her condition?
▶ What kidney disease may be present in this patient?

ANSWERS TO CASE 49:

Intracerebral Hemorrhage and
Increased Intracranial Pressure

Summary: A 47-year-old woman has a history of severe headache, hypertension, neck stiffness, and lethargy. She is scheduled for clipping of an aneurysm.

- **Most likely diagnosis**: Subarachnoid hemorrhage.

- **Most common cause**: Ruptured berry aneurysm of the cerebral arteries (circle of Willis).

- **Causes/risk factors:** Head trauma, ruptured aneurysm, ruptured arteriovenous malformation, hypertension, polycystic kidney disease, coarctation of the aorta, fibromuscular dysplasia, female gender, tobacco use, cocaine abuse, and oral contraceptives.

- **Possible kidney disease:** Autosomal dominant polycystic kidney disease may be present in patients with saccular aneurysms. Patients with polycystic kidney disease may also have cysts in other parts of the body, including the circle of Willis.

CLINICAL CORRELATION

The symptoms of the worst headache of one's life, neck stiffness, and/or neurological deficits are highly suggestive of subarachnoid hemorrhage. The most common reason is the rupture of an aneurysm in the circle of Willis. Patients with autosomal dominant polycystic kidney disease may have saccular defects of the cerebral vessels.

The prevalence of intracranial aneurysms is about 5% in the general population. Intracranial aneurysm is the most common cause of intracranial hemorrhage and accounts for 75% of all subarachnoid hemorrhages. One-third of patients do not survive the initial acute bleed. The risk factors are female gender, genetic predisposition, hypertension, pregnancy, polycystic kidney disease, cocaine abuse, and vascular abnormalities (ie, arteriovenous malformation). The most common location for an aneurysm is the junction of the anterior cerebral artery and the anterior communicating artery. The size ranges from small (<12 mm) to large (12-24 mm) and giant (>24 mm). About 80% of aneurysms are classified as small, and the risk of rupture ranges from 0.14% to 1.10% per year for smaller aneurysms. The risk of rupture increases proportionally with the size of the aneurysm. The pathophysiology of a ruptured aneurysm is that the arterial blood released raises the intracranial pressure (ICP), which lowers the cerebral perfusion pressure (CPP) and cerebral blood flow (CBF), and ultimately causes the patient to lose consciousness.

The three main causes of mortality and morbidity are the initial bleed, a rebleed, and vasospasm. Other factors include surgical complications, hydrocephalus, and neurological dysfunction.

Severe and sudden onset of a headache, which occurs in 90% of patients, is the most common presenting symptom. A brief loss of consciousness may occur. The symptoms are similar to meningitis; for instance, there may be neck stiffness, nausea and vomiting, or photophobia. In addition, both sensory and motor deficits can accompany the other signs and symptoms. Medical imaging by CT or MRI is used to confirm the diagnosis. The clinical approach includes using imaging or following clinical signs to determine if there is ongoing arterial bleeding from the ruptured aneurysm.

The most common surgical intervention is "clipping" or ligation of the aneurysm. A coordinated approach between the neurosurgeon and the anesthesiologist is necessary to provide the safest care for the patient, and various anesthetic techniques can be employed to optimize conditions for the surgeon. The challenge in the anesthetic management of this surgery is in ensuring adequate cerebral blood flow. Understanding the neurophysiology, the surgical procedure, and the effects of anesthesia is essential to the optimal management of the clipping of an aneurysm. Relatively recently, in 1995, the FDA approved the endovascular embolization of intracranial aneurysms with metal coils, but this is not a treatment option in all aneurysms.

In neonates, intraventricular hemorrhage is a common complication of prematurity, most commonly occurring in infants born prior to 32 weeks' gestation. The source of the intraventricular hemorrhage is the germinal matrix, which is prone to bleed due to its thin-walled blood vessels.

APPROACH TO:

Intracerebral Hemorrhage and Increased Intracranial Pressure

OBJECTIVES

1. Describe the determinants of cerebral blood flow (CBF).

2. Know how to decrease intracranial pressure (ICP).

DEFINITIONS

CEREBRAL PERFUSION PRESSURE: Calculated by the formula:

$$CPP = MAP - ICP$$

where MAP is the mean arterial pressure and ICP is the intracranial pressure,

ANESTHETIC INDUCTION: The period between the entry of the patient to the operating room and the surgical incision. Induction consists of the administration of medications to anesthetize the patient, intubation of the trachea, and placement of any invasive monitors.

OSMOTIC DIURESIS: The administration of mannitol to reduce extracellular volume, which improves surgical conditions by producing a "relaxed" brain.

DISCUSSION

Cerebral blood flow (CBF) averages 40 to 50 mL per 100 g brain tissue per minute. The brain receives 15% of the total cardiac output because of the relatively high cerebral metabolic rate of oxygen ($CMRO_2$) and the inability of the brain to store oxygen. The gray matter has a greater blood flow than the white matter. Cerebral blood flow is determined by a number of factors: the $CMRO_2$, the CPP, the arterial carbon dioxide ($PaCO_2$) and oxygen (PaO_2) levels, and anesthetics (both intravenous and inhalational). It is autoregulated; between a MAP of 50 and 150 mm Hg the CBF is constant, regardless of the CPP. Inhalational (volatile) anesthetics cause cerebral vasodilation and an increase in CBF. On the other hand, intravenous anesthetic agents (eg, propofol, etomidate, and barbiturates) cause cerebral vasoconstriction, which reduces CBF. Opioids have a minimal effect on CBF. Cerebral blood flow can also be disrupted by cerebral ischemia, tumors, hypoxia, hypercarbia, and edema.

The cerebral metabolic rate of oxygen ($CMRO_2$) is a critical factor in determining cerebral ischemia. The rate is 3.0 to 3.8 mL O_2 per 100 g brain tissue per minute. The brain has a very high energy requirement, accounting for 20% of the body's oxygen consumption. If oxygen consumption is high and blood perfusion is low, the risk of ischemia is high. The $CMRO_2$ is directly coupled with CBF and changes are proportional. Intravenous anesthetic agents (except ketamine) and inhalational (volatile) anesthetics both decrease $CMRO_2$. However, intravenous agents further decrease CBF by vasoconstriction, whereas the volatile anesthetics increase CBF by vasodilation. Temperature also affects the $CMRO_2$ by decreasing CBF by 7% for every 1°C decrease.

Cerebral perfusion pressure (CPP) is the difference between the MAP and the ICP, and is normally 50 mm Hg. There is autoregulation to maintain cerebral blood flow (CBF) over a wide range of CPP (from 50-150 mm Hg). The response time to adjust to changes to CPP is about 1-3 minutes.

The curve shifts left during anesthesia (cerebral blood flow maintained over lower CPP's), which allows a patient to tolerate hypotension better during anesthesia. Chronic hypertension shifts the curve to the right, which makes patients prone to cerebral ischemia at a higher than normal CPP. Autoregulation of CBF is altered by head trauma and intracranial pathology such as a brain mass.

Intracranial pressure is normally 10 mm Hg. A sustained ICP of greater than 15 mm Hg is considered increased ICP. As ICP increases there is a decrease in CPP, which decreases CBF and can result in regional and/or global cerebral ischemia. The brain is located in the fixed confines of the skull, and the intracranial cavity comprises 80% brain tissue, 10% blood, and 10% cerebrospinal fluid (CSF) by volume. If the volume of any one component of the intracranial cavity is increased, then the volume of other two must be reduced to prevent an increase in ICP.

In this case, the bleeding from the ruptured aneurysm results in an increase in cerebral blood volume (CBV). Initially, as ICP increases the cerebrospinal fluid relocates to the spinal canal to accommodate for the increase in pressure. As the spinal canal is filled to capacity, a small change in intracranial volume causes an exponential increase in ICP leading to ischemia and edema. Either maintenance or reduction of ICP is desirable in neuroanesthesia because it is a key component in determining CPP.

Carbon dioxide affects the brain in a potent manner. Between 20 and 80 mm Hg, as the $PaCO_2$ increases the CBF increases. For every 1 mm Hg increase, there is an increase in CBF of 1 mL/100 g/min. Hypercarbia causes vasodilation and an increase in CBV by the effect it has on the pH of the CSF. Hypocarbia decreases CBF resulting in a decrease in CBV and ICP. Mild hyperventilation to a $PaCO_2$ of 30-35 mm Hg is effective in decreasing the ICP. Below 30 mm Hg, the risks of systemic effects of hypocarbia outweighs the benefit. The effects of hyperventilation decrease after 6 hours. The ability of the anesthesiologist to manipulate $PaCO_2$ by changing ventilation is an important aspect of intraoperative management that can directly affect the surgeon's operative conditions.

Hypoxia (PaO_2 <50 mm Hg) results in an increase in CBF and CBV secondary to abrupt vasodilation.

Anesthetics influence CBV. All volatile (inhalational) anesthetics increase CBF and decrease $CMRO_2$. Intravenous induction agents, with the exception of ketamine, decrease CBF, $CMRO_2$, and ICP. Opioids decrease CBF and have little effect on ICP, but can cause respiratory depression and hypercarbia, which is a potent cerebral vasodilator. Neuromuscular blocking agents have no effect on ICP, with the possible exception of succinylcholine, which theoretically increases ICP.

Ways to decrease ICP:

- Elevation of the head
- Hyperventilation
- CSF drainage
- Osmotic diuresis
- Corticosteroids
- Reduce CBV (by the use of barbiturates or propofol)
- Avoid cerebral vasodilation (avoid volatile inhalational anesthetics)

Signs and symptoms of raised ICP:

- Nausea/vomiting
- Hypertension
- Bradycardia
- Personality changes
- Altered consciousness
- Altered breathing
- Papilledema

CASE CORRELATES

- See Cases 1 (glioblastoma), Case 35 (uncal herniation), Case 39 (medulloblastoma), Case 41 (brain laterality), Case 42 (visual perception), Case 43
  (spatial cognition), Case 44 (language disorders), Case 45 (memory), Case 46
  (disconnection syndrome), Case 47(executive function), and Case 48 (consciousness).

COMPREHENSION QUESTIONS

49.1 A 38-year-old woman suddenly has the worst headache of her life and then
     develops dizziness, neck stiffness, and vomiting. CT imaging shows enhance-
     ment of the subarachnoid space, consistent with subarachnoid hemorrhage.
     Which of the following is the most likely etiology?

   A. Congenital thrombocytopenia

   B. Coagulopathy

   C. Weakness at branch points of cerebral arteries

   D. Hypertensive disease

49.2 A 62-year-old man develops accelerated hypertension, and the concern is
     excessive cerebral blood flow. However, autoregulation of the cerebral blood
     flow maintains constant blood flow to the brain. Which of the following is
     known to cause vasoconstriction and reduce cerebral blood flow?

   A. Elevated serum sodium concentration

   B. Low arterial oxygen level

   C. Low arterial carbon dioxide level

   D. Low serum glucose levels

49.3 A 70-year-old woman presents with the sudden onset of right eye blindness
     and severe jaw pain. The patient did not have any other pain or neurological
     deficits. The blindness lasts for 20 minutes and then resolves. Laboratories
     reveal a haemoglobin of 12 g/dL, platelet count of $150 \times 10^3/\mu L$, and an ESR of
     110 mm/h. Which of the following is the most likely to establish a diagnosis?

   A. CT of the head

   B. MRI of the head

   C. Antinuclear antibody test

   D. Temporal artery biopsy

   E. Angiography of cerebral arteries

ANSWERS

49.1 **C.** The most common cause of subarachnoid hemorrhage is a ruptured berry (or saccular) aneurysm. These aneurysms occur due to weakness at the bifurcation of cerebral arteries; cerebral arteries are at increased risk because they have a very thin adventitia and no external elastic lamina. Berry aneurysms usually occur at the branching sites of the large cerebral arteries in the anterior portion of the circle of Willis.

49.2 **C.** High arterial oxygen levels, or low carbon dioxide levels, lead to cerebral vessel vasoconstriction and decreased cerebral blood flow. This phenomenon is used to promote hyperventilation in stroke patients, which decreases intracranial pressure and therefore reduces the risk of further brain damage.

49.3 **D.** This patient likely has temporal arteritis. The symptoms of jaw pain, especially beginning with a meal, unilateral painless blindness in the absence of other neurological findings, and an extremely high ESR (typically exceeding 90 mm/h) are almost diagnostic. A temporal artery biopsy confirms the diagnosis. Treatment with corticosteroids can prevent permanent blindness.

NEUROSCIENCE PEARLS

▶ The most common cause of subarachnoid hemorrhage is a ruptured berry aneurysm.

▶ A patient with a subarachnoid hemorrhage often complains of "the worst headache of their life."

▶ Hyperventilation and hypocarbia causes vasoconstriction and decreased intracranial pressure.

▶ Intraventricular hemorrhage is a common complication in extremely premature infants, originating from the germinal matrix.

REFERENCES

International Study of Unruptured Intracranial Aneurysms investigators. Unruptured intracranial aneurysms—risk of rupture and risks of surgical intervention. *N Engl J Med.* 1998;339:1725-1733.

Matta BF, Mayberg TS, Lam AM. Direct cerebrovasodilatory effects of halothane, isoflurane, and desflurane during propofol-induced isoelectric electroencephalogram in humans. *Anesthesiology.*1995;83(5): 980-985.

The Brain Tumor Foundation. The American Association of Neurological Surgeons. The Joint Section on Neurotrauma and Critical Care: Hyperventilation. *J Neurotrauma.* 2000;17(6-7):513-520.

Listing of Cases

Listing by Case Number

Listing by Disorder (Alphabetical)

INDEX

Note: Page numbers followed by *f* or *t* indicate figures or tables, respectively.